Damage and Cracking of Concrete Structures

To my son Sébastien
who died prematurely while this book was being written

Jacky Mazars

Series Editor
Gilles Pijaudier-Cabot

Damage and Cracking of Concrete Structures

From Theory to Practice

Jacky Mazars
Stéphane Grange

iSTE

WILEY

First published 2022 in Great Britain and the United States by ISTE Ltd and John Wiley & Sons, Inc.

ISTE Ltd
27-37 St George's Road
London SW19 4EU
UK

www.iste.co.uk

John Wiley & Sons, Inc.
111 River Street
Hoboken, NJ 07030
USA

www.wiley.com

Library of Congress Control Number: 2022939173

British Library Cataloguing-in-Publication Data
A CIP record for this book is available from the British Library
ISBN 978-1-78630-699-9

Contents

Foreword by Franz-Josef Ulm

Since the inception of mechanics of materials as a core discipline of civil engineering, the concept of damage has been central for engineers. It always looms deep in our engineering conscience as a fear of the inceptive failure of materials and structures built in ever more challenging environments, affecting the well-being and sustainability of our societies at large. But when damage is rationalized in the form of engineering mechanics models and put to work, it turns into a science-enabled tool of empowerment for engineers to quantify, predict and avoid premature deterioration and failure of materials and structures. This empowering message is at the core of this textbook dedicated to damage mechanics of concrete and concrete structures.

Damage mechanics condenses discrete fracture processes into a tractable observable, the degradation of elastic stiffness. Set within a continuum framework, such deterioration processes are best described within the context of the thermodynamics of irreversible processes. This is the starting point of this intellectual journey the authors take us on. It builds upon Jacky Mazars' groundbreaking discovery in the late 1980s that damage mechanics applied to concrete needs to break away from the dogma of yield-design-based approaches that dominated concrete mechanics at the time, while rationalizing strain-based approaches for concrete cracking into the larger framework of thermodynamics-based damage mechanics. After more than three decades of science-enabled engineering research, the fruits of Mazars' endeavor are now in our hands as a powerful toolbox for engineers, which this book opens up to us.

From the onset, Mazars' damage mechanics of concrete departs from the premise that the choice of the appropriate driving force of damage must be both rooted in fundamental principles and tractable for engineers. An ingenious stroke is the choice of linking damage to the positive eigenvalue of the strain tensor, which all engineers immediately recognize as a potential threat of fracture due to tension. But when derived from fundamental principles, this engineering quantity is readily expressed in terms of the continuum counterpart of the energy release rate of fracture mechanics. Henceforth, it becomes possible to link fundamental principles to experimental observations, such as stress–strain curves, and translate these observations into thermodynamically sound evolution laws of damage for a variety of relevant loading situations: monotone, cyclic, confinement, etc.

The second pillar of Mazars' damage mechanics is the sound computational mechanics approach, which is a condition sine qua non for damage mechanics to be adopted by engineers. As a continuum theory, damage mechanics has no intrinsic length scale. Yet, when considered consistently as a bulk mean-field representation of a surface fracture process, a length scale is readily derived, which needs to be taken into account in the computational implementation. The result, known as non-local damage theory, achieves just this: ensuring a sound finite element realization of damage mechanics with model parameters of clear physical significance and accessible by (almost) standard engineering tests.

The litmus test of all engineering mechanics models is its application for a large range of problems engineers need to address today; ranging from classical concrete design solutions to earthquake design, account of rate effects risk of early age concrete damage caused by the competition of maturation (stiffening) and thermal and chemical eigenstress development. It is through these applications that Mazars' damage mechanics achieves lasting engineering impact.

More generally, as the perils of global warming force us engineers to think ahead, and design structures not only for strength, but for durability and resilience in day-to-day operations and under extreme events, this book stands out: developed by two eminent educators and engineering scientists in the field, it delivers the key elements of damage mechanics of concrete and concrete structures for both instructors and students in a pedagogically fine-tuned way, and establishes damage mechanics at the heart of contemporary education of mechanics of materials in civil engineering with lasting impact.

Franz-Josef ULM
July 2022

Foreword by Pierre Labbé

Some events, such as the tragic ruin of the Polcevera Viaduct in Genoa, reveal to the general public and periodically remind engineers and scientists of the considerable societal and economic stakes associated with damage to concrete structures. In this context, among the many problems that can affect concrete, cracking is certainly the one that has mobilized the most knowledge and know-how of engineers. Indeed, as soon as the concept of limit state is introduced in engineering practices, there are formulas for spacing and opening cracks in French regulation called BAEL 68.

In the recent past, scientific and technical knowledge related to the lifespan of structures has traditionally been less invested than that which presides over design and construction and is regarded as less prestigious. It must be said that we do not invite the press to inaugurate a maintenance site! Fortunately, things are changing as a result of environmental concerns and a growing interest in the sustainability of structures and heritage conservation.

The book by Jacky Mazars and Stéphane Grange is therefore fully in line with the issues that will be at the heart of the profession of the next generations of civil engineers. This book reflects the personality of the authors who are both experienced scientists, attached to the development and progress of engineering sciences, as well as practitioners attentive to their operational implementation by design offices. As a result, both researchers and engineers wishing to perfect their knowledge of the physics of cracking concrete structures and its modeling can benefit from reading this book.

Without a doubt, Mazars and Grange are scientists and researchers, for example, in the care they have taken to develop a theoretical framework adapted to the formulation of their model in the three-dimensional continuous medium or in the presentation of the foundations of the behavioral laws they develop and use, paying attention to their compatibility with the main principles of thermodynamics. Being scientists, they are also in the spectrum of situations for which they question the appearance and development of cracks. This is how they turn their attention to the behavior of concrete at early ages, whose crucial role in the appearance of cracks at the maturity of materials and structures is now increasingly understood. They look with the same acuity, and no doubt with the same passion, at the very slow phenomena of creep and those of fast dynamics, and show that the models they propose adapt to these different situations.

Alongside the arduous developments that the presentation of their scientific approach requires, and taking the same care that will be appreciated by engineers, Mazars and Grange present the practical implementation of the models they propose. Beneath its apparent banality, the subtitle, "from theory to practice", has rarely been more relevant. Here, we especially commend the efforts of simplification that the authors brought to their initial approach by retaining a model with isotropic damage under monotonic load, which they then extended to cyclic situations. It should also be noted that they handle tools dear to engineers with ease, such as Timoshenko's beam theory. Even if the behavior models they present obviously require more parameters than what engineers are accustomed to, the authors have made it a point of honor that the setting of these parameters is possible from conventional laboratory tests. They even provide numerical values that can be adopted by default and have gone so far as to ensure implementation in finite element software for industrial situations requiring large meshes, as in the example presented of a containment structure mock-up.

To conclude on a more personal note, I would like to add that after having shared several exciting research adventures with Mazars, ranging from the shake-table studies of the CAMUS program to the national CEOS.fr project, including the SAFE, SMART and VeRCoRs mentioned in this work, I think I can be the spokesperson for a large community by saying that I have always appreciated his lofty view, scientific and intellectual rigor, and unshakeable probity. After he trained many engineers and scientists, including Grange, I have no doubt that the new generations will also find material to train well when reading this book.

Pierre LABBÉ
July 2022

List of Notations

Barring exceptions, each notation appears only once in this table. Unspecified units are either without unit (wu) or without computational interest.

Chapter	Symbol	Description	Unit
1	f_c, f_t	Uniaxial compression and tensile strengths, respectively, measured at 28 days	N/m^2
	f_{ck}, f_{ctk}	Characteristic uniaxial compression and tensile strengths, respectively	N/m^2
	f_{cm}, f_{ctm}	Mean uniaxial compression and tensile strengths, respectively	N/m^2
	σ_i	Stress measured in direction i	N/m^2
	ε_i	Strain measured in direction i	(wu)
	σ_m	Hydrostatic stress (= p)	N/m^2
	$\dot{\varepsilon}$	Strain rate (= $d\varepsilon/dt$)	s^{-1}
	E	Modulus of elasticity or Young's modulus	N/m^2
	K	Bulk modulus	N/m^2
	ν	Contraction coefficient or Poisson's ratio	(wu)
2	D	Damage variable	(wu)
	$\tilde{S}$, $\tilde{\sigma}$	Effective section, effective stress	m^2, N/m^2
	U, u	Internal energy, specific internal energy	
	K, T	Kinetic energy, temperature	
	P_i, P_e	Power of inner and outer forces	
	Q, $\vec{q}$	Heat received, current vector of heat	
	ρ, r	Density, internal heat production volume density	

Chapter	Symbol	Description	Unit
2	σ, ε	Stress and strain tensors	
	Λ	Matrix of the elastic coefficients of the material	
	λ, μ	Lamé coefficients	
	ψ	Specific free energy	
3	ε_{eq}	Equivalent strain (Mazars model)	(wu)
	$\langle \varepsilon_i \rangle_+$	Positive part of the i^e principal strain	"
	ε_0, k_D	Initial damage threshold (Mazars model) and damage threshold during loading	"
	D_t, D_c	Tensile damage, compression damage	"
	A_t, B_t	Material parameters for tensile damage	"
	A_c, B_c	Material parameters for compression damage	"
	α_t, α_c	Partitioning coefficient in the damage of the tensile part and the compression part	"
	β	Power applied to previous settings	"
	ε_t, ε_c	Equivalent strain (μ model) under tension and compression (the threshold is indexed 0)	(wu)
	Y_t, Y_c	Maximum ε_t, ε_c values obtained on the load path	"
	Y, d	Pilot variable of the effective damage d (μ model)	"
	R	Stress triaxiality factor	"
	A, B	Parameters of evolution of the damage law	"
	K	Asymptotic shear calibration	"
	C	Coefficient to be attributed to Y_c for low and medium confinement	"
	d_{hc}	Consolidation damage for high confinement	"
	I_ε, $I_{\varepsilon 0}$, $I_{\varepsilon 1}$	First invariant of the strain tensor (or trace), lower threshold, upper threshold for d_{hc}	"
4	$[N]$	Nodal displacements interpolation matrices – displacement within a finite element	
	$[B]$	Nodal displacements interpolation matrices – strains within a finite element	
	$[k]^e$, $[K]$	Elementary and overall stiffness matrix	
	$\{F\}$, $\{U\}$	Vector of external forces and nodal displacements	

Chapter	Symbol	Description	Unit
4	F_r, P_r	Force at failure (maximum value) of a specimen	N
	m	Weibull coefficient	(wu)
	V_i, V_{eq}	Specimen volume i, equivalent volume of a structure	m^3
	G_F	Concrete tensile fracture energy	N/m
	G_f	Intrinsic tensile fracture energy	N/m
	δ, w	Displacement at mid-span, crack opening	m
	E_t	Post peak tangent stiffness (tensile curve)	N/m^2
	h	Dimension of a band of finite elements	m
	$g(x-s)$	Weight function at $x-s$ coordinate point	(wu)
	l_c	Characteristic length	m
	$\bar{\varepsilon}_{eq}$	Non-local equivalent strain	(wu)
5	s_c, w_c	Average spacing between cracks, mean crack opening	m
	σ_a	Damping stress	N/m^2
	β_1, β_2	Damping coefficients without and with damage	(wu)
	σ_{fc}, ε_{fc} σ_{ft}, ε_{ft}	Parameters to build the uniaxial behavior curve with hysteresis and permanent strains	N/m^2, (wu)
	ε_e, ε_p, ε_s	Elastic, plastic strain of steel and strain resulting from concrete-reinforcement slip	(wu)
	f_e	Yield strength of rebars	N/m^2
6	ε_{t0}^{s}, ε_{t0}^{d}	Static and dynamic tensile damage threshold	(wu)
	R_t	Dynamic threshold amplification factor	(wu)
	a_t, b_t, c_t, d_t, M_t	Damage threshold amplification factor calibration parameters	(wu)
	T, T_{ext}	Temperature, external temperature	°K
	C	Volumetric heat capacity	Jm^3K^{-1}
	k	Thermal conductivity	$Wm^{-1}K^{-1}$
	L	Latent heat of hydration	Jm^{-3}
	ξ	Degree of hydration	(wu)
	$A(\xi)$	Chemical affinity	(wu)
	M	Maturity of the cementitious mixture	(wu)
	φ	Heat flux	Wm^{-2}

Chapter	Symbol	Description	Unit
6	h	Exchange coefficient	$Wm^{-2} K^{-1}$
	ε_{bc}, ε_{au}, ε_{th}	Creep, autogenous shrinkage, and thermal strain	(wu)
	α, θ	Coefficient of thermal expansion, temperature change	K^{-1},°K
	κ_∞	Value of long-term autogenous shrinkage	µm/m
	k_{bc}, η_{bc}	Value of stiffness and viscosity in the Kelvin–Voigt model	N/m^2, $N/m^2/s$
	F_c, F_e, F_s	Forces supported in a section by: concrete, struts, rebars	N
7	$[B]$	Projection matrix of degrees of freedom at the interface between 2D and 1D finite elements	
	$[M]$, $[C]$, $[K]$	Mass, damping and stiffness matrices	
8	M, M_r, M_f	Bending moment, failure moment, cracking moment	N.m
	P, F	Force applied to a structure	N
	θ, δ	Angular variation, displacement	d°, m
	A_s, A_c	Steel area, concrete area in a reinforced concrete section	m^2
	σ_s, σ_c	Stress in steel and in concrete	N/m^2
	ρ	Steel to concrete ratio in a section	(wu)
	E_s, E_c	Modulus of elasticity, for steel, for concrete	N/m^2
	σ_{ch}, ε_h, d_h	Diffuse scheme: stress, strain, damage in concrete	N/m^2, (wu), (wu)
	x_r	Stress transfer length between reinforcement and concrete from a crack	m
	ε_r	Failure strain of concrete ($\sigma = 0$) in bilinear behavior	(wu)
	ε_m, ε_M	Strains at the top of the parabolas describing the behavior under tension and compression concrete in the diffuse scheme	(wu)
9	β	Ratio between lateral and vertical stress in an axisymmetric triaxial loading	(wu)
	s	Threshold applied on the lateral stress in two-phase triaxial loading	N/m^2
	$I_{\varepsilon+}$	Sum of positive principal strains	(wu)

Chapter	Symbol	Description	Unit
9	$I_{\varepsilon 0}$	Projection of $I_{\varepsilon +}$ on the nearest plane $\sigma_i = 0$ (we set the weakest stress to 0)	(wu)
	R	Restraint degree	%
Appendix	$\sigma_{ij}, \epsilon_{ij}$	Projections on the direction i of the stress and strain tensor terms relating to the normal j	MPa, (wu)
	σ_i, ϵ_i	Principal stresses and strains in direction i	"
	s_{ij}, e_{ij}	Projections according to the direction i of the terms of tensors of the deviators of stresses and the deviator of strains relative to the normal j	"
	s_i, e_i	Principal components of stress and strain tensor deviators	"
	U_r	Radial displacement in axisymmetric conditions	
	$\partial\Omega_u, \partial\Omega_f$	Edges of the field of study for the application of the boundary conditions under displacement and under force	
	U^*	Virtual kinematic field	
	$\hat{A}$	Writing of quantities in the space of overall degrees of freedom	
	γ, β	Newmark time integration model parameters	
	Φ_i	Modal shape of mode i	
	ω_i	Pulse of mode i	
	Γ_i	Mode i participation factor	
	$\widetilde{M}_i$	Effective modal mass of mode i	
	ν_i	Participation coefficient of mode i	

Introduction

Every year, almost as many cubic meters of concrete are used worldwide as there are people on the planet. It is by far the most widely used material, in particular to construct all kinds of buildings and infrastructures. In a world where sustainable development has become an unavoidable subject, where the increase in world population and the consequences of climate change constitute enormous challenges for future generations, engineers and civil engineering researchers have a duty and responsibility to ensure the durability and safety of all of these constructions.

Schematically, we can say that the damage to constructions is related to:

– the aging of materials under the effect of time, climatic variations and/or an "aggressive" environment;

– the repetition of heavy loads and the effects of excessive loads due to accidental (natural or technological risks), or even intentional causes (terrorist attacks).

Having a good control of these different aspects requires both a good command of the design of the structures, the manufacture of the materials and an in-depth knowledge of the way in which the materials and the structures react in the event of mechanical or chemical attacks. In all cases, these "attacks" result in more or less intense cracking, which themselves become the vector of environmental attacks (by water or certain corrosive products).

The understanding of cracking, initially approached by empirical concepts, then by fracture mechanics, has seen, during the last four decades, the emergence of approaches linked to the concept of damage to describe the degradation of concrete by cracking. The use of this concept and the development of tools for small-scale observation of the phenomena that underlie them led to the development of behavioral models that are now widely used in universities and the world of business.

The authors of this book have done a lot of research on these topics and their ideas and contributions have been sought by the construction industry on issues including:

– major energy-producing structures, in particular dams and power stations;

– pre- and post-seismic analyses in France, Japan, or as part of the development of European construction codes;

– innovations in the design and analysis of buildings and engineering structures, aimed in particular at better control of safety margins.

In this context, they believe that the know-how and the developments acquired make it possible to offer a presentation of the work carried out that is both complete and concise, by showing that it is sufficiently accomplished for a transition to engineering.

Thus, the following chapters aim to enable those who are in training or an apprenticeship, whether in research or engineering, or even as an operational engineer, to find the foundations, tools and applications that the work carried out by the authors on the damage of concrete structures allows. This work deliberately stands out from other more complete ones which, taking the international works in the field more broadly, act as a collection of know-how on the subject, but require the reader to have expert knowledge and a huge work of assimilation.

The idea here was to produce a small book, that is as self-sufficient as possible (the necessary prerequisites in solid mechanics and on finite element digitization are in the Appendix); some training exercises are also proposed and a wide variety of applications are given to allow those who must “do” to be guided in their work. Also, it is not necessary to practice a linear reading of the book; readers are advised to first immerse themselves in the contents and then to seek out the chapters that best suit their needs.

The subtitle of the work “from theory to practice” indicates that the modeling established was done so within a framework respecting the main principles of the thermodynamics of irreversible processes (Chapters 2 and 3) and are implemented by satisfying, at best, the state of the art on current numerical procedures (Chapters 4 and 5), but with an ever-present objective: “to make it as simple as possible while aiming for good performance”. Chapter 8 is a good example of this. Chapters 5–8 precisely describe how to approach a wide variety of applications, such as the cyclic behavior and seismic response of structures, the description of fast dynamic situations (shocks in particular), or the effects due to the maturation of concrete at an early age in massive structures. Chapter 9 is a collection of exercises and supplements to perfect the practice of the models presented. Finally, a references list covering the whole of the field makes it possible for those who wish to do so to deepen their knowledge in certain subjects that have been covered.

1

Mechanisms of Deformation and Failure of Concrete

1.1. Concrete: a material that is both widespread and misunderstood

Concrete, the most widely used manufactured material in the world (1 m^3 per year and per inhabitant of the planet), is at the heart of the construction of terrestrial infrastructures. Whether it be residential buildings, industrial buildings, energy-producing works, transport infrastructure, etc., concrete, combined with metal to build reinforced concrete structures or prestressed concrete, is present everywhere.

The Romans knew how to make concrete. They discovered that to make a hydraulic binder that would set under water, it was necessary to mix the waste from the manufacture of tiles and bricks or volcanic ash (coming in particular from Pozzuoli, in the bay of Naples, which gave its name to pozzolana, a volcanic rock) with lime. This knowledge, for example, made it possible for them to build harbors protected by concrete piers that set under water, unlike lime. Their knowledge was then lost in the Middle Ages. It was the perfecting and development of the production of modern artificial cements, in particular by Louis Vicat (1817), which allowed the development of current concrete construction. Reinforced concrete boats (1848) by Joseph-Louis Lambot were followed by the creation of entrepreneurs who developed reinforced concrete "systems": François Coignet (1814–1888), who perfected concrete blocks; Joseph Monier, whose patents of 1877 and 1878 were exploited in Germany; François Hennebique (1842–1921), whose company built more than 7,000 structures, most of which still exist; Armand Considere (1841–1914), who invented confined concrete (1901), and many others. In 1906, this first phase ended with the publication of the *Instructions relatives à l'emploi du béton armé*, the first real French regulation for the calculation of reinforced concrete structures, which would lead to the construction of the audacious tower by the architect Auguste Perret in Grenoble in 1924 (see Figure 1.2).

a) b)

Figure 1.1. *a) The first cast cement bridge (Louis Vicat, Grenoble, 1853); b) the Glacières Viaduct (France, 1988), truss structure in prestressed concrete (source: P. Paultre and J. Mazars). For a color version of this figure, see www.iste.co.uk/mazars/damage.zip*

a) b)

Figure 1.2. *a) The first reinforced concrete tower (Perret Tower, Grenoble, 1924); b) Architectural audacity of the dancing house (Prague, 1996) (source: P. Paultre and J. Mazars). For a color version of this figure, see www.iste.co.uk/mazars/damage.zip*

However, it was the invention of prestressed concrete by Eugène Freyssinet (1879–1962) that opened up new horizons for concrete material. Prestressing consists of keeping the concrete in compression under load by applying a prior stress (prestress) by means of external jacks or cables under tension installed within the structure itself. It makes it possible to produce original concrete structures in terms of their shape, lightness and mechanical audacity (see Figure 1.1).

Admittedly, concrete exists in several categories, ranging from the most rustic – concrete used for self-construction – to the most elaborate – very high performance concretes resulting from mixtures of ingredients and specific manufacturing processes. However, in any case, getting it to hold up well under load and over time is a tricky problem to address. Figure 1.3 shows all the complexity, from the moment of pouring until the long term: a varied number of phenomena related to the maturation of the concrete, the application of the loads or the aging intervene.

At the heart of these problems, there is the progressive degradation of the material of a thermohydromechanical nature or under the effects of time. The purpose of this book is to propose modeling tools based in particular on fundamentals in solid-state physics and damage mechanics, making it possible to help engineers solve a number of questions mentioned in Figure 1.3.

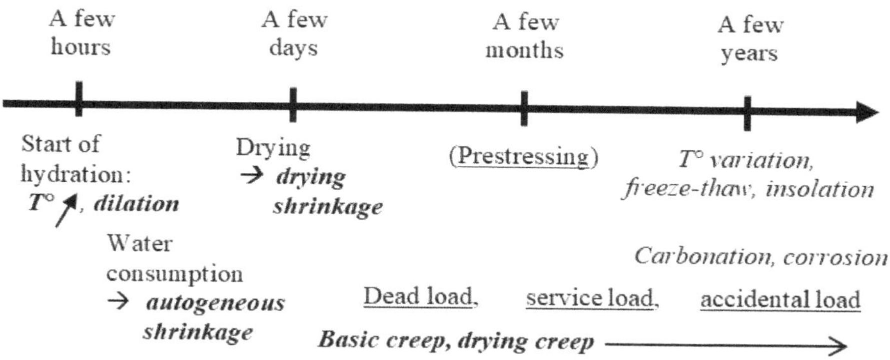

Figure 1.3. *Phenomena acting on concrete during the life of a structure; in bold italics, the physicochemical effects; underlined, the loads applied; in italics, climate and environmental actions (Lavinia et al. 2010)*

This summary presentation could advantageously be supplemented by the following documents: Pliskin (1995), Merceron-Vicat (1999) and Association Freyssinet (2004).

1.2. Composition and behavior of concrete at an early age

Concrete results from the mixture of the following main ingredients:

– cement, which is a hydraulic binder (hardening under the action of water) obtained by firing at 1,450°C a mixture of limestone and clay (clinker) that is finely ground with additions, giving it specific characteristics;

– aggregates, which include sand (grains up to 5 mm) and gravel (grains from 5 to 25 mm for common concretes);

– water, any additives used to control or promote certain properties (delay or acceleration of setting, plasticity of fresh concrete, etc.), as well as air occluded during placement in molds.

The proportions for common concretes are given in Table 1.1. The whole of this is introduced into a mixer adapted to the volume to be processed to constitute a mixture that is as homogeneous as possible.

In addition, for certain applications, it is possible, by playing on the quality of the different ingredients, to obtain special concretes: light concretes or heavy concretes by the use of suitable aggregates (expanded clay, barite or magnetite aggregates), high-performance concrete (additions of ultrafine silica fumes), fiber-reinforced concrete (additions of metallic or organic fibers), etc.

	Water	**Air**	**Cement**	**Aggregates**
Volume (in %)	14–22	1–6	7–14	60–78
Weight (in %)	5–9		9–18	65–85

Table 1.1. *Order of magnitude of the proportions of the constituents for common concrete (according to CimBéton (2013))*

Figure 1.4 shows, at three different scales, the main aspects of the microstructure of hardened concrete:

– decimetric scale (that of the volume element, in the mechanical sense of the term), showing the different sizes of aggregates and the cementitious matrix;

– centimeter scale, showing the cracking of the matrix between grains, which may result from effects linked to maturation or from premature mechanical loading;

– micrometric scale, for which the microstructure of the hydrated cement paste appears.

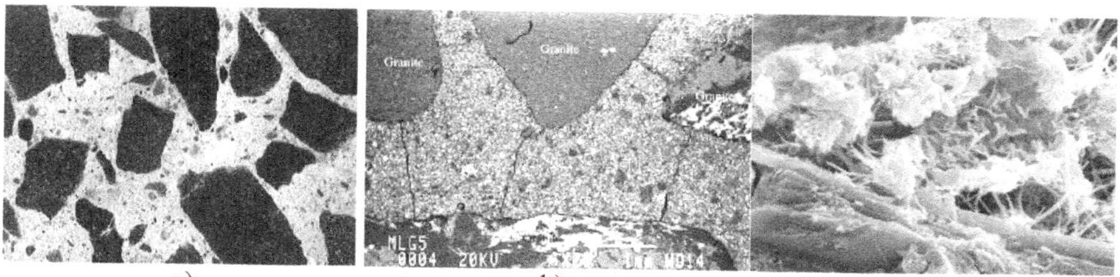

a) b) c)

Figure 1.4. *Three levels of concrete microstructure: a) volume element (in the mechanical sense), b) grain scale and c) microstructure of the hydrated cement, which is a tangle of C-S-H and ettringite needles (from Aftes (2012) and CimBéton (2013))*

1.2.1. *Concrete curing*

The major elements contained in the cement that intervene during hydration are the silicates. In contact with water, the tricalcium silicates (Ca_3SiO_5) and the dicalcium silicates (Ca_2SiO_4) contained in the cement dissolve in the form of ions, which interact with each other and form hydrated calcium silicates ($(CaO)x(SiO_4)y(H_2O)z$), C-S-H in simplified form, and portlandite ($Ca(OH)_2$). These reactions are exothermic.

C-S-H are non-crystalline compounds. They have the characteristics of a gel and have a variable composition; their average composition is close to $(CaO)_{1.7}(SiO_4)(H_2O)_4$. The entanglement of the C-S-H gel gives cement its solidity: it develops on the surface of the non-hydrated cement grains and progressively fills in the capillary pores between the grains (see Figure 1.5).

After a few hours, the layer of hydrates that coats the cement grains becomes thick enough to slow down the diffusion of ions and water toward the anhydrous components of the system. Hydration slows down more and more but can continue for months, or even years.

However, tricalcium aluminate is also present, which is the cement compound that is the most reactive with water. This is because the hydration of aluminates is so rapid that cement manufacturers add gypsum to the clinker to control their reactions. Aluminates react with gypsum sulfate to form ettringite TSA (hydrated calcium trisulfoaluminate). This reaction is highly exothermic. A layer of hydrates surrounds the aluminates, which then form a protective layer around the grain of cement. When the gypsum is exhausted, the sulfate concentration of the solution drops; the ettringite then becomes unstable and dissolves to form hydrated calcium monosulfate. After a few weeks, in general, the ettringite is completely transformed into monosulfoaluminate (called AFm).

As we have just seen, the hydration of Portland cement involves the reaction of the various mineral phases that compose it. As cement is mainly composed of tricalcium silicates, it is the hydration of the C3S that controls the overall kinematics of hydration. However, there is a synergy between the different reactions. Figure 1.5 indicates the main phases of hydration, which, from the start of setting, lead to the formation of a solid that undergoes heating then cooling at the same time as it hardens (improvement of mechanical properties).

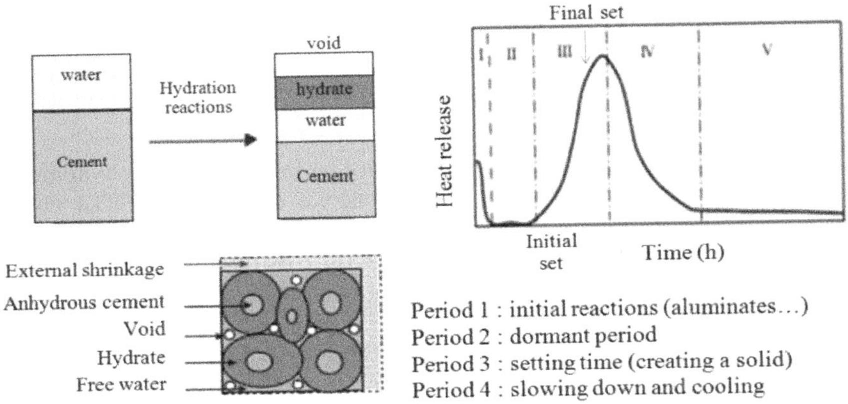

Figure 1.5. *Cement hydration mechanisms showing the exothermicity of the reaction and the autogenous shrinkage it induces*

However, concrete includes, in addition to water and cement, aggregates: sand and gravel. The hydration of the cement occurs in this highly heterogeneous medium, which leads to thermomechanical reactions during the chemical process. Indeed, aggregates are generally inert materials, but are sensitive to heating and shrinkage of the cementitious matrix. This is how they will play a dual role during the maturation phases:

– guiding the development of hydration products creating a different microstructure near the aggregates (particularly the larger ones), thus enclosing a zone called the "transition zone" likely to create local mechanical weakness;

– constitute hard points during the retraction of the material (see Figure 1.5) showing that the final volume of the hydrated products is lower than the initial volume, in particular due to the consumption of water during the chemical reaction. The grains oppose this shrinkage, which generates internal stresses within the matrix and risks of damage, as shown in Figure 1.4(b).

On the scale of a structural element, these cumulative dilation-shrinkage phenomena can lead, when these movements are hampered by specific boundary conditions, to cracking (see Figure 1.6).

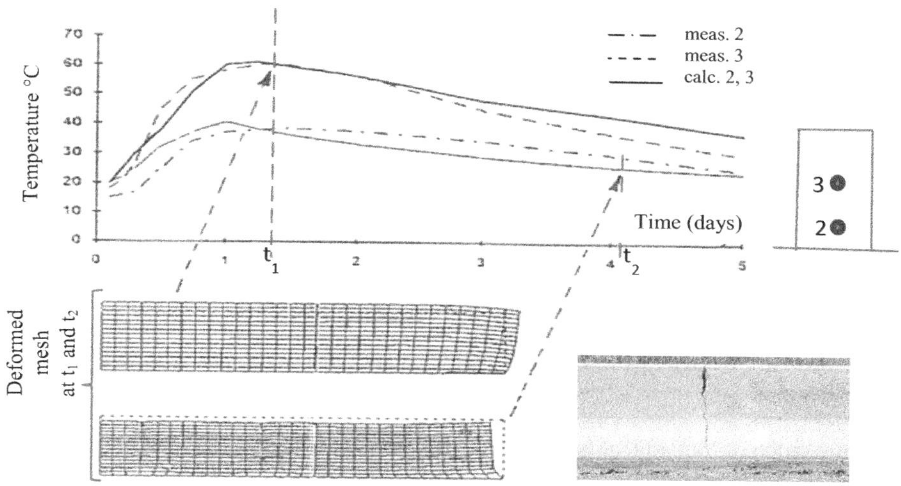

Figure 1.6. *Cumulative consequences of dilation-shrinkage effects in a concrete element whose movement is hampered by its adhesion to the supporting soil; the strains are amplified (according to Mazars and Bournazel (1998))*

SUMMARY.– *The hydration of cement, which leads to the maturation of concrete, is accompanied by physicochemical phenomena that generate:*

– on the barely solid material (beginning of setting), a dilation created by the exotherm of the reaction (see Figure 1.8(a));

– on the material at already advanced maturity (end of setting), the accumulation of cooling and shrinkage (autogenous shrinkage created in particular by the reduction in the overall volume of the products entering the hydration reaction (see Figure 1.5)), which cause structural stresses internal to the material likely to cause them at the level of the structural elements according to the hamper to the strain brought by the boundary conditions.

It is important to note that the development of stresses that apply to a material of viscous nature leads to the generation of relaxation (lowering of the stress at imposed strain) and/or creep (increase of the strain at imposed stress). At an early age, it is generally relaxation that prevails, and its effect is mechanically positive by limiting the risk of damage.

1.2.2. *Consequences of curing and phenomena related to the aging of concrete*

The lessons from section 1.2.1 allow us to consider that two zones of importance appear in the microstructuring of common concrete:

– the cementitious matrix-aggregate bond where a less resistant zone is observed due to:

- the specific structuring of hydrates;

- the existence of microvoids from air bubbles or water droplets trapped in particular under the grains during placement in the molds;

- any microcracks resulting from the hard spot effects of the aggregates during the shrinkage of the matrix (see Figure 1.4(b));

– the matrix outside this bond is dotted with capillaries of micrometric dimension, vestiges of those existing in fresh concrete and micropores, linked to the very formation of hydrates, the size of which is of the order of a few tens of Angstroms (see Figure 1.7(a)).

In general, as shown in Figure 1.7(b), the strength of concrete (here, under compression) decreases with the porosity of the material, which will mainly depend on the water/cement (W/C) ratio of the initial mix. The greater the W/C is, the more excess water there is in the concrete after maturing, which increases the porosity accordingly (the capillaries are filled with air or water). This greater porosity has two consequences:

– the material is less compact and less resistant;

– the capillaries are often interconnected (unlike the pores of hydrates), which increases the permeability of the material.

This open porosity promotes the evaporation of water from the capillaries, which generates a retraction of the material: drying shrinkage or desiccation shrinkage, which is added to the autogenous shrinkage due to hydration (see Figure 1.8(b)). In addition, this desiccation acts on the creep behavior (strain under imposed stress) by generating, in addition to the basic creep linked to maturation, a drying creep which, like the drying shrinkage, acts in the long term (Ollivier et al. 2012).

All of this information leads to a better understanding of Figure 1.3, confirming the different physical phenomena that accompany the maturation of concrete at an early age and the aging of mature concrete. To all this, it is now necessary to add the effects of mechanical loadings; this is what we propose for analysis in section 1.3.

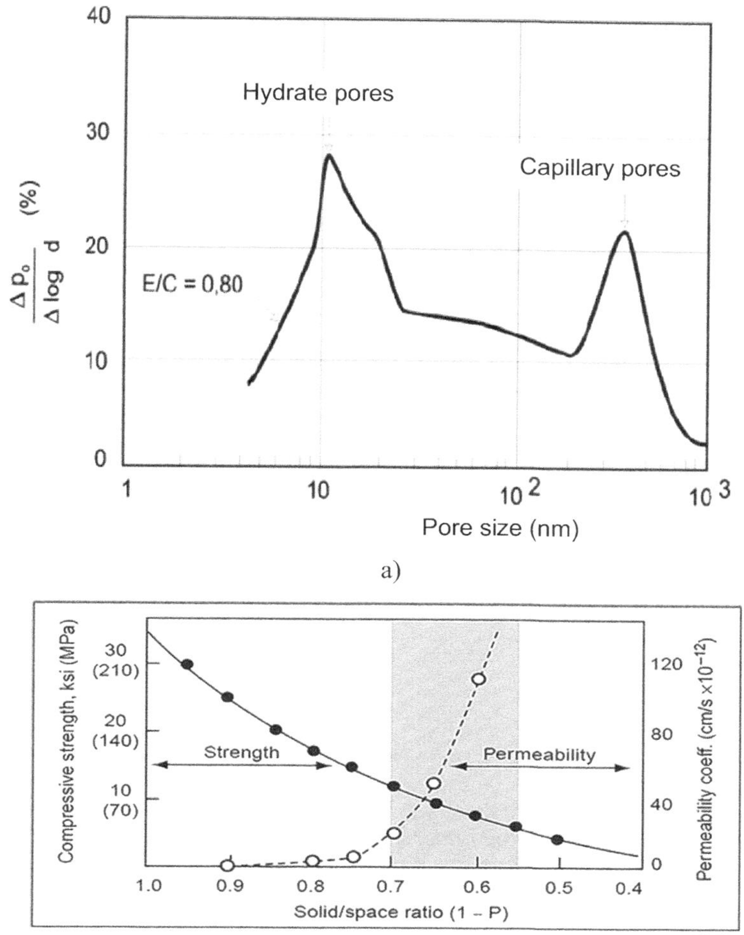

a)

b)

Figure 1.7. *Porosity and strength. a) The two categories of pores (Verbeck et al. 1968); b) decrease in the compressive strength and increase in the permeability of concrete with the increase in its porosity P (according to Hilal (2014))*

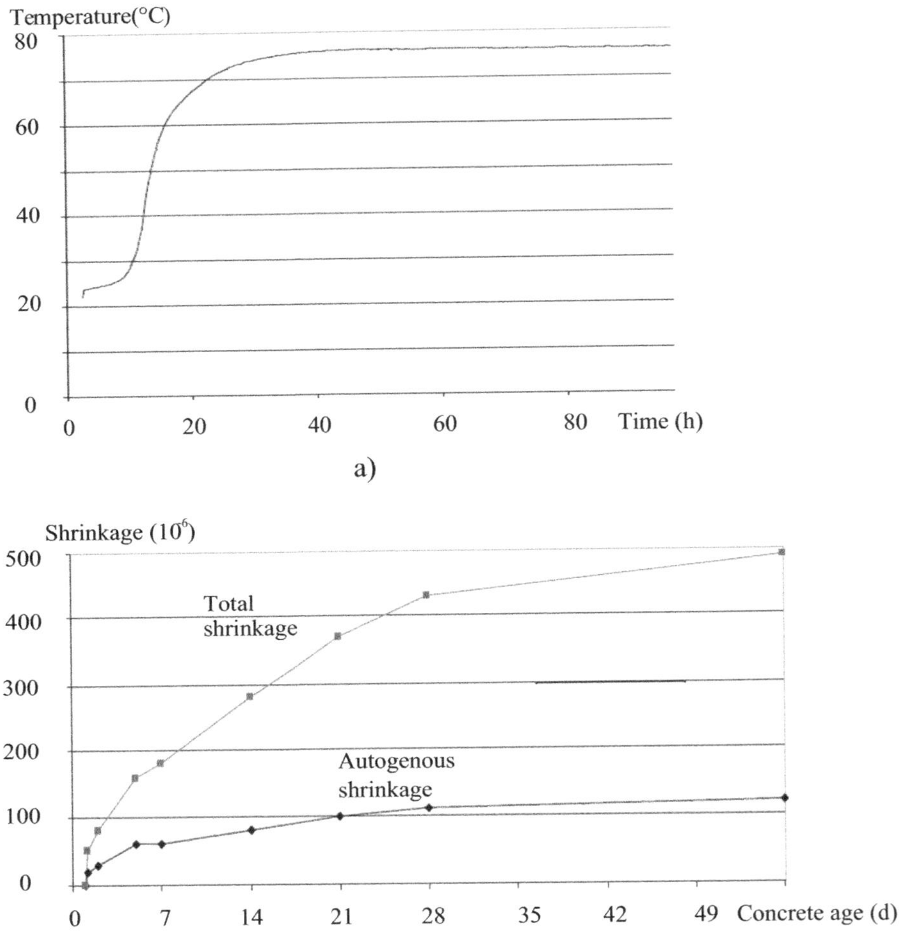

Figure 1.8. *Phenomena of early age of C50/60 concrete. a) Temperature rise during hydration measured under adiabatic conditions; b) autogenous and total shrinkage: autogenous + desiccation (according to CEOS (2017))*

1.3. Main aspects of the mechanical behavior of concrete

1.3.1. *Concrete under uniaxial loading*

1.3.1.1. *Uniaxial compression*

The failure properties of the material under compression are at the center of the characteristics used by the designers of concrete structures who rely on standards and codes to work. First, concretes are designated by their resistance class: for example, C 30/37 corresponds to a characteristic compressive strength of 30 MPa on a cylinder and 37 MPa on a cube. These strengths are characteristic strengths f_{ck} (defined at the fractile 5%) of hardened concrete at 28 days.

Cylinder strength occurs on specimens twice as high as they are wide, at least 110 mm in diameter (in France, cylinders are 160 mm in diameter and 320 mm in height). The cube strength f_{ck} is higher because, due to the interactions between the press platens and the specimen (friction locally generating a lateral pressure), the incidence is felt up to the middle of the cube; the stress state then has a triaxial character, which artificially improves its strength. Having a cylinder of height that is twice the diameter (or more) makes it possible to provide, in the center of the specimen, an area in quasi-uniaxial compression (see Figure 1.9(a)). From this f_{ck} value at the fractile 5%, which means that in a batch of specimens tested, 95% of them have a strength greater than f_{ck}, a number of characteristics are deduced according to Eurocode 2 (2004):

– the average value of compressive strength: $f_{cm} = f_{ck} + 8$ MPa;

– the calculation strength: $f_{cd} = f_{ck}/\gamma_c$ ($\gamma_c = 1.2$ for accidental situations, 1.3 for earthquakes, 1.5 for long-term or transient situations);

– the "instantaneous" modulus of elasticity: $E_{cm} = 2{,}200\ (f_{cm}/10)^{0.3}$ MPa;

– the "delayed" modulus (taking creep into account): $E_{c,eff} = E_{cm}(t_0)/(1 + \varphi(t_0))$, t_0 age at first loading and $\varphi(t_0)$ creep coefficient (current case, $\varphi(t_0) = 2$);

– the average tensile strength: $f_{ctm} = 0.3(f_{ck})^{2/3}$ MPa, if $f_{ck} < 50$ MPa and $f_{ctm} = 2.12\ \ln(1 + f_{cm}/10)$ MPa if $f_{ck} > 50$ MPa, from which we deduce the characteristic strength $f_{ctk,0.05} = 0.7\ f_{ctm}$ and the calculation strength $f_{ctd} = f_{ctk}/\gamma_c$ (γ_c as defined for the compression).

Thus, the designer has an arsenal of values that make it possible to approach a project for new structures within the framework of the corpus of rules established by the Eurocode and its national appendices. Eurocode 2 (2004) also provides relationships for estimating strength to early age, shrinkage, creep, etc.

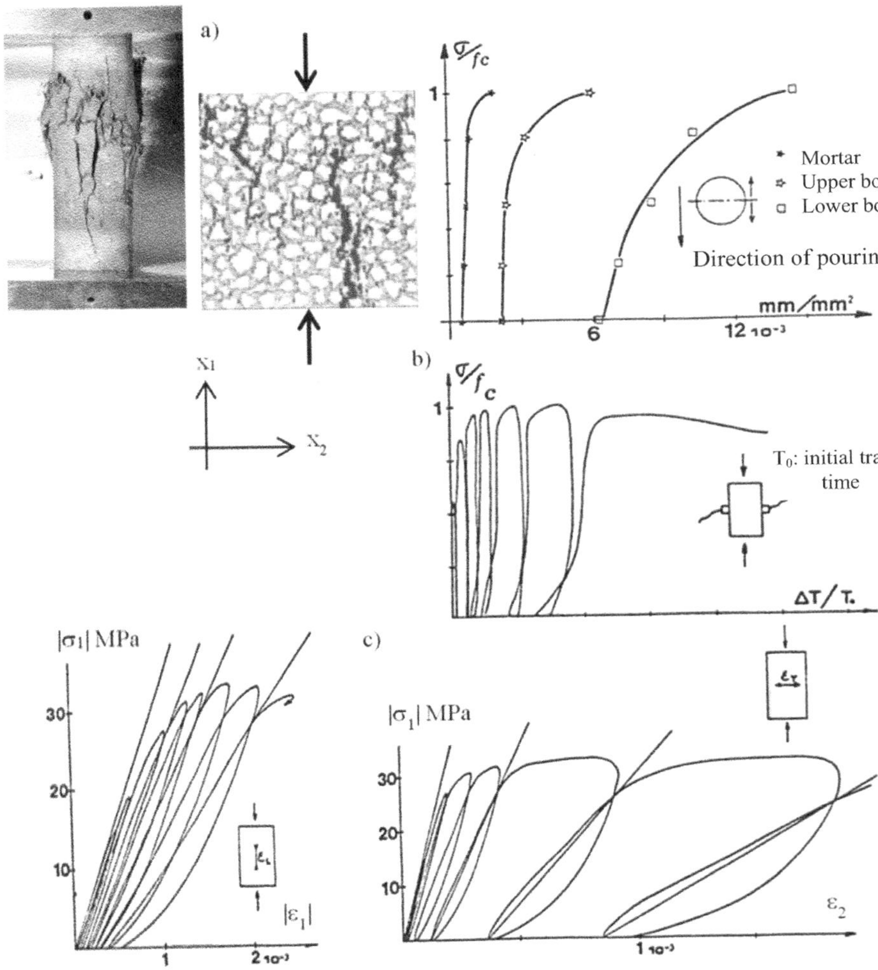

Figure 1.9. *Behavior of concrete under compression. a) Experimental cylinder failure and fracture facies on a meso-model; b) progression of microcracks within the mixture and evolution of ultrasonic waves in the transverse direction (x_2) of the sample; c) $|\sigma_l|-|\varepsilon_l|$ and $|\sigma_l|-\varepsilon_2$ curves, with discharges, showing the evolution of stiffness (from Mazars (1984))*

The use of the modeling strategies proposed in this book is outside the corpus of Eurocodes, but when certain test results are not available at the time of calculations, data from previous relationships may be needed. In this case, *it is the average strengths that are to be considered* and not the characteristic strengths that are values including a certain safety, and whose use would distance the results obtained from the reality that the present models must approach as closely as possible to carry out a relevant structural analysis.

With regard to the uniaxial compression behavior of ordinary hardened concrete, Figure 1.9 provides information on the overall microstructural and mechanical response. Indeed, we note that:

– the failure is obtained by multicracking with the orientation of the cracks being globally parallel to the direction of the load (see Figure 1.9(a));

– the monitoring of microcracking during the test leads to the observation of cracking present before loading, mainly in the lower bond of the largest grains (compared to the direction of pouring) and that it is around the grains that the cracking first progresses, before spreading in the matrix as it approaches failure (see Figure 1.9(b));

– the curve of evolution of the velocity of ultrasonic waves in the direction X_2 is characteristic of the opening of cracks, especially approaching the maximum stress and beyond, where a runaway phenomenon appears at the same time as failure (see Figure 1.9(a)).

The global curves $|\sigma_1| - |\varepsilon_1|$ and $|\sigma_1| - \varepsilon_2$ exhibit three stages (see Figure 1.9(c)):

– a quasi-elastic zone, 30–40% f_c (f_c = compressive strength), a period during which the initial microcracking does not progress; the modulus of elasticity E is constant, as is Poisson's ratio ν;

– a nonlinear zone with increasing stress (pre-peak), the cracks progress around the aggregates, and the modulus of elasticity E of the material decreases (see the slopes of the load-unload loops in Figure 1.9(c)), but the Poisson's ratio barely changes;

– a nonlinear zone with decreasing stress (post-peak), the cracks develop more and more in the matrix and open (see evolution of ultrasonic velocities in Figure 1.9(b)), transverse strains ε_2 are amplified and the apparent Poisson's ratio greatly increases: this is the failure (see Figure 1.9(a)).

The load–discharge curves (see Figure 1.9(c)) show, in addition to the evolution of the stiffness, the presence of hysteresis loops, which are characteristic of energy dissipation, mainly due to friction between the lips of cracks that are close. Also present are permanent strains at zero stress; the explanation that can be given is that they are the manifestation of the release of internal stress created during the maturation of the material (see section 1.2). Once these stresses have been released, the microcracked medium does not return to its reference state at zero stress, which generally results in residual strains.

NOTE 1.1.– The determination of the modulus of elasticity and Poisson's ratio is carried out from curves such as those presented in Figure 1.9(c). The details of obtaining them are given in the exercises part of the book (see section 9.1).

1.3.1.2. *Uniaxial tension*

Overall, the tensile strength of ordinary concrete is an order of magnitude lower than its compressive strength. For the design engineer, this is still a useful property, but is of less importance than compression.

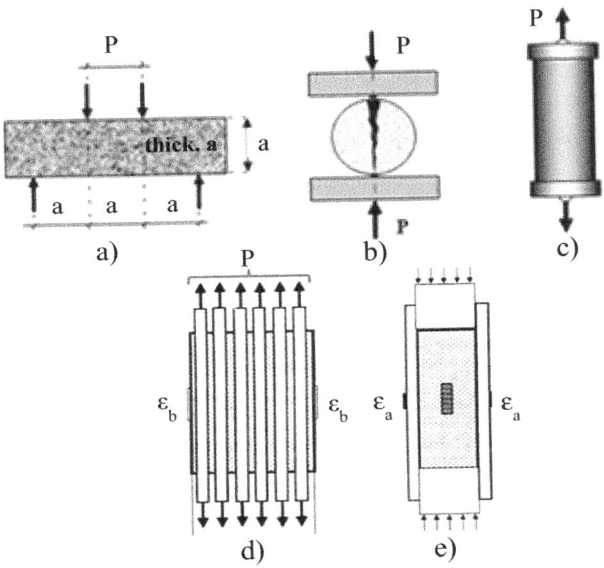

Figure 1.10. *Different types of tests for analyzing behavior. a) Tension by bending; b) tension by splitting; c) direct tension; d) and e) direct tension-compression to multicrack and analysis of the unilateral character of cracked concrete*

This is for several reasons:

– apart from a few structures such as dams, concrete is rarely used alone; it is either reinforced (the reinforcements being mainly present to overcome the deficiency of concrete tensile strength) or prestressed (the structure is placed in a state of permanent compression to avoid weaknesses due to tension);

– testing concrete under tension is less easy and therefore less widespread than the compression test. This led the coder, as indicated in section 1.3.1.1, to define the tensile properties according to the results of the compression tests.

This being said, the cracking of structures is totally controlled by the tensile properties and this is throughout the total life of the structure (from an early age to a long time). This has major consequences on the durability of structures and some specific properties, such as sealing for reservoirs and containment structures.

Figure 1.10 shows the variety of tests that are performed to test concrete under tension. From left to right:

– the tension by bending test, easy to perform on a prismatic specimen (in France, dimension a is 100 mm); the concrete cracks on the lower side by failure under tension, we suppose that the behavior is linear until the failure and, for the maximum load, we deduce the tensile stress at failure;

– the tension by splitting test, a compressive load applied on a diameter of the test cylinder, leads to a distribution of tensile stresses perpendicular to the diameter and, as before, assuming that everything happens linearly until failure, the tensile stress at failure is deducted from the maximum load;

– the direct tension test, generally carried out on a cylinder of constant or variable section, to locate the failure, which always appears as a crack perpendicular to the direction of the load;

– the test making it possible to cause multicracking to obtain more information on post-peak behavior. This same test makes it possible to exert, following tension (see Figure 1.10(e)), compression to analyze the unilateral nature of the concrete behavior under alternating loading (restoration of the stiffness with the closing of the cracks).

In the rest of this section, we will only analyze the behavior under direct tension of the last two types of tests, but we will return to the others (tension by bending and splitting) in Chapter 4 in regard to comparing the results of these tests, which must be done with correctness, within the framework of the identification of the parameters of the damage models.

Figure 1.11 gives the essential elements that make it possible to understand the particularities of concrete under direct tension, which will be useful to model its behavior.

Two main phases appear as follows:

– a quasi-linear phase up to the peak (see Figure 1.11(b)), where the material remains quasi-elastic linear;

– a post-peak phase, resulting from the instability created by the formation of a localization of microcracks that coalesce into a main crack, which causes a sudden drop in stress as the strains continue to grow, as well as a significant stiffness loss (see Figure 1.11(b)).

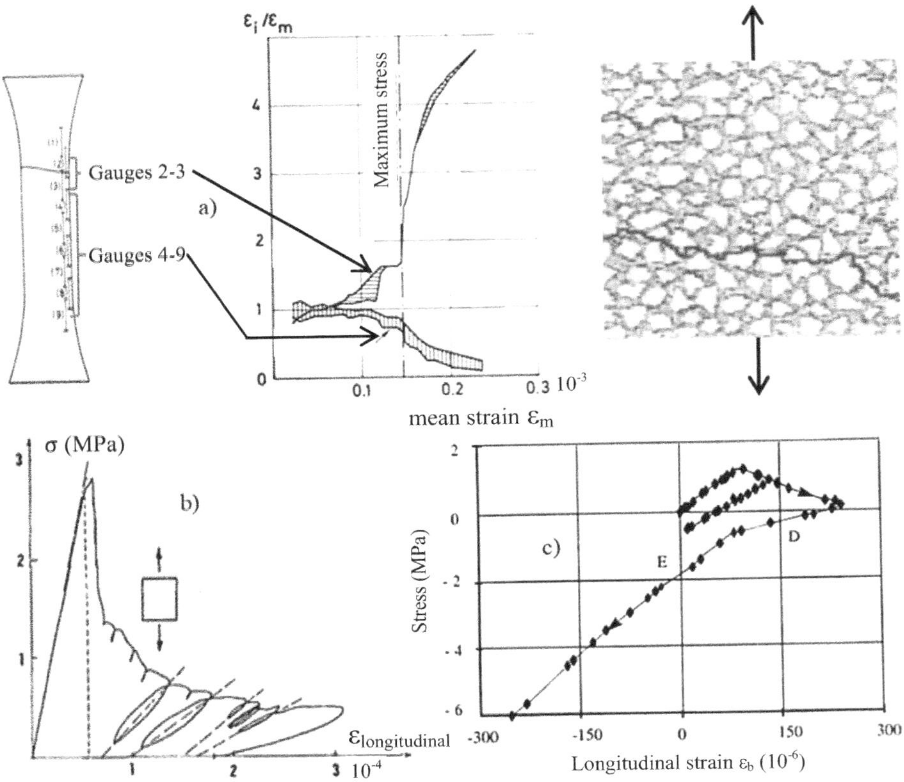

Figure 1.11. *Behavior under direct tension. a) Localization of strains and final cracking (according to Heilmann et al. (1969)); b) overall mechanical behavior (according to (Terrien 1980)); c) particularity of the behavior resulting from the test creating a multicracking and allowing alternate loading (according to Mazars et al. (1989))*

The crack is generally intergranular and globally oriented perpendicular to the direction of the force. The measurements made by Heilmann et al. (1969) (see Figure 1.11(a)) provide a better understanding of the localization process because of the measurement of strains by gauges at and around the crack. While the strains of gauges 2 and 3, through which the crack passes, progress compared to the average strain of the specimen ε_m, the strains measured by gauges 1, 4 and 9, which are around this zone, regress. This redistribution of the deformations inside the specimen is characteristic of the appearance of the localization, which is noticeable before the peak. The specimen is no longer in a uniform field (stress and strain) and becomes a structure that the models will have to know how to simulate.

In the so-called multicracking test (see Figures 1.10(d) and 1.10(e)), the tensile force is exerted on the parallelepipedal sample by means of aluminum rods glued to the large sides of this sample. Thus, these rods apply a state of uncontrolled strain on either side of the specimen, which prevents any premature localization. The material here is microconcrete, and at the end of the tensile sequence, multicracking appears (four to five cracks barely visible to the naked eye), indicating that the degradation is widely distributed inside the material; we will discuss diffuse damage here. A second sequence is then carried out by freeing the rods of all forces and applying compression directly to the two extremities of the specimen. The curve (see Figure 1.11(c)) results from a post-processing of the global force-strain curve concerning the assembly (ε_1 measured at the center of the specimen), from which the forces passing through the bars of the assembly must be deducted (these remain elastic throughout the entire test).

From these results, it can be seen that:

– we find the phases of tensile behavior already seen in the direct tension test: quasi-elasticity up to the peak, then beyond that a drop in stiffness and stress, less brutal than in the direct tension test;

– the subsequent compression phase highlights a restoration of the stiffness due to the closure of the cracks created under tension, which leads to almost regaining the initial stiffness and reducing the permanent strains resulting from the tension phase. This phenomenon is called the “unilateral effect” (Mazars et al. 1990); it plays a major role in the behavior of concrete structures under alternating loads, such as earthquakes.

In fact, this test is representative of the tensile effects on concrete in the presence of reinforcement, the role of which is played here by the bonded aluminum bars.

NOTE 1.2.– The compressive and tensile properties for a given concrete are dependent on the loading rate measured in strain rate. We will take up this point in detail in Chapter 6 during the studies on the simulation of rate effects. It is shown that the phenomenon is especially marked under tension for high velocity (the resistance is multiplied by 2, 3, 4, respectively, for $\dot{\varepsilon} = d\varepsilon/dt = 10^{-1}$/s, 1/s, 10^2/s) (below $\dot{\varepsilon} = 10^{-2}$/s, the increase is not significant and may not be taken into account). Under compression, the effects are much less marked (approximately 1.3 for $\dot{\varepsilon} = 1$/s). In Chapter 6, we show how these aspects must be considered to address situations of rapid dynamics and, in particular, of impact on structures.

NOTE 1.3.– Among the different types of concrete exists the so-called high-performance concrete (HPC), which is a highly compact material obtained by the addition of elements allowing for the deflocculation of the cement grains and by extension of the granular mixture toward very thin elements (limestone fillers, silica fume, etc.), which, in addition to better compactness, makes it possible to limit the quantity of water necessary for implementation. The resulting hardened material has significantly improved properties in strength and durability (drastic reduction in open porosity). Figure 1.12(a) shows an example of the behavior under compression of this type of concrete ($f_c = 105$ MPa, E = 51 GPa), which shows quasi-elastic behavior up to failure and very brittle behavior beyond that. The same concrete has a tensile strength, measured in the splitting test, of 6.5 MPa.

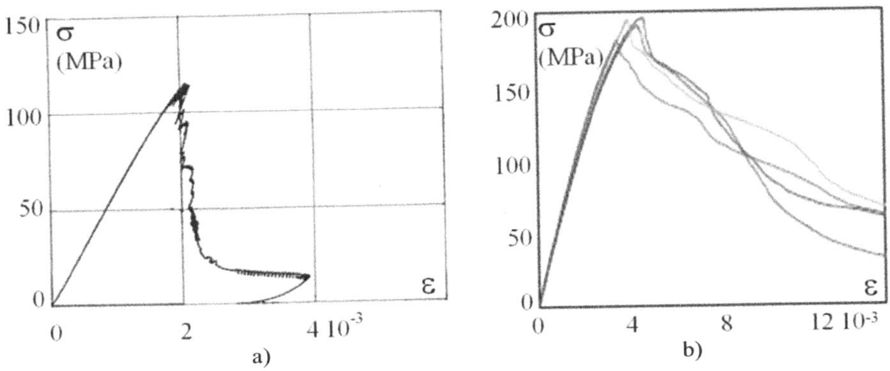

Figure 1.12. *Behavior in compression of special kinds of concrete: a) high-performance concrete (HPC); b) ultra-high-performance fiber-reinforced concrete (UHPFRC) (according to Jungwirth and Muttoni (2004)). For a color version of this figure, see www.iste.co.uk/mazars/damage.zip*

Figure 1.12(b) gives an example of another category of concrete: ultra-high-performance fiber reinforced concrete (UHPFRC), whose composition ensures good compactness and includes fibers that are generally, but not exclusively, metallic. Due to a very controlled manufacture, the compressive strengths reach 200 MPa and higher.

The behavior is characterized by:

– a range of elastic strain limited by the strength at extension of the cementitious matrix;

– a post-cracking domain characterized by the tensile strength of the fiber material obtained after the cracking of the matrix.

This leads to a very softened post-peak, which settles on large strains and gives it a kind of ductility.

The models presented in the following chapters focus on so-called "ordinary" kinds of concrete with an average strength of up to 60–70 MPa. Beyond that, we fall into the category of special kinds of concrete, and it goes without saying that the modeling of these concrete types must be adjusted in relation to those presented in this book.

1.3.2. *Concrete under multiaxial loading*

From the analyses made with respect to uniaxial loading, we remember in particular that the directions of cracking are:

– under tension, perpendicular to the direction of the force;

– under compression, parallel to the direction of the force.

The common denominator of these two situations is that the cracks are globally perpendicular to the directions of extensions ($\varepsilon > 0$). This crack opening mode is called "mode I".

Mode I promotes unstable behavior (present under compression beyond the peak and very marked under tension). The presence of a stress in one or two complementary directions changes the response of the material.

The addition of one or two tensile directions will favor the development of mode I in their respective orthogonal direction, which tends to weaken the strength in the initial direction, whether loaded under compression or under tension.

With the addition of compression, the cracking facies obtained are illustrated in Figure 1.13a (Kupfer et al. 1969):

– if the original stress is tension, this compression will add extensions in the direction of the tension, favor mode I and consequently weaken the corresponding strength;

– if the original stress is compression, this new compression will hamper the development of mode I in an orthogonal direction, which creates a kind of confinement in the bicompression plane that is as strong as this stress is intense, and the strength is increased. Figure 1.13(b) shows the effect of this additional compression on the variation in volume of the material.

With the addition of two compressions:

– if the original stress is tension, the observation is the same as that expressed for the addition of one compression, and the strength under tension is weakened;

– if the original stress is compression, depending on the relative intensity of the three compressions, the development of mode I is increasingly hampered, and we are then in the most favorable cases of strength. We will discuss this particular situation in detail in section 1.3.2.2 by considering two cases: where the extensions remain possible and where they are no longer, for which new modes of cracking in shear (mode II) will develop.

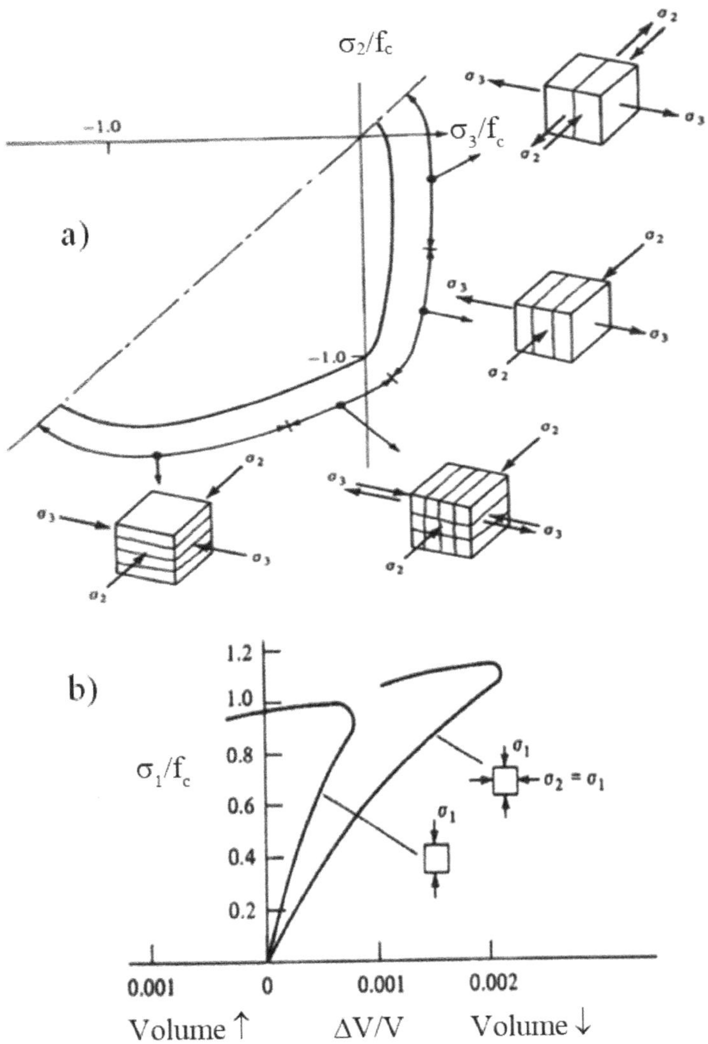

Figure 1.13. *Biaxial loading (σ_3 = 0). a) Facies of degradation under biaxial loading; b) influence of confinement on the volume variation ($\Delta v/v = \varepsilon_1 + \varepsilon_2 + \varepsilon_3$)*

1.3.2.1. *Behavior under biaxial loading*

A series of authoritative experimental results on this subject are those of Kupfer and Gerstle (1973). Performed on parallelepipedal specimens (200 × 200 × 50 mm) for three kinds of concrete of different uniaxial compressive strengths (f_c = 19.1 – 32.8 – 59.4 MPa). They show that the shape of the failure surface in the plane σ_3 = 0 (normalized by the compressive strength f_c) is very close, which is major information for establishing a model (see Figure 1.14(a)).

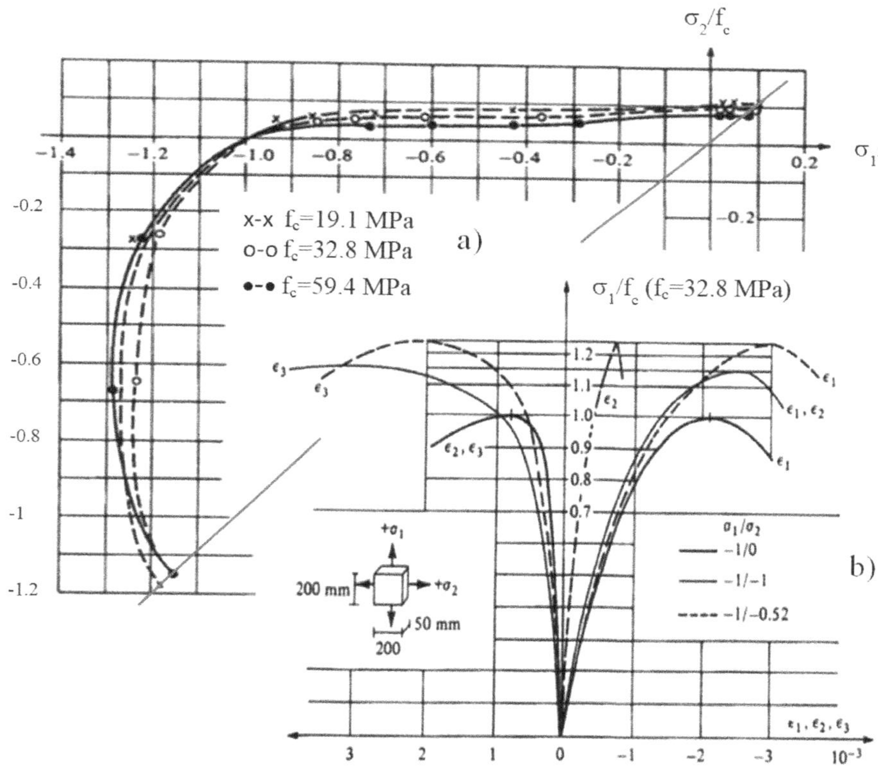

Figure 1.14. *Experimental results under biaxial loading (source: Kupfer et al. 1973): a) trace of the failure surface in the plane σ_3 = 0, for three different kinds of concrete; b) comparison of behavior for concrete f_c = 32.8 MPa and for three ratios σ_1/σ_2 (–1/0, –1/–1, –1/–0.52); we can note the increase in strength with the intensity of confinement. For a color version of this figure, see www.iste.co.uk/mazars/damage.zip*

This failure surface confirms the observations made on the incidence of the addition of orthogonal loading to uniaxial loading in the three sectors:

tension-tension (little difference with the uniaxial strength), tension-compression (weakening tensile strength) and compression-compression (increase in compressive strength due to the confinement created).

On the compression-compression sector, Figure 1.14(b) shows, for concrete f_c = 32.8 MPa, the behavior for three different bicompression ratios ((σ_1/σ_2 = −1/0, −1/−1, −1/−0.52).

It can be seen that it is for the ratio $\sigma_1/\sigma_2 = -1/-0.52$ that the strength is the greatest, approximately 1.25 times the uniaxial compressive strength, while it is only 1.15 when $\sigma_1 = \sigma_2$.

It is also noted that the failure processes are of the same nature for bicompression as for uniaxial compression: linearity up to approximately one-third of the strength, progressive nonlinearity with increasing stress up to the peak and then an amplification of the strains of extension (ε_3), which translates the formation and the opening of cracks in mode I in direction 3, which quickly leads to the collapse of the sample in layers parallel to the plane of loading.

1.3.2.2. *Behavior under triaxial loading of compression*

Figure 1.15 shows the two main types of triaxial tests:

– the test using a prism on which the three orthogonal stresses are exerted; this is done by a system of jacks and interface plates, which must allow all the strains during the test, which is not simple; it is said to be "true" triaxial because all of the combinations of compressions are possible on the three axes (see Figure 1.15(a));

– the test in a triaxial cell (see Figure 1.15(b)); inherited from soil mechanics, it is carried out on a cylinder that receives a lateral pressure under the effect of a pressurized fluid (sealed protection is necessary at the interface between the fluid and the specimen) and a vertical stress via an axial jack, which leads to an axisymmetric stress state ($\sigma_1 \neq \sigma_2 = \sigma_3$).

In both cases, the experimental system is complex and expensive. It requires real expertise for its use (the realization of measurements often poses problems). Triaxial cells with high confinement exist in some laboratories around the world and make it possible to carry out tests whose results can be used to simulate impacts on structures (the pressures under impactors are indeed highly triaxial and very intense).

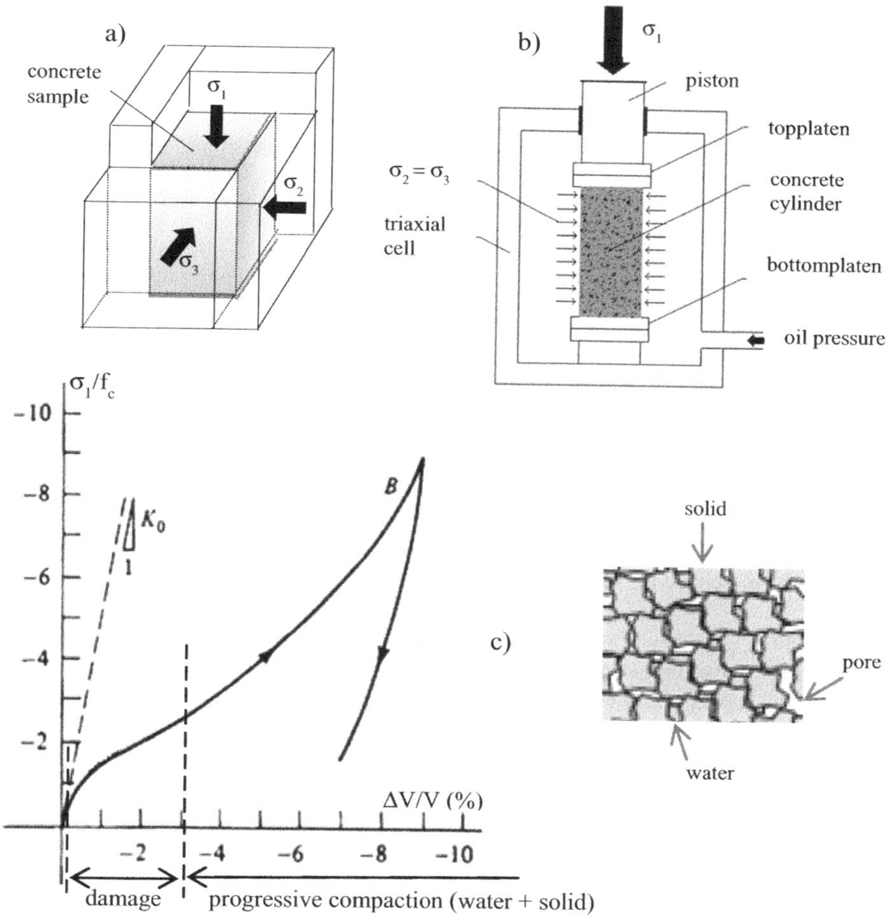

Figure 1.15. *Triaxial tests for concrete: a) "true" triaxial ($\sigma_1 \neq \sigma_2 \neq \sigma_3$); b) confinement cell ($\sigma_1 \neq \sigma_2 = \sigma_3$); c) response of a concrete specimen to a hydrostatic load ($\sigma_1 = \sigma_2 = \sigma_3$) showing, after the collapse of the microporous structure, the compaction of water and solid (according to Rouquand et al. (2007)). For a color version of this figure, see www.iste.co.uk/mazars/damage.zip*

1.3.2.2.1. Triaxial cell: hydrostatic test and axisymmetric triaxial test

With a high-performance triaxial cell, it is possible to perform a test such as ($\sigma_1 = \sigma_2 = \sigma_3$). Figure 1.15(c) shows an example of a response curve obtained presented in the coordinate system benchmark (σ_m, $\Delta V/V$). σ_m is the hydrostatic stress (or pressure) and $\Delta V/V = \varepsilon_1 + \varepsilon_2 + \varepsilon_3$ is the volume variation. The stress is normalized by the compressive strength.

The resulting behavior has three phases:

– a quasi-elastic phase; at the scale of the curve, it is small, but it makes it possible to show the bulk modulus:

$$K = \frac{(\sigma_1+\sigma_2+\sigma_3)}{3(\varepsilon_1+\varepsilon_2+\varepsilon_3)} = \frac{1}{3}\frac{E}{(1-2\vartheta)} \quad [1.1]$$

– a phase of progressive collapse of the microporous matrix that generates a nonlinearity with increasing stress and a degradation of the medium that we will call consolidation (in the sense of soil mechanics), a mode very different from that related to the creation of a crack;

– a phase after collapse, gradually making the porosity filled with air disappear, it is then the compaction of the water, initially present in some pores, and that of the solid phase (grains + matrix). As a result, the behavior is very much influenced by the water content of the medium: the greater it is, the more prematurely the inflection of the curve upward occurs (Pontiroli et al. 2010).

In Figure 1.16, we show the results of a series of tests carried out with the triaxial cell of the Giga machine (3SR Grenoble, France), whose performance is unique in the world of concrete research (Gabet et al. 2008).

Figure 1.16(a) shows the scheme of the confinement system (possibly up to 1 GPa), the axial jack having a capacity of 10 MN. On a specimen with 70 mm diameter and 140 mm height, the possible axial pressures are of the order of 2.5 GPa. The concrete considered is ordinary concrete; its failure in uniaxial compression appears for $f_c = 34$ MPa and a peak deformation of approximately 0.2%.

The principle of the test presented is as follows. Two loading phases are applied:

– a hydrostatic phase up to a given pressure value p ($= \sigma_m$);

– a phase with a constant lateral pressure $\sigma_2 = \sigma_3$ (= p) and axial charge σ_x increasing until failure (see Figure 1.16(b)).

This was done for six values of p, from 0 (single compression) to p = 650 MPa. Figure 1.16(b) gives the corresponding loading paths in the coordinate system deviator (σ_x– p) average pressure (p); in the first phase, the deviator is zero, and in the second phase, it progresses at the same time as the axial load (σ_x) progresses.

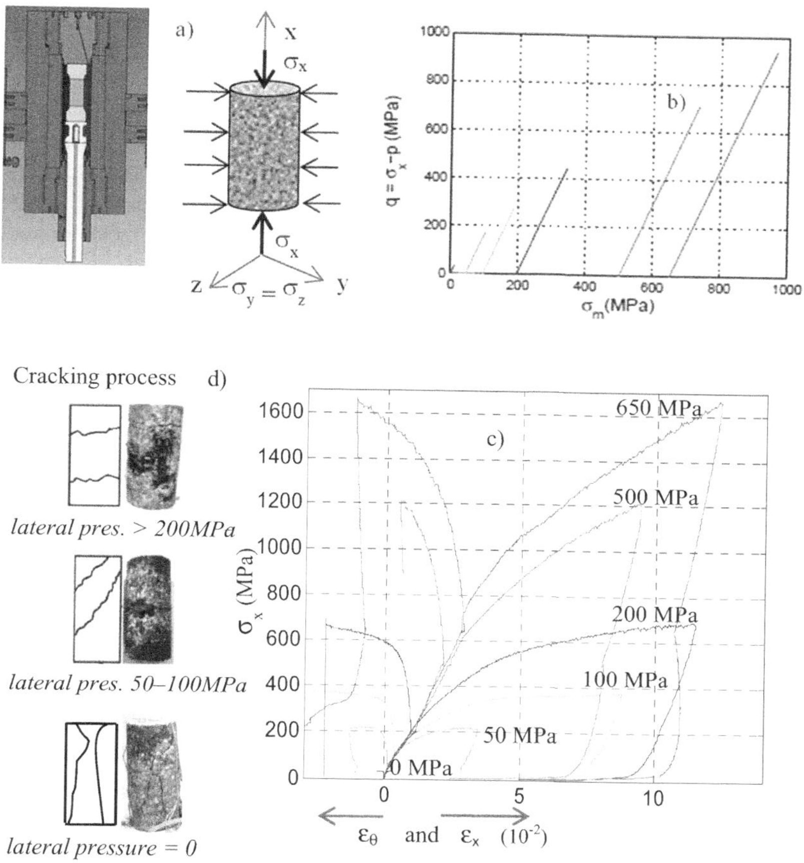

Figure 1.16. *Concrete response to triaxial loading. For a color version of this figure, see www.iste.co.uk/mazars/damage.zip*

COMMENTS ON FIGURE 1.16.– *(a) Scheme of the Giga confinement apparatus; (b) different loading paths: from a given hydrostatic pressure p, evolution of the deviator, $\sigma_x - p$, until failure (the lateral pressure $\sigma_y = \sigma_z$ being kept constant); (c) results for different values of the lateral pressure p from 0 to 650 MPa; (d) evolution of the crack facies from vertical in uniaxial compression to horizontal with a lateral stress greater than 200 MPa (from Gabet (2006)). ε_θ is the radial strain measured on the circumference of the specimen.*

The finding is that the value of the confinement pressure (p) has a major influence on the behavior obtained. Beyond 100 MPa of confinement, axial strains (ε_x) reach or exceed 10% and maximum stresses exceed 500 MPa.

These results show that confinement, by preventing any extension within the material and, consequently, any development of cracking in mode I, substantially increases the strength of concrete.

The analysis of the rupture is interesting. Figure 1.16(d) shows pictures of the crack facies obtained as a function of the intensity of the confinement. Starting from a globally vertical cracking for low confinements, we come to inclined cracks for medium confinements (50–200 MPa), which are the results of cracks in mode II (shear).

In this mode, due to confinement, the movements between the lips of cracks generate intense friction, and the energy to be spent to make them progress increases with p, giving the material a form of "ductility" (quasi-plateau with large strains).

Beyond a confinement to 200 MPa, we return to the observations made in the case of the hydrostatic test. The average stress σ_m reaches values that intensely activate the consolidation; bands where this consolidation is located are then created inside the specimen in the weakest areas and perpendicular to the strongest stress (ε_x).

Additionally, noteworthy in these phases of failure is the incursion of the transverse (or radial ε_θ) strain in the field of extensions, which favors the failure of the sample.

It is conceivable that, under these conditions, the modeling of such behaviors will have to include a number of specificities related in particular to degradation modes (mode I, mode II, consolidation) in relation to the possibilities or not of extensions and the existence of strong average stresses. Chapter 2 shows how these problems can be approached.

Loading type		**Local damage mode**	**Behavioral peculiarities**
A	– Uniaxial – Biaxial – Triaxial, with tension or low compression in at least one direction	Microcracking, first around the grains and then in the matrix, mainly in mode I; orthogonal cracks to the directions of extensions 	Existence of extensions in at least one direction. Behavior with peak and post-peak, modest strain value at failure (approximately 10^{-4} to 10^{-3})

B	Triaxial, superimposing hydrostatic pressure on a deviatoric pressure	Microcracking mainly in mode II with some confinement 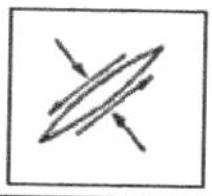	Extension not allowed in all three directions. Ductile behaviour with significatant strain at failure (approximately 10^{-2})
C	Hydrostatic pressure creating a strong confinement	Collapse of microporosity generating consolidation 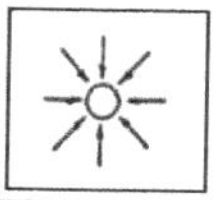	Several phases of behavior: softening first then stiffening when the liquid (water) and solid (matrix + grains) phases are under pressure
C'	Triaxial combining strong confinement and deviator	Combination of type B and C damage modes and, under strong deviator, appearance of compression band localization	Like type B, ductile tendency behavior with high strain

Table 1.2. *Summary of the different modes of damage according to the type of load and associated behavioral characteristics*

SUMMARY.– *Based on all the observations made throughout section 1.3, Table 1.2 proposes a classification of behaviors into four classes according to the type of load applied to a representative volume of concrete. These classes form the basis of the hypotheses and choices formulated to establish the behavioral models proposed in Chapter 3.*

2

Damage Concept and Its Applicability to Concrete

Damage to a material or structure expresses the gradual or sudden deterioration of the medium that forms them. Corresponding to a decohesion within the material or a local failure within a structure, the cause is usually local excessive loading. Causing nonlinearities of behavior, damage can lead to a gradual loss of strength or stability if redistribution of local or generalized stresses is no longer possible. This is why it is important for the modeling to have access to the local or global security reserve of a structure or system.

2.1. Damage concept

Catastrophic breaks in structures are spectacular and devastating. The Malpasset dam failure in December 1959 (see Figure 2.1(b)) destroyed a valley and killed 423 people; it was a concrete structure supported by a rocky massif that suddenly gave way locally, modifying the boundary conditions which are, as we will see in the chapters that follow, a major point in the holding of concrete structures. If we understand the logic of the cracking of concrete, that of metals, a ductile material if there is one, appears less trivial. Figure 2.1(a) shows a type of failure that appeared on *Liberty Ships* during World War II; more recently, in 1988, as a result of the sudden development of cracks, a Boeing 737 lost part of its cabin mid-flight, generating a sudden depressurization that did not prevent it from landing and saving all passengers. In general, it is found that, in metals, "persistent" loads (fatigue, creep) generate, under certain conditions, local decohesions that can, by coalescence, turn into cracks and cause failure. From these observations, scientists sought to analyze the phenomena and propose concepts and models to address simulations likely to predict the behavior of materials to avoid these sudden failures.

a)

b)

Figure 2.1. *a) Sudden failure of a Liberty ship during the Second World War (source: unknown); b) failure of the Malpasset dam (France, December 2, 1959), following a failure of a rock support (source: J. Mazars). For a color version of this figure, see www.iste.co.uk/mazars/damage.zip*

2.1.1. *Miner cumulative damage law*

Prior to the prediction of the fatigue failure of metal structures, tests are carried out on specimens to which sinusoidal loads are applied. Three parameters matter: the amplitude of the applied stress, the average stress and the number of cycles to failure N_F. Figure 2.2 proposes a test with alternating loading (zero mean stress) and the application of two series of cycles of different amplitudes, $2\sigma_1$ on N_1 cycles and $2\sigma_2$ on N_2 cycles.

Previous tests have shown that under an amplitude of $2\sigma_1$, failure appears at the end of the N_{F1} cycles, while for an amplitude of $2\sigma_2$, failure appears for the N_{F2} cycles. In such a context, Miner (1945) envisioned the following concept:

– to characterize the evolution of the degradation of the material until failure, he proposed the use of the damage variable D, which is 0 when the material is healthy and 1 at failure;

– for a given fatigue test, of which we know the number of cycles to failure N_F with amplitude and fixed average stress, the damage evolves according to the following rule: each cycle brings an increment of damage $\delta D = 1/N_F$, as well as for a given number of cycles $N : D = N/N_F$;

– he proposed, in addition, a linear cumulation rule which, for the case shown in Figure 2.2, leads after the first series of N_1 cycles with amplitude $2\sigma_1$, to $D_1 = N_1/N_{F1}$ and after N_2 supplementary cycles with amplitude $2\sigma_2$, to $D = N_1/N_{F1} + N_2/N_{F2}$; the failure ($D = 1$) occurring for $N_2 = N_{2R}$, that is, for a total number of cycles at failure, $N_F = N_1 + N_{2R}$.

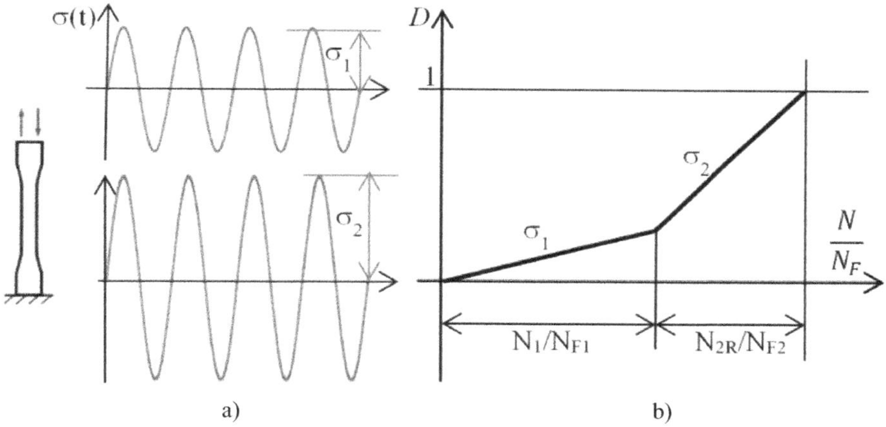

Figure 2.2. *Fatigue test on metal: a) application of two series of alternating cycles (first series: N_1 half-amplitude cycles σ_1; second series: N_2 half-amplitude cycles σ_2); b) evolution of damage during the two series of cycles (Miner 1945)*

Thus, the rule of linear cumulation is written as:

$$D = \sum_{i=1}^{p} N_i / N_{Fi} \qquad [2.1]$$

N_i is the number of cycles of sequence i, and failure appears when $D = 1$.

A criticism of this law is that it does not take into account the order of the sequences or the value of the average stress, which in certain cases appears necessary, particularly when there is a large variation of these parameters. To this end, nonlinear cumulation laws have been proposed (see Lemaitre and Chaboche (1978) or Jardin et al. (2009)).

2.1.2. Katchanov's progressive damage law

Creep is a phenomenon observed in most materials when loading stresses persist over time. At constant stress, the material deforms, which is the creep strain, and depending on the loading conditions, this may stabilize itself and sometimes cause failure.

Figure 2.3 shows the results obtained on a metal at high temperature. The curves in Figure 2.3(a) show the logarithmic evolution for two different heating temperatures. In both cases, three phases appear:

– phase I, which is primary creep: the strain rate decreases over time;

– phase II, which is secondary creep: the strain rate is constant;

– phase III, which is tertiary creep: the strain rate increases more or less suddenly, and this phase can lead to failure.

Phases I and II correspond to metals to dislocation activation with temperature and tertiary creep to the creation of progressive decohesions at grain boundaries (see Figure 2.3(b)). This decohesion progresses into microcracking and by coalescence into cracking, which causes failure.

As early as 1929, Norton proposed a law to describe secondary creep: $d\varepsilon/dt = a(\sigma)^n$, where a and n are two characteristic coefficients of the material depending on the temperature conditions of the test. Taking up Norton's law, Kachanov (1958) proposed introducing the concept of damage into this law to describe tertiary creep:

$$d\varepsilon/dt = A(\sigma/(1-D))^n \quad [2.2]$$

$(1 - D)$ is a term that should be understood as being representative of the reduction in the active surface of the sample cross-section due to microcracking: if S is the cross-sectional area of the element tested, then this, reduced by the effects of cavities and microcracks, is equal to $\tilde{S}$ (effective resistant area or section) and:

$$\tilde{S} = S(1-D) \quad [2.3]$$

$D = 0$ when the surface is totally active and therefore undamaged and $D = 1$ when the surface is totally broken.

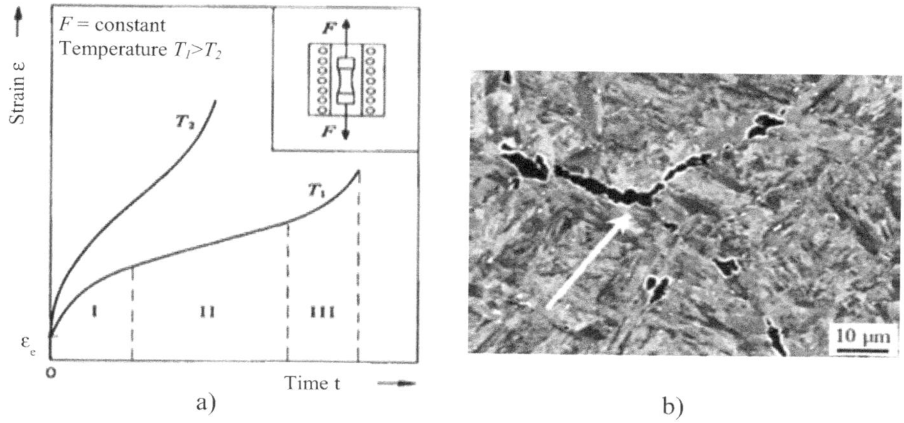

Figure 2.3. *a) Creep tests and curves on metal for two different temperatures; b) microcracking observed during tertiary creep (source: Centre des matériaux de MINES ParisTech 2010)*

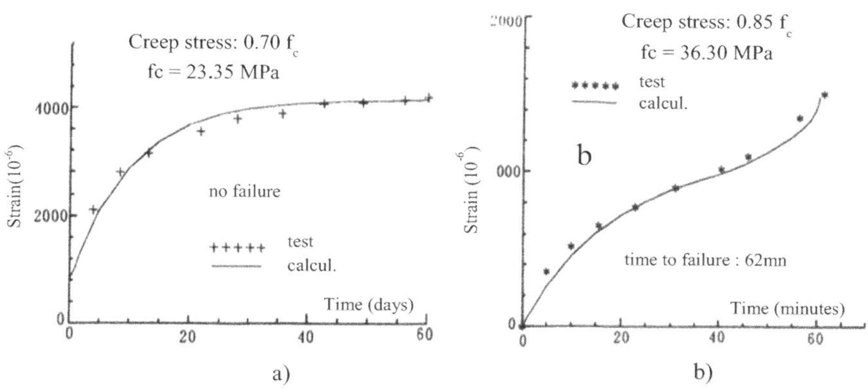

Figure 2.4. *Creep of an ordinary kind of concrete under compression: a) under a stress $\sigma = 0.7fc$, beyond 40 days, the creep evolves into a plateau; b) under a stress $\sigma = 0.85fc$, tertiary creep appears at approximately 40 min (according to Smadi et al. (1985))*

At a given moment, the stress in the effective section is worth:

$$\tilde{\sigma} = F/\tilde{S}, \text{ that is, } \tilde{\sigma} = F/S(1 - D) \qquad [2.4]$$

$\tilde{\sigma}$ is called the effective stress. The relationship that binds the usual stress: σ ($= F/S$) and the effective stress is then written as:

$$\tilde{\sigma} = \sigma/(1 - D) \quad [2.5]$$

It is therefore this concept that was introduced by Kachanov (see equation [2.2]) in the law previously proposed by Norton to describe tertiary creep.

NOTE 2.1.– Concrete, as we saw in Chapter 1, is a viscous material and therefore sensitive to creep. However, the mechanisms behind it are of a different nature than those associated with metals. If the same tendencies are found for primary creep, tertiary creep can only appear for loads very close to that of static failure (Figure 2.4(b) indicates this for 85%). Under conventional load conditions, secondary creep is established following a power curve with an asymptotic tendency (the phenomenon saturates in the long term (see Figure 3.3(a))).

2.1.3. *Elasticity-damage coupling*

2.1.3.1. *Analysis of the 1D case*

After the Russian school (Kachanov 1958; Rabotnov 1969), many authors around the world have worked on these damage concepts. Here, we have borrowed, in particular from Lemaitre and Chaboche (1990), certain points of the presentations that follow in this chapter.

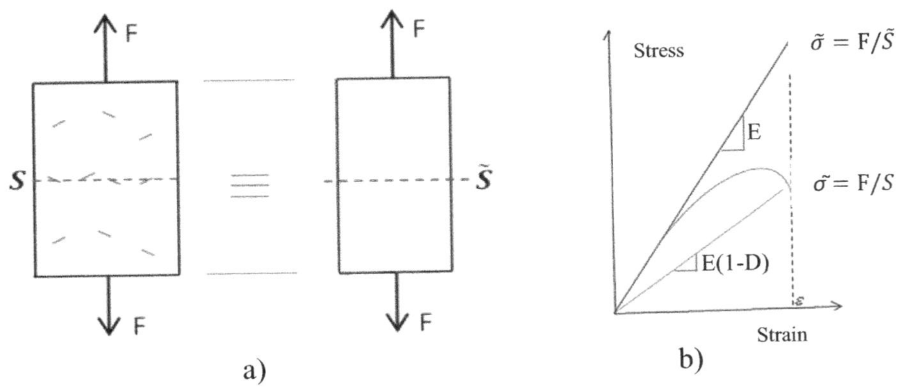

Figure 2.5. *Elastic-damage behavior: a) equivalence in strain; b) associated behavior curve*

In a material with a brittle tendency, such as rocks, ceramics and concrete, the nonlinearities under load are, as we saw in Chapter 1, associated with local degradations of the micro- or macrocrack type. These nonlinearities are thus good candidates for a description using the damage concept.

Consider a specimen of this type of material and assume that it is exclusively elastic and damageable. The cross-sectional area of the original specimen is S and it is subjected to tensile loading.

Let us return to the concept of effective section $\tilde{S}$ seen in section 2.1.2 and consider in parallel to the specimen of section S (which is elastic, damageable), an equivalent specimen, in the sense of the strains, of section $\tilde{S}$, which is only elastic, but with an evolving section ($\tilde{S}$ decreases with D). This concept is called the principle of equivalence in strain.

If at a given moment of loading the damage is D, then the effective section, that of the specimen that remains elastic, is $\tilde{S} = S(1 - D)$. The two specimens have the same strain at all times, then:

– for the elastic specimen, the effective stress is expressed by:

$$\tilde{\sigma} = F/\tilde{S} = E\varepsilon \qquad [2.6]$$

– for the elastic-damageable specimen, the usual stress is written as:

$$\sigma = F/S = F(1 - D)/\tilde{S} = E(1 - D)\varepsilon \qquad [2.7]$$

Thus, everything happens as if the initial modulus of elasticity E was affected by the factor $(1 - D)$. This modulus evolves between E ($D = 0$, healthy material) and 0 ($D = 1$, broken material). This leads to nonlinear evolving behavior (see Figure 2.5(b)). On this curve, the effective stress (on the initial elasticity line) and the usual stress (at the intersection of the nonlinear curve and the elasticity line associated with $E(1 - D)$) have been positioned. The behavior described is thus an elastic behavior with variable elasticity.

2.1.3.2. *Generalization to the 3D case*

This uniaxial description must be extended to a three-dimensional description.

If we rely on the evolutions noted in Chapter 1 for the concrete under various loadings, the deterioration (the damage) is of anisotropic nature. Take the example of uniaxial compression (see section 1.3.1.1), which generates microcracking mainly oriented parallel to the direction of the load.

Figure 2.6 summarizes this aspect by considering an orthotropic description of the damage (anisotropy with isotropic properties in plane 2-3). Such a state can be described using two damage variables, D_1 characterizing the effects in direction 1 and D_2 characterizing the effects in directions 2 and 3 (see Figure 2.6(a)). The corresponding variations of D_1 and D_2, resulting from experiments on cylindrical specimens, are given for an ordinary kind of concrete (see Figure 2.6(b)).

However, describing all mono-, bi-, or three-dimensional situations involving traction and compression would require a complete anisotropic description (21 independent elastic characteristics). Under these conditions, the damage becomes tensorial, and the effective stress is described via an operator that is a tensor of order 4:

$$\tilde{\sigma} = \mathbf{M_D} : \sigma \qquad [2.8]$$

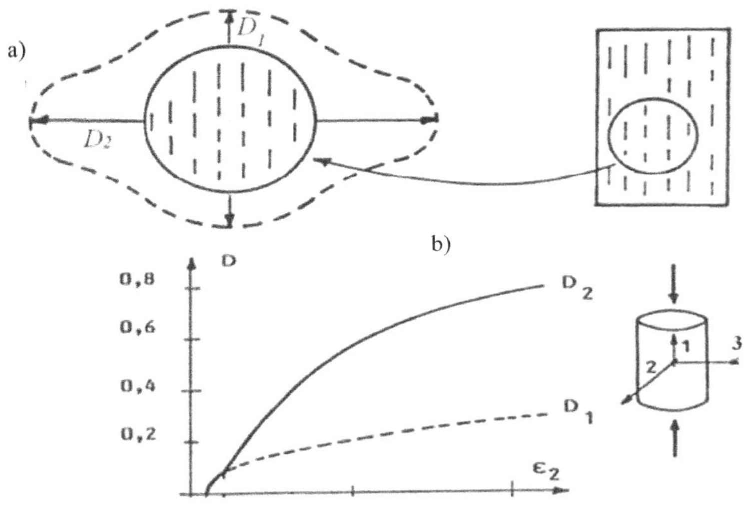

Figure 2.6. *Anisotropy of the damage of concrete in uniaxial compression: a) bulb of damage around an area; b) orthotropic description of the damage including two variables D_1 (direction 1) and D_2 (directions 2 and 3)*

Many proposals for anisotropic damage models can be found in the literature (Halm and Dragon 1998; Lemaitre and Desmorat 2005; Murakami 2014).

In addition, as we saw in Chapter 1, the damage generates secondary effects (creation of permanent deformations, consolidation due to a strong confinement depending on the water content, etc.). To do this, some authors have proposed models coupling damage and plasticity (Jason et al. 2006). Everything is possible, but the price to pay is the complexity that increases with the desired details of phenomenological description.

The purpose of this book is to provide a simplified and operational description of concrete damage. Thus, for this description, we will favor several fundamental hypotheses:

– the material is considered initially isotropic, which is acceptable for concrete on a macroscopic scale (volume element of a decimetric order);

– the damage is considered isotropic, which means that the damage is likened to spheres that grow within the material; this assumption makes it possible to preserve the properties of isotropy;

– the material is considered initially linear elastic, and the damage conserves this type of property.

We will see that, under certain conditions (uniaxial description, for example), it will be possible to override some of these assumptions, but this will remain the exception.

Thus, for a three-dimensional situation but with isotropic damage, the ratio $S/\tilde{S} = (1 - D)$ does not depend on the orientation, and the relationship between effective stress tensor and "usual" stress tensor is of the same nature as that seen in equation [2.5] for the one-dimensional case. In this case, the operator, seen in equation [2.8], reduces to the scalar $1/(1 - D)$ and:

$$\tilde{\boldsymbol{\sigma}} = \boldsymbol{\sigma}/(1 - D) \quad [2.9]$$

2.2. Theoretical bases of damage mechanics

Building a model in mechanics requires that it respects some principles, at the risk of violating certain laws of physics and therefore making it unsuitable for good simulation. In this section, the presentation aims to present the framework of the thermodynamics of irreversible processes (science that deals with energy transfers between two thermomechanical states), which is based on two major principles, one relating to the conservation of energy and the other to the production of entropy.

2.2.1. *Elasticity-damage coupling*

At a given instant, the state of a material is described by state variables, which are classified into two categories: observable state variables (the strains ε, the temperature T) and the internal state variables (in our case, the damage D). Each of them has its associated variable (with the strains, the stresses σ; with the temperature, the entropy s; with the damage, the rate of energy restitution Y). Entropy is a term that accounts for the dispersion of energy within a system.

The first principle of thermodynamics translates the conservation of energy and connects, for a given body subjected to thermomechanical loading, the internal energy of the material U, the kinetic energy of the body if there is movement K, with the power of the external forces P_e and the rate of heat received Q:

$$\frac{d(U+K)}{dt} = P_e + Q \qquad [2.10]$$

The axiom of the equilibrium of the powers involved states that, in a dynamic equilibrium, the power of the quantities of acceleration (P_a) is equal to the sum of the power of the internal forces ($P_i = -\int \boldsymbol{\sigma}:\dot{\boldsymbol{\varepsilon}} dV$) and the power of the external forces (P_e). By noting (Lemaitre and Chaboche 1990) that $P_a = dK/dt$ and that the rate of heat received Q by the volume considered is the sum of the heat created by external actions (volume density of internal production r) and the heat received by conduction across the boundary of this volume (characterized by the current vector of heat $\vec{q}$), this relationship transformed into a local expression (at a point of the body considered) becomes:

$$\rho\frac{du}{dt} = \boldsymbol{\sigma}:\dot{\boldsymbol{\varepsilon}} + r - div\vec{q} \qquad [2.11]$$

$\dot{\boldsymbol{\varepsilon}} = \frac{d\varepsilon}{dt}$, ρ is the density and u is the specific internal energy.

The second law of thermodynamics postulates that the rate of entropy production is always greater than or equal to the rate of heat received, divided by the temperature. From the definitions seen for the first principle, this one is written in local form as:

$$\rho\frac{ds}{dt} + div\frac{\vec{q}}{T} - \frac{r}{T} \geq 0 \qquad [2.12]$$

By deferring the value of r from equation [2.11], we obtain:

$$\rho\frac{ds}{dt} + div\frac{\vec{q}}{T} - \frac{1}{T}\left(\rho\frac{du}{dt} - \boldsymbol{\sigma}:\dot{\boldsymbol{\varepsilon}} + div\vec{q}\right) \geq 0 \qquad [2.13]$$

By introducing the specific free energy, or Helmotz energy ($U - TS$), which at the local scale is expressed as $\Psi = u - Ts$, we obtain the following by derivation:

$$\frac{d\Psi}{dt} = \frac{du}{dt} - T\frac{ds}{dt} - s\frac{dT}{dt}, \text{ that is, } T\frac{ds}{dt} - \frac{du}{dt} = -\left(\frac{d\Psi}{dt} + s\frac{dT}{dt}\right) \qquad [2.14]$$

By transferring to equation [2.13] and using the mathematical identity, $div\left(\frac{\vec{q}}{T}\right) = \frac{div\vec{q}}{T} - \frac{\vec{q}\,\overrightarrow{grad}\,T}{T^2}$, this result leads to:

$$\boldsymbol{\sigma}: \dot{\boldsymbol{\varepsilon}} - \rho(\frac{d\Psi}{dt} + s\frac{dT}{dt} - \vec{q}\,\frac{\overrightarrow{grad}\,T}{T}) \geq 0 \quad [2.15]$$

This relationship condenses the two principles of thermodynamics. It is called the "Clausius-Duhem inequality". Respecting it in an irreversible process of change of state in the mechanics of materials, such as damage, is necessary.

To help with its resolution, intrinsic dissipation is generally decoupled from thermal dissipation by imposing:

$$\boldsymbol{\sigma}: \dot{\boldsymbol{\varepsilon}} - \rho(\frac{d\Psi}{dt}) \geq 0 \text{ and } s\frac{dT}{dt} - \vec{q}\,\frac{\overrightarrow{grad}\,T}{T} \geq 0 \quad [2.16]$$

2.2.1.1. *Application to linear elasticity*

The Helmotz free energy is a state function whose variation makes it possible to obtain the useful work likely to be provided by a closed thermodynamic system. The specific free energy (per unit mass) depends on the state variables (here, $\boldsymbol{\varepsilon}$).

The conditions of isotropy and linearity require that the specific free energy (or thermodynamic potential) is a positive definite quadratic function of the components of the strain tensor (Lemaitre and Chaboche 1990):

$$\Psi = \frac{1}{2\rho}\boldsymbol{\Lambda}: \boldsymbol{\varepsilon}: \boldsymbol{\varepsilon} \quad [2.17]$$

Ψ thus defined, a complementary law appears, called the state law, which connects the observable state variable $\boldsymbol{\varepsilon}$ and its associated variable σ:

$$\boldsymbol{\sigma} = \rho\frac{d\Psi}{d\varepsilon} = \boldsymbol{\Lambda}: \boldsymbol{\varepsilon} \quad [2.18]$$

This relationship is the behavior law of the linear elastic material, and $\boldsymbol{\Lambda}$ is the matrix of the elastic coefficients, which, as presented in section A.1.1, for an isotropic material, is written as:

$$\boldsymbol{\sigma} = \lambda Tr(\boldsymbol{\varepsilon})\mathbf{1} + 2\mu\boldsymbol{\varepsilon} \quad [2.19]$$

λ and μ are Lamé coefficients: $\lambda = \frac{\vartheta E}{(1+\vartheta)(1-2\vartheta)}$, $\mu = \frac{E}{2(1+\vartheta)}$ and **1** is the unit tensor.

It is noted that by considering equations [2.16] and [2.18], the case of linear elasticity verifies the equation of Clausius–Duhem without any problem because:

$$\boldsymbol{\sigma}:\dot{\boldsymbol{\varepsilon}} - \rho\left(\frac{d\Psi}{dt}\right) = \boldsymbol{\sigma}:\dot{\boldsymbol{\varepsilon}} - \rho\left(\frac{d\Psi}{d\varepsilon}\right):\dot{\boldsymbol{\varepsilon}} = 0 \qquad [2.20]$$

2.2.2. *Isotropic damage theory*

Starting from the corpus of the previous equations, let us now see what this implies for the damage. To simplify the matter, in this section we only consider the situation announced beforehand, that of the isotropic behaviors coupling elasticity and damage. Moreover, we will only consider processes that occur at constant temperature.

We have seen previously that, in the case of isotropic damage, the relationship between the effective stress $\tilde{\boldsymbol{\sigma}}$ and the usual stress $\boldsymbol{\sigma}$ involves the $1/(1-D)$ ratio. By analogy with the thermodynamic presentation of linear elasticity, let us choose the following thermodynamic potential:

$$\Psi = \frac{1}{2\rho}\Lambda(1-D):\boldsymbol{\varepsilon}:\boldsymbol{\varepsilon} \qquad [2.21]$$

The state law that links the variable $\boldsymbol{\varepsilon}$ to its associated variable $\boldsymbol{\sigma}$ then leads to:

$$\boldsymbol{\sigma} = \rho\frac{\partial\Psi}{\partial\varepsilon} = \Lambda(1-D):\boldsymbol{\varepsilon} \qquad [2.22]$$

This relationship is in coherence with the concept of effective stress concerning the equivalent linear elastic material (see Figure 2.5) and whose behavior is given in equation [2.18]:

$$\tilde{\boldsymbol{\sigma}} = \frac{\sigma}{(1-D)} = \Lambda:\boldsymbol{\varepsilon} \qquad [2.23]$$

The damage variable is also a state variable because it is associated with the state of the material at a given time. It is an internal variable, unlike strain, which is an observable variable because it is measurable.

By analogy with the strain, whose associated variable is the stress, we associate with the internal variable D, an associated variable Y defined by:

$$Y = \rho\frac{\partial\Psi}{\partial D} = -\frac{1}{2}\Lambda:\boldsymbol{\varepsilon}:\boldsymbol{\varepsilon} \qquad [2.24]$$

Y is, to the nearest sign, the variation in elastic energy generated by a variation in damage with constant stress. Y is the rate of restitution of elastic energy.

Moreover, using the Clausius–Duhem inequality for the intrinsic part (see equation [2.16]), we obtain:

$$\left(\boldsymbol{\sigma} - \rho\frac{\partial \Psi}{\partial \varepsilon}\right) : \dot{\boldsymbol{\varepsilon}} - \rho\frac{\partial \Psi}{\partial D}\dot{D} \geq 0 \qquad [2.25]$$

By taking into account the equations of state [2.22] and [2.24], we finally obtain the relationship that must be verified so that the two fundamental principles of thermodynamics are respected:

$$-Y\dot{D} \geq 0 \qquad [2.26]$$

SUMMARY.– *What can we learn from all these results?*

– *From equations [2.19] and [2.22], we obtain:*

$$\boldsymbol{\sigma} = \lambda Tr\big((1-D)\boldsymbol{\varepsilon}\big)\mathbf{1} + 2\mu(1-D)\boldsymbol{\varepsilon}$$

$$= \lambda(1-D)Tr(\boldsymbol{\varepsilon})\mathbf{1} + 2\mu(1-D)\boldsymbol{\varepsilon} \qquad [2.27]$$

– *However,* $\lambda = \frac{\vartheta E}{(1+\vartheta)(1-2\vartheta)}$ *and* $\mu = \frac{E}{2(1+\vartheta)}$. *Thus, at fixed D, everything happens as if we had used an elastic material whose modulus of elasticity would be E(1 – D) and the Poisson's ratio* ν.

– *Since –Y, given by equation [2.24], is a positive definite quadratic form of* $\boldsymbol{\varepsilon}$, *the relationship that translates the two principles of thermodynamics dictates*:

$$\dot{D} \geq 0 \qquad [2.28]$$

As defined, the damage can only increase or be constant. This fact is a guideline to be respected in regard to defining the damage evolution laws.

NOTE 2.2.– As we will see in Chapter 3, other definitions are possible for damage. In a more general case, to describe it, we can deal with several internal state variables (V_i). This leads to a thermodynamic potential that takes the form $\Psi = \Psi(\boldsymbol{\varepsilon}, V_i)$, from which we deduce, by the state laws, the corresponding associated variables:

$$\boldsymbol{\sigma} = \rho\frac{\partial \Psi}{\partial \varepsilon} \text{ and } A_i = \rho\frac{\partial \Psi}{\partial V_i} \qquad [2.29]$$

For the Clausius–Duhem relationship (see equation [2.16]), this leads to:

$$\left(\sigma - \rho\frac{\partial \Psi}{\partial \varepsilon}\right) : \dot{\varepsilon} - \sum_i \rho \frac{\partial \Psi}{\partial V_i}\dot{V}_i \geq 0, \text{ that is, } -\sum_i \frac{\partial \Psi}{\partial V_i}\dot{V}_i \geq 0 \qquad [2.30]$$

The concept is therefore the same; it simply completes the conditions that the laws of evolution of the chosen internal variables must respect.

NOTE 2.3.– In Chapter 7, we will discuss the damage to concrete at an early age. As seen in Chapter 1, the maturation of concrete is underpinned by an exothermic hydration reaction. Consequently, the temperature evolution that appears during the process requires completing the thermodynamic potential given by equation [2.21] to take this fact into account.

2.2.3. *Damage threshold and notion of loading surface*

In an initially elastic material, the damage will evolve beyond a threshold, and the condition that the evolution beyond this threshold must respect is given to us by equation [2.27]. Thus:

– In the case of uniaxial loading in compression σ_1 in the direction X_1 (see Figure 2.7(a)), damage appears when $\varepsilon_1 = \varepsilon_{c0}$. In the same way, under tension, the damage begins when $\varepsilon_1 = \varepsilon_{t0}$. ε_{c0} and ε_{t0} are the thresholds of damage under compression and under tension, respectively, expressed here in strain.

– In a multiaxial case, a single component does not make it possible to define the threshold; also, one generally calls upon a scalar variable, which is a function of the state of loading of the material, expressed starting from the components of the tensor of the strains (it is what we chose for the concrete, but it can also be the stress tensor). Beyond the threshold, this variable, which we will call $p(\boldsymbol{\varepsilon})$, will be the pilot variable of the evolution of the damage, $D = g(p, p_0)$.

Let p_0 be the value of p at the initial damage threshold ($p_0 = \varepsilon_{c0}$ or ε_{t0} in the uniaxial case of Figure 2.7(a)). In the space of strains, the function that defines the initial threshold: $f(p, p_0) = p - p_0 = 0$ is a surface. The surface that we imagine for the moment should be arbitrary (see Figure 2.7(b)), but must include the thresholds ε_{c0} and ε_{t0} of the uniaxial situation.

Thus, during loading, three cases arise:

– if $p < p_0$, the loading point is inside the surface $f(p, p_0) = 0$, then the material remains elastic;

– if $p = p_0$, the loading point is on the surface $f(p, p_0) = 0$, then the damage is able to appear;

– if the loading continues, then the point of loading involves the surface with it and creates a new threshold, which is written as $f(p, k_D) = p - k_D = 0$. k_D is the maximum value of p reached during the entire loading phase, and the function $f(p, k_D) = 0$ is called the loading surface.

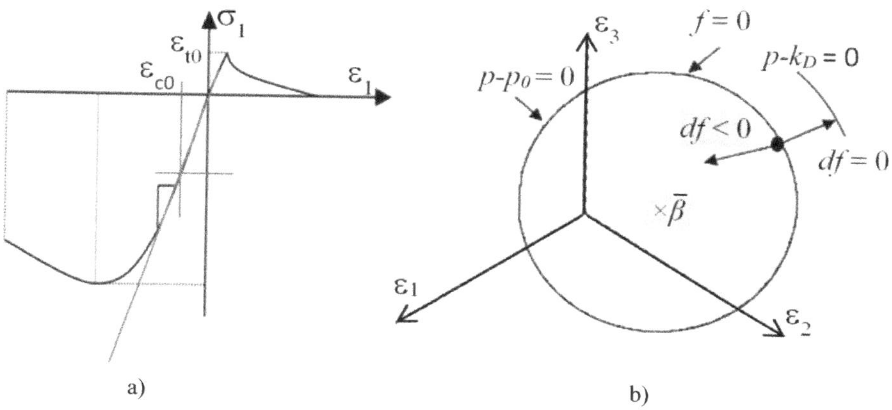

Figure 2.7. *Initial threshold of damage and its evolution: a) uniaxial situation; b) initial threshold and loading surface in strain space*

In this context, the charge–discharge criterion is expressed mathematically as follows (see Figure 2.7(b)):

– $f(p, k_D) < 0$, the point is inside the surface, then $\dot{D} = \frac{dD}{dt} = 0$ (D does not progress);

– $f(p, k_D) = 0$ and df < 0, the point is on the surface and comes back inside (discharge) then $\dot{D} = 0$;

– $f(p, k_D) = 0$ and df $= 0$, the point is on the surface and drags the surface with it and $\dot{D} > 0$ (D increases).

For the models presented later in the book, the pilot variable is the equivalent strain ε_{eq} (Mazars 1984) or a combination of equivalent strains when there are several (Mazars et al. 2015). The equivalent strains considered are scalar functions of the current values of the components of the strain tensor, as we will show in Chapter 3.

3

Damage Modeling

The literature offers many models, including damage dedicated to cementitious materials (Lubliner et al. 1989; Mazars and Pijaudier-Cabot 1989; La Borderie et al. 1994; Li and Mróz 1994; Jiràsek 2004). In this chapter, we limit ourselves to the presentation of two damage models designed for concrete, and for each of them, we show their applicability to the main loading situations encountered in civil engineering structures: monotonous, repeated, cyclical, and dynamic loadings.

3.1. Mazars models for monotonous loadings

Developed within the framework of the thermodynamics of irreversible processes, this model was built to deal with the case of situations where the loading is monotonic and where the damage that it creates at the local level is in mode I, corresponding to the type load A according to the classification made in Table 1.2. This type of loading is most commonly found in civil engineering structures (beams, columns, walls, etc.), so this model has wide areas of application. Its seniority (Mazars 1984, 1986) places it among the first models of its kind and it is still widely used, both for educational purposes and industrial applications.

3.1.1. *Constitutive equations*

Here, let us again take the framework of isotropic damage mechanics, as presented in section 2.2.1. As shown in equation [2.22], the stress $\boldsymbol{\sigma}$ is defined by the following tensor equation:

$$\boldsymbol{\sigma} = \boldsymbol{\Lambda}(1 - D) : \boldsymbol{\varepsilon} \qquad [3.1]$$

where $\boldsymbol{\Lambda}$ is the stiffness matrix of the undamaged material and $\boldsymbol{\varepsilon}$ is the stress tensor.

This law coupling elasticity and damage is said to be elastic-damageable. The plasticity and other dissipative phenomena are not taken into account. D is the internal variable that describes the state of damage within the meaning defined in Chapter 2. Its value is between 0 for undamaged material and 1 for broken material.

In section 2.2.3, we introduced the concept of the pilot variable for damage, which is a scalar function of the strain tensor components. To take into account the fact that the damage linked to the formation of microcracks in mode I is associated with the local existence of tension strain (see Table 1.2), Mazars (1984) introduced the notion of equivalent strain:

$$\varepsilon_{eq} = \sqrt{\langle \varepsilon_i \rangle_+^2} \qquad [3.2]$$

The symbol $< >_+$ defines the positive part so that:

$$\langle \varepsilon_i \rangle_+ = \varepsilon_i \text{ if } \varepsilon_i > 0$$

$$\langle \varepsilon_i \rangle_+ = 0 \text{ otherwise} \qquad [3.3]$$

ε_i is the i[e] principal strain. ε_{eq} is an indicator of the state of tension within the material; this state of tension is responsible for the damage.

This quantity makes it possible to define the threshold of damage, as well as its evolution via the loading surface f, which is such that:

$$f = \varepsilon_{eq} - k_D = 0 \qquad [3.4]$$

where $k_D = \varepsilon_0$ if $D = 0$ and ε_0 is the initial strain threshold of the damage (see Figure 3.1).

Figure 3.1(a) shows that, as desired, this surface limits the space of the extensions ($\varepsilon_i > 0$) and not that of the confinements ($\varepsilon_i < 0$), a domain in which the material remains linear elastic. The μ model presented in section 3.2 is more stringent on this point.

This being the case, we note despite everything, and this is a strong point of the model, that it respects the tension–compression asymmetry, the initial thresholds of damage being in a ratio $-\vartheta\sqrt{2}$ (i.e., approximately –3.5). This feature will be detailed in section 3.1.2.

As shown in 2.2.3, when in the strain space the point of loading is on the surface $f = 0$ and loading continues ($df = 0$), then the point drags the surface with it and the damage increases.

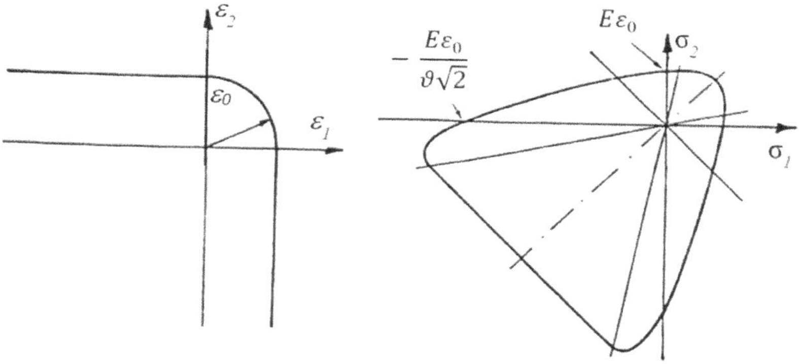

a) b)

Figure 3.1. *Mazars model, initial surface threshold of damage (D = 0): a) plot in the plane $\varepsilon_3 = 0$ of strain space; b) plot in plane $\sigma_3 = 0$ of the stress space*

Taking the considerations made in section 2.1.3, the physical reality is that concrete damage is not isotropic. It is therefore through compromise that we have moved from this observation to a model that retains the isotropy of the material by integrating two major ideas:

1) The first idea is that the model must satisfy a good representation of nonlinearities under tension and uniaxial compression. Figures 1.8 and 1.10 show that the two types of nonlinearities are different: sudden after the peak under tension and much more progressive under compression. To meet this requirement, two different damage laws are proposed for the two types of behavior: $D_t = f_t(\varepsilon_{eq})$ under tension and $D_c = f_c(\varepsilon_{eq})$ under compression, each having to comply with the Clausius–Duhem law, $\dot{D} \geq 0$, whatever the loading increment ($\dot{\varepsilon}_{eq} \geq 0$).

2) The second idea is that, in the case of a combination of compression or tension loads on several of the three axes of the stress space, the resulting damage retains the concept of isotropy, but remains marked by the relative prevalence of tension and compression.

Thus, in the general case of loading on the three axes, D is defined as a combination of the two damage modes defined by D_t and D_c, each varying between 0 and 1 depending on the associated state of damage. The relationship between these two variables is as follows:

$$D = \alpha_t D_t + \alpha_c D_c \qquad [3.5]$$

The coefficients α_t and α_c ensure this prevalence between tension and compression damage and must meet the following requirements:

– when only tension is activated (in one or more directions) $\alpha_t = 1$ while $\alpha_c = 0$;

– when only compression is activated then $\alpha_c = 1$ and $\alpha_t = 0$;

– in addition, in all cases $\alpha_t + \alpha_c = 1$.

The coefficients α_t and α_c are functions, at each moment of loading, of the relative intensity of tension and compression. In this context, the approach adopted to determine them is as follows.

A stress and strain partition is produced in the principal coordinate system, such as:

$$\boldsymbol{\sigma} = \langle\boldsymbol{\sigma}\rangle_+ + \langle\boldsymbol{\sigma}\rangle_- \text{ and } \boldsymbol{\varepsilon} = \boldsymbol{\varepsilon}_t + \boldsymbol{\varepsilon}_c \quad [3.6]$$

where:

$$\boldsymbol{\varepsilon}_t = \frac{1+\vartheta}{E}\langle\boldsymbol{\sigma}\rangle_+ - \frac{\vartheta}{E}\mathrm{Tr}\langle\boldsymbol{\sigma}\rangle_+ \text{ and } \boldsymbol{\varepsilon}_c = \frac{1+\vartheta}{E}\langle\boldsymbol{\sigma}\rangle_- - \frac{\vartheta}{E}\mathrm{Tr}\langle\boldsymbol{\sigma}\rangle_- \quad [3.7]$$

The symbols $< >_+$, $< >_-$ define the positive part and the negative part, respectively, of the stress (only the positive and negative principal components are retained).

This leads, in the principal coordinate system, with summation on the indices (i = 1, 2, 3), to the following expressions:

$$\alpha_t = \sum_i \mathrm{H_i}\frac{\varepsilon_{ti}(\varepsilon_{ti}+\varepsilon_{ci})}{\varepsilon_{eq}^2} \text{ and } \alpha_c = \sum_i \mathrm{H_i}\frac{\varepsilon_{ci}(\varepsilon_{ti}+\varepsilon_{ci})}{\varepsilon_{eq}^2} \quad [3.8]$$

$\mathrm{H_i}$ is the Heaviside symbol; it is 1 if $\varepsilon_i > 0$ and 0 otherwise. By noting that $\varepsilon_{eq}^2 = \sum_i \mathrm{H_i}\,(\varepsilon_{ti} + \varepsilon_{ci})^2$, it is verified that $\alpha_t + \alpha_c = 1$. Moreover, when $\langle\boldsymbol{\sigma}\rangle_- = 0$, then $\varepsilon_{ci} = 0$; therefore, $\alpha_c = 0$ and $\alpha_t = 1$. Conversely, when $\langle\boldsymbol{\sigma}\rangle_+ = 0$ then $\varepsilon_{ti} = 0$ and therefore $\alpha_t = 0$ and $\alpha_c = 1$. Thus, these expressions clearly verify the objectives set.

A particular feature of this model is its explicit writing because at a given loading stage, D_t and D_c depend solely on the maximum value reached of the equivalent strain, k_D, and not on its history, as is generally the case in plasticity models. This implies that all of the variables are calculated directly, which presents an advantage in numerical resolutions.

To limit complexity, for the two laws of evolution of the damage D_t and D_c, a single and same expression has been kept, but with a different set of parameters:

$$D_t = 1 - \frac{(1-A_t)\varepsilon_0}{k_D} - A_t \exp\left(-B_t(k_D - \varepsilon_0)\right) \quad [3.9]$$

$$D_c = 1 - \frac{(1-A_c)\varepsilon_0}{k_D} - A_c \exp\left(-B_c(k_D - \varepsilon_0)\right) \quad [3.10]$$

$k_D = \sup\ (\varepsilon_0, \max(\varepsilon_{eq}))$ and A_t, A_c, B_t, and B_c are material parameters that allow the shape of the post-peak curve to be modulated. They are easily identified from uniaxial tests, as we will see in Chapter 4.

3.1.2. Using the model (uniaxial case)

The uniaxial state is defined by a one-dimensional state under stress but a two-dimensional state under strain. The material and damage are assumed to be isotropic. Thus, in the principal coordinate system (X_1, X_2, X_3) and from equation [A.3], the behavior is written as:

$$\varepsilon_i = \frac{\sigma_i}{E(1-D)}\left(1 - \vartheta\left(\sigma_j + \sigma_k\right)\right), \text{ i, j, k permuting on values 1, 2, 3} \quad [3.11]$$

For uniaxial loading σ_1 in the X_1 direction, these relationships lead to the following results:

$$\varepsilon_1 = \frac{\sigma_1}{E(1-D)},\ \varepsilon_2 = \frac{-\vartheta\sigma_1}{E(1-D)},\ \varepsilon_3 = \frac{-\vartheta\sigma_1}{E(1-D)} \quad [3.12]$$

If $\sigma_1 > 0$ (tension), then $\varepsilon_1 > 0$ and $\varepsilon_2 = \varepsilon_3 < 0$. The equivalent strain given by equations [3.2] and [3.3] is:

$$\varepsilon_{eq} = \varepsilon_1 \quad [3.13]$$

For a given ε_1, the damage D is therefore equal according to equations [3.5] and [3.6], $D = D_t$ or:

$$D = 1 - \frac{(1-A_t)\varepsilon_0}{\varepsilon_1} - A_t \exp\left[-B_t(\varepsilon_1 - \varepsilon_0)\right] \quad [3.14]$$

The complete curve of the behavior under tension is then written as:

$$\text{if } \varepsilon_1 < \varepsilon_0 : \sigma_1 = E\varepsilon_1$$

$$\text{if } \varepsilon_1 > \varepsilon_0 : \sigma_1 = E(1-D)\varepsilon_1 \quad [3.15]$$

If σ_1 is < 0 (compression), then $\varepsilon_1 < 0$ and $\varepsilon_2 = \varepsilon_3 > 0$. The equivalent strain given by equations [3.2] and [3.3] is then written as:

$$\varepsilon_{eq} = \sqrt{\varepsilon_2^2 + \varepsilon_3^2} = \frac{-\vartheta\sigma_1}{E(1-D)}\sqrt{2} = -\vartheta\varepsilon_1\sqrt{2} \qquad [3.16]$$

At a given ε_1, the damage D is therefore, according to [3.5] and [3.6], $D = D_c$, that is:

$$D = 1 - \frac{(1-A_c)\varepsilon_0}{-\vartheta\varepsilon_1\sqrt{2}} - A_c \exp\left[-B_c\left(-\vartheta\varepsilon_1\sqrt{2} - \varepsilon_0\right)\right] \qquad [3.17]$$

It is found that the damage under compression develops less rapidly, as a function of the axial strain ε_1, than under tension. This point and the adjustment of the material parameters A_c and B_c (different from A_t and B_t) lead to the curves shown in Figure 3.2, which correctly reproduce the behavioral asymmetry stated for concrete.

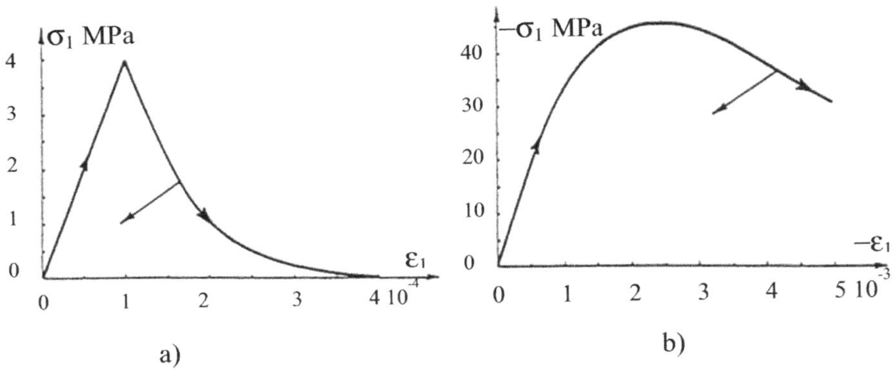

Figure 3.2. *Response of the Mazars model: a) uniaxial tension; b) uniaxial compression*

This model has been widely used to simulate the behavior of structures under monotonous loading. Without going into detail, in Figure 3.3 we give the example of a concrete beam (not reinforced) tested with a 4-point bending set up (length between supports = 1.40 m, between loads = 0.60 m, height = 0.22 m, width = 0.15 m, f_c = 32 MPa).

The failure occurs by single cracking appearing on the lower face at the level of gauge no. 1 and close to gauge no. 2. After identifying the parameters of the model, the simulation shows that it is capable of grasping the overall behavior (force–displacement), as well as the local behavior by creating a strain localization in the tensile area between the applied forces. This leads to describing very correctly, in the force–strain diagram, the sudden increase in the strain at the time of localization that corresponds (gauge 1) to the formation and opening of the crack and in its vicinity (gauge 2) to the local shrinkage of material when the crack opens.

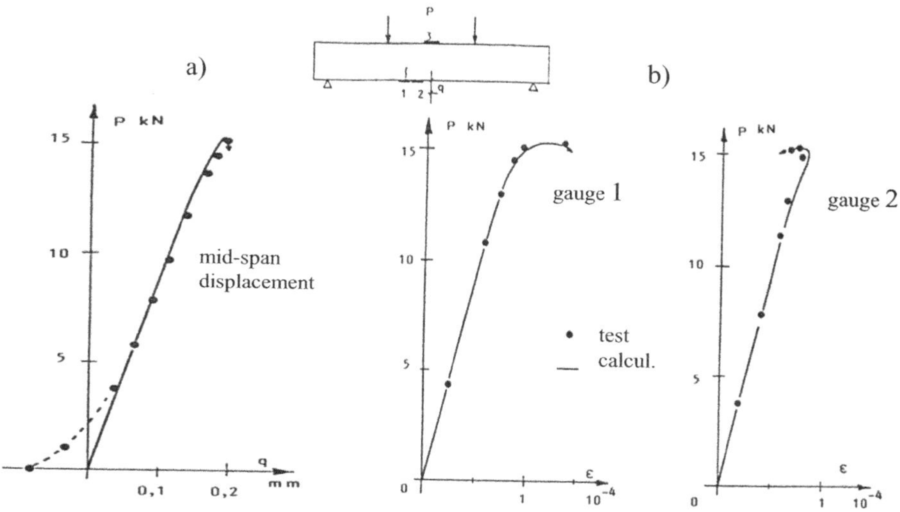

Figure 3.3. *Plain concrete beam tested with a 4-point bending set up, test-calculation comparison. a) Overall force–displacement behavior; b) local effort–strain behavior in the environment of a crack (from Mazars 1984)*

NOTE 3.1.– To further develop the use of this model, an example of calculation for multiaxial loading situations is given in section 9.2.1.

3.1.3. *Strengths and weaknesses of the Mazars model*

This model is easy to use, has good robustness in numerical calculations and gives, as we have just seen, good results for monotonous loading situations for common structural elements.

However, some weaknesses were noted, including the following:

– in multicompression loading situations when the loading point is located on the compression side in the vicinity of the bisecting plane of the stress space; the original model significantly underestimates the strength of the concrete in bicompression; an improvement can be made to this weakness by incorporating a correction for the equivalent strain:

$$\varepsilon_{eq}^{c} = \gamma \varepsilon_{eq}$$

where:

- $\gamma = -\frac{\sqrt{\sum_i \langle \tilde{\sigma}_i \rangle_-^2}}{\sum_i \langle \tilde{\sigma}_i \rangle_-}$ if at least one principal stress is negative;

- $\gamma = 1$ otherwise; [3.18]

– in loads under pure shear; this point can be troublesome for the simulation of the concrete-rebars connection in reinforced concrete; in fact, the damage is concentrated around the rebar, and the latter loses its adhesion prematurely to the concrete; to limit this problem, a power β has been introduced (Pijaudier-Cabot et al. 1991) in the expression, combining tension and compression damage, which gives:

$$D = \alpha_t^{\beta} D_t + \alpha_c^{\beta} D_c \qquad [3.19]$$

Figure 3.4(a) shows the effects of these corrections. It shows the impact of power β on the shear response of the model. The idea is to obtain for the major distortions, not a shear which progressively reduces ($\beta = 1$), but a shear which is maintained with a residual value capable of simulating the slip effects of the reinforcements which occurs with high friction. Typically, the β value is set at 1.06 (Davenne et al. 1989).

Figure 3.4(b) shows how, thanks to the corrected equivalent strain, the fracture surface is "inflated" on the bicompression side in the plane $\sigma_3 = 0$ of the stress space, thus approaching the experimental results obtained by various authors (Kupfer and Gerstle 1973; Bascoul and Maso 1974).

Finally, more recently, a so-called "reviewed" version of the model incorporating these two points, with another approach to the laws of evolution, has been proposed. It can be found in the Code_Aster manual (2020), where it has been implemented.

Despite its interest, this first model was designed only for monotonous loads. For cyclic loads, it is necessary to take into account not only the gradual opening of the cracks (increase in damage), but also their gradual closing when the stress changes from tension to compression. This phenomenon, shown in the results presented in Figure 1.11(c), is called the "unilateral effect"; it is very present in the case of cyclic loading (under the effect of an earthquake, for example) and its consequences are important.

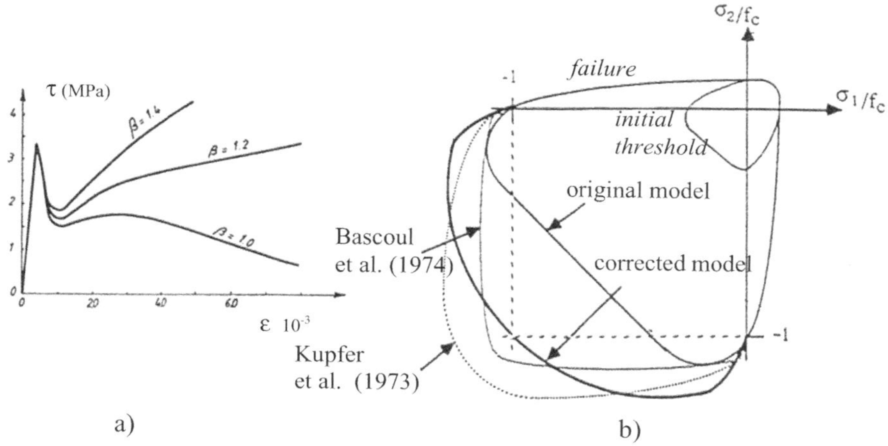

Figure 3.4. *Corrected model. a) Impact of power β on shear behavior; b) plane $\sigma_3 = 0$, comparison of failure surfaces of original Mazars model, corrected model and experimental curves*

To this end, in section 3.2 we present the μ model (mu, for Mazars unilateral (Mazars et al. 2015)), which was designed to improve the weaknesses of the original model and manage the effects of unilaterality as simply as possible.

3.2. Model for cyclic loadings: the μ model

3.2.1. *Concept of effective damage variable*

In Figure 1.10(c), we report the results of an experiment on a concrete prism loaded in two phases: first, by tension that causes, by virtue of the system for applying the loads by bonded metal bars, multifracturing, and then compression loading showing that, when the cracks are closed, there is a restoration of the stiffness of the specimen that regains its initial rigidity (see Figure 3.4(a)).

Thus, the damage created by a "tension" leads to a reduction in the modulus of elasticity of the material when the load is of the same sign as that which created it. On the other hand, the change from a "tension" to a "contraction" leads to the restoration of this modulus. In such a mechanism, the relationship of damage/loss of stiffness is therefore not systematic.

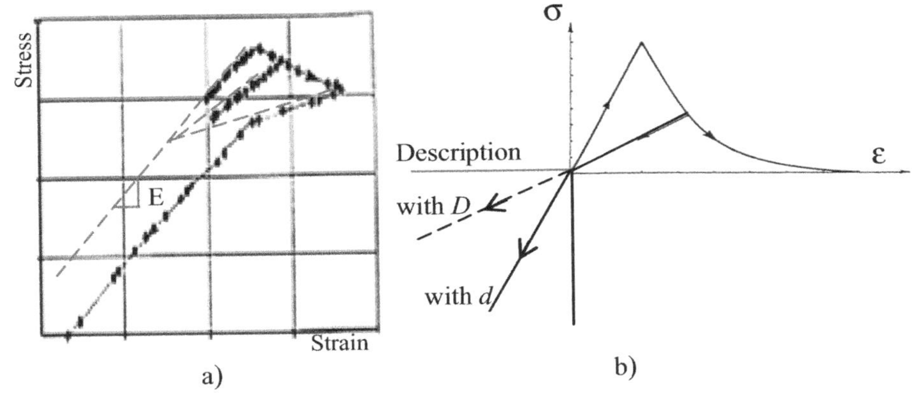

Figure 3.5. *Unilateral effect*

COMMENT ON FIGURE 3.5.– *(a) Experimental behavior showing the stiffness of the material during tensile damage and its gradual restoration during the transition to compression (according to Mazars et al. 1990); (b) modeled curve showing the non-representation of this phenomenon by the "state" damage variable D and the way in which the "effective" damage d can describe it in the context of elasticity-damage coupling.*

At this stage, we distinguish the damage of state D, characteristic of the state of degradation of the material, which is irreversible, and the mechanical effect of this damage, which acts on the modulus of elasticity or not depending on whether the load acts on the opening or closing of the cracks. To describe this mechanical effect, we propose the introduction of the concept of "effective" damage. The effective damage is activated ($d > 0$) when the stress state is of the same sign as that which created it (the cracks then remain open) and is not activated when this state changes sign and the cracks are closed (then $d = 0$).

The mechanical effect thus described is called the "unilateral effect". As shown in Figure 3.5(b), the use of the state variable D used in the Mazars model does not make it possible to describe this effect: the damage created under tension remains the same when going under compression. However, as defined, the effective damage variable, d, which is deactivated during the tension–compression transition, will allow for this description within the framework of the μ model presented below (Mazars et al. 2015). As mentioned before, the context is always that of elastic-damageable materials and permanent strains are neglected (see Figure 3.5(b)).

3.2.2. Constitutive equations

As mentioned before, we return to the framework of the mechanics of isotropic damage, as presented in section 2.2.1. As shown in equation [2.22], the stress $\boldsymbol{\sigma}$ is derived from the state law defined by the following tensor equation:

$$\sigma = \rho \frac{d\Psi}{d\varepsilon} = (1-d)\Lambda : \varepsilon \quad [3.20]$$

Ψ is the specific free energy, $\boldsymbol{\Lambda}$ is the stiffness matrix of the undamaged material, $\boldsymbol{\varepsilon}$ is the strain tensor and d is the effective damage.

As with the Mazars model, the concept of equivalent strain is taken into consideration. However, to distinguish the effects of tension loads from those of compression and therefore to manage the unilateral effects, two equivalent strains are used. That relating to *cracking* under tension, ε_t, and that relating to *crushing* under compression, ε_c:

$$\varepsilon_t = \frac{I_\varepsilon}{2(1-2\vartheta)} + \frac{\sqrt{J_\varepsilon}}{2(1+\vartheta)} \quad [3.21]$$

$$\varepsilon_c = \frac{I_\varepsilon}{5(1-2\vartheta)} + \frac{6\sqrt{J_\varepsilon}}{5(1+\vartheta)} \quad [3.22]$$

$$I_\varepsilon = \varepsilon_1 + \varepsilon_2 + \varepsilon_3 \text{ and } J_\varepsilon = 0.5[(\varepsilon_1 - \varepsilon_2)^2 + (\varepsilon_2 - \varepsilon_3)^2 + (\varepsilon_3 - \varepsilon_1)^2] \quad [3.23]$$

I_ε and J_ε are, to within one coefficient, the first invariant of the strain tensor and the second invariant of the strain deviator tensor (see equations [A.8] and [A.9]). Thus, ε_t and ε_c are independent of the chosen coordinate system.

Two independent load surfaces are associated with them:

$$f_t = \varepsilon_t - Y_t \text{ and } f_c = \varepsilon_c - Y_c \quad [3.24]$$

As presented in section 2.2.3, when, in the strain space, the point of loading is on the surface $f_{t,c} = 0$ and the loading continues ($df_{t,c} = 0$), the point drags the surface with it and the damage progresses (see Figure 2.7(b)). Y_t and Y_c define the maximum value of the equivalent strains ε_t and ε_c, obtained over the entire loading path:

$$Y_t = Sup[\varepsilon_{t0}, max\varepsilon_t] \text{ and } Y_C = Sup[\varepsilon_{c0}, max\varepsilon_c] \quad [3.25]$$

ε_{t0} and ε_{c0} are the initial thresholds of ε_t and ε_c, respectively.

A few additions on this subject can be found in section 9.3.1.

3.2.2.1. *Evolution of damage*

As in the Mazars model, the existence of both modes of damage, cracking in tension or crushing in compression, is considered. Y_t and Y_c are the pilot variables for cracking and crushing, respectively.

In the same spirit as for the Mazars model, it is assumed that the case of any loading is a combination of cracking and crushing. However, instead of performing the combination at the level of the damage variables, it is proposed here at the level of the pilot variables of damage. Thus, in the most general case, the pilot variable is, for the μ model, a linear combination of the two preceding variables.

It gives us:

$$Y = r\,Y_t + (1-r)Y_c \text{ where: } r = \frac{\sum_i \langle \widetilde{\sigma_i} \rangle_+}{\sum_i |\widetilde{\sigma_i}|} \quad [3.26]$$

r is the triaxiality factor (Lee and Fenves 1998), which varies in the stress space from 0 for the compression zone to 1 for the tension zone (see section 9.3.2). As we have seen in equation [2.9], $\widetilde{\boldsymbol{\sigma}} = \frac{\sigma}{(1-d)} = \boldsymbol{\Lambda}:\boldsymbol{\varepsilon}$ is the effective stress; $\langle x \rangle_+$ and $|x|$ are the positive part and the absolute value of x, respectively. Thus, r does not depend on the damage and can be determined at each calculation step without iteration.

In the original model, the two damage modes are each described by a damage evolution law, $D_t = f_t(k_D, A_t, B_t)$ and $D_c = f_c(k_D, A_c, B_c)$. To find the same laws for these two modes, the law of evolution proposed here for effective damage is defined by:

$$d = 1 - \frac{(1-Y)Y_0}{Y} - A\exp(-B(Y - Y_0)), \text{ where } Y_0 = r\varepsilon_{0t} + (1-r)\varepsilon_{0c} \quad [3.27]$$

Y_0 is the initial threshold for Y. Variables A and B determine the shape of the laws of evolution of the effective damage and, consequently, that of the behavior curves. $[A, B]$ changes from $[A_t, B_t]$ for cracking curves to $[A_c, B_c]$ for crushing curves.

The proposals for variables A and B are as follows:

$$A = A_t\big(2r^2(1-2k) - r(1-4k)\big) + A_c(2r^2 - 3r + 1)$$

$$B = r^{(r^2-2r+2)}B_t + (1 - r^{(r^2-2r+2)})B_c \quad [3.28]$$

Thus:

– when the damage is purely due to tension, $r = 1$, $Y = Y_t$, which induces $A = A_t$, $B = B_t$: the law of evolution is the same as that of D_t of the previous model (see equation [3.9]);

– when the damage is purely due to compression, $r = 0$, $Y = Y_c$, which induces $A = A_c$, $B = B_c$: the law of evolution is the same as that of D_c of the previous model (see equation [3.10]).

A_t, B_t, A_c and B_c are material parameters, as previously identified from uniaxial tests.

k is introduced to calibrate asymptotic stresses for large shear strains (useful for representing rebars-concrete slips. For example, in pure shear, $\sigma_1 = -\sigma_2 > 0$ and $\sigma_3 = 0$; then, according to equation [3.26]:

$$r = \sigma_1/(\sigma_1 + |\sigma_2|) = 0.5$$

which leads, according to equation [3.28], to:

$$k = A_{r=0.5}/A_t \tag{3.29}$$

Figure 3.9 shows what this gives in the case where $A_t = 1$ and $k = 0.7$ (the usual value used).

3.2.2.2. *The μ model within the thermodynamic framework*

As has just been seen, the effective damage d is directly associated with the thermodynamic variables Y_t and Y_c, and its intensity depends directly on their value.

At the thermodynamic level, unlike D, the effective damage, d, no longer has the status of a state variable. Indeed, for the same given state of damage, when the cracks pass from the open to the closed situation, its value changes from a positive value to 0, and the Clausius–Duhem equation, which stipulates $\dot{D} \geq 0$, must be reconsidered. In equation [2.16], we saw that in the absence of temperature variation, the Clausius–Duhem equation was reduced to:

$$\boldsymbol{\sigma}:\dot{\boldsymbol{\varepsilon}} - \rho\left(\frac{d\Psi}{dt}\right) \geq 0 \tag{3.30}$$

We have seen in section 2.2.1 that:

$$\Psi = \frac{1}{2\rho}\boldsymbol{\Lambda}(1-d):\boldsymbol{\varepsilon}:\boldsymbol{\varepsilon} \tag{3.31}$$

However, in the μ model, $d = f(Y_t, Y_c)$ and therefore $\Psi = (\boldsymbol{\varepsilon}, Y_t, Y_c)$, an expression that falls within the scope of remark 2.2 (see section 2.2.1).

In this context, equation [3.30] becomes:

$$\left(\boldsymbol{\sigma}-\rho\frac{\partial\Psi}{\partial\varepsilon}\right):\dot{\boldsymbol{\varepsilon}}-\rho\frac{\partial\Psi}{\partial Y_t}\dot{Y}_t-\rho\frac{\partial\Psi}{\partial Y_c}\dot{Y}_c\geq 0 \qquad [3.32]$$

Equation [3.20] leads to the first term being cancelled, and the other two terms can be written as:

$$-\rho\frac{\partial\Psi}{\partial d}\left(\frac{\partial d}{\partial Y_t}\dot{Y}_t+\frac{\partial d}{\partial Y_c}\dot{Y_c}\right)\geq 0 \qquad [3.33]$$

However, as seen in section 2.2.1, $-\rho\frac{\partial\Psi}{\partial d}=\frac{1}{2}\boldsymbol{\Lambda}:\boldsymbol{\varepsilon}:\boldsymbol{\varepsilon}$ is a positive defined quadratic form of $\boldsymbol{\varepsilon}$, therefore, the Clausius–Duhem equation will be satisfied if:

$$\frac{\partial d}{\partial Y_t}\dot{Y}_t+\frac{\partial d}{\partial Y_c}\dot{Y}_c\geq 0 \qquad [3.34]$$

Since $Y = rY_t + (1-r)Y_c$, in the case of radial loading (r = const):

$$\frac{\partial d}{\partial Y_t}\dot{Y}_t=\frac{\partial d}{\partial Y}\frac{\partial Y}{\partial Y_t}\dot{Y}_t=\frac{\partial d}{\partial Y}r\dot{Y}_t\geq 0$$

$$\text{and }\frac{\partial d}{\partial Y_c}\dot{Y}_c=\frac{\partial d}{\partial Y}\frac{\partial Y}{\partial Y_c}\dot{Y}_c=\frac{\partial d}{\partial Y}(1-r)\dot{Y}_c\geq 0 \qquad [3.35]$$

r and $(1 - r)$ are by definition non-negative, and $\dot{Y}_t$, $\dot{Y}_c$, are always positive or zero during loading according to equation [3.25].

Thus, when r is constant, the relationship of equation [3.34] is always satisfied. When r is variable, the partial derivatives $\frac{\partial Y}{\partial Y_t}$ and $\frac{\partial Y}{\partial Y_c}$ involve $\dot{r}$, thus making the problem more difficult to solve. However, by numerically verifying the positivity of the relationship of equation [3.34], it has been shown that the principles underlying the Clausius–Duhem inequality are generally respected, even for complex loads.

3.2.3. *Response of the μ model under various types of loading*

3.2.3.1. *Uniaxial loading*

On the same basis as those used for the Mazars model (see section 3.1.2), it is easy to deduce the response of the μ model in tension–compression. The first step is the determination of equivalent strains. From equation [3.23], for a load in direction X_1, we obtain:

$$I_\varepsilon=\varepsilon_1(1-2\vartheta),\ J_\varepsilon=\varepsilon_1^2\ (1+\vartheta)^{\ 2}\text{ and }\sqrt{J_\varepsilon}=|\varepsilon_1|(1+\vartheta) \qquad [3.36]$$

Hence, according to equations [3.21] and [3.22]:

$$\varepsilon_t = \frac{\varepsilon_1(1-2\vartheta)}{2(1-2\vartheta)} + \frac{|\varepsilon_1|(1+\vartheta)}{2(1+\vartheta)} = \frac{\varepsilon_1}{2} + \frac{|\varepsilon_1|}{2}$$

$$\text{and } \varepsilon_c = \frac{\varepsilon_1(1-2\vartheta)}{5(1-2\vartheta)} + \frac{6|\varepsilon_1|(1+\vartheta)}{5(1+\vartheta)} = \frac{\varepsilon_1}{5} + \frac{6|\varepsilon_1|}{5} \quad [3.37]$$

However, we have seen in equations [3.25] and [3.26] that:

$$Y_t = \sup(\varepsilon_{t0}, \max\varepsilon_t),\ Y_c = \sup(\varepsilon_{c0}, \max\varepsilon_c) \text{ and } r = \frac{\sum_i\langle\widetilde{\sigma_i}\rangle_+}{\sum_i|\widetilde{\sigma_i}|} \quad [3.38]$$

Thus:

– if $\sigma_I > 0$ (tension), then $r = 1$ and $Y = Y_t = \max\,\varepsilon_t = \max\,\varepsilon_I$;

– if $\sigma_I < 0$ (compression), then $r = 0$ and $Y = Y_c = \max\,\varepsilon_c = \max|\varepsilon_I|$.

Furthermore, from the evolution laws of parameters A and B (see equation [3.28]), it is deduced that:

– when $r = 1$ (tension), then $A = A_t$ and $B = B_t$;

– when $r = 0$ (compression), then $A = A_c$ and $B = B_c$.

The combination of these elements leads, in both cases, to the deduction from equation [3.27] of the damage values and the behavioral laws as follows:

– under tension:

$$\sigma_1 = E(1-d)\sigma_1$$

where:

$$d = d_t = 1 - \frac{(1-A_t)\varepsilon_{0t}}{Y_t} - A_t\exp(-B_t(Y_t - \varepsilon_{0t})) \quad [3.39]$$

– under compression:

$$\sigma_1 = E(1-d)\sigma_1$$

where:

$$d = d_c = 1 - \frac{(1-A_t)\varepsilon_{0c}}{Y_c} - A_c\exp(-B_c(Y_c - \varepsilon_{0c})) \quad [3.40]$$

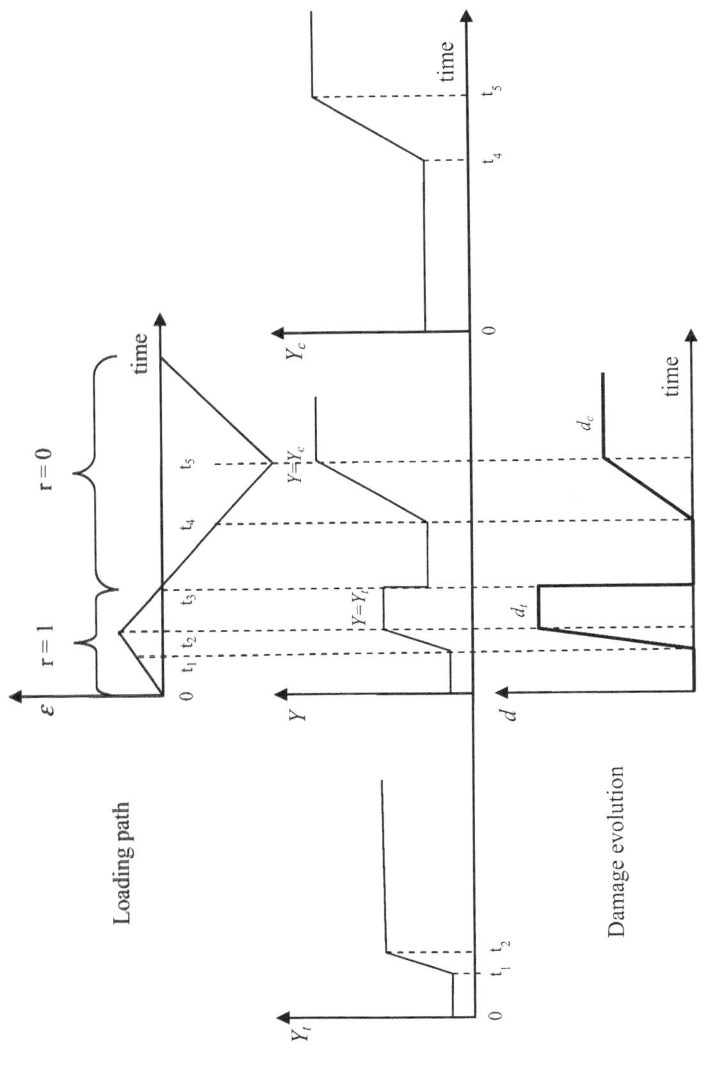

Figure 3.6. *Evolution of the variables Y_t and Y_c, their combination Y and the damage d, during an alternate loading of tension–compression*

Particularized in d_t and d_c, these variables describe tension and compression damage, respectively (cracking and crushing). In this context, Figure 3.6 shows how the μ model works during its response to alternating tension–compression loading. This figure shows the incidence of the triaxiality factor of the stresses, r, to manage the combination $Y = f(Y_t, Y_c)$ and the activation or deactivation of the damage d depending on whether the cracks are open or closed.

NOTE 3.2.– The expressions in equations [3.39] and [3.40] are the same as those presented in equations [3.9] and [3.10] for the Mazars model, with a particularity related to the thresholds:

– in the original model, the two thresholds are linked to the equivalent strain ε_{eq} (see equations [3.13] and [3.16]), which leads to the damage thresholds $k_D = max\varepsilon_1$ under tension and $k_D = -\vartheta max\varepsilon_1\sqrt{2}$ under compression, respectively;

– in the μ model, the two thresholds Y_t and Y_c are completely independent. We find $Y_t = \max\varepsilon_1$ under tension, but it is $Y_c = \max|\varepsilon_1|$ under compression.

Thus, for the same concrete (mechanical characteristics E and ν identical), to rediscover the same uniaxial responses of the two models, it will be necessary to:

– apply the same thresholds: $\varepsilon_{t0} = \varepsilon_0$ under tension, $\varepsilon_{c0} = -\vartheta\varepsilon_0\sqrt{2}$ under compression;

– use the same parameters A_t and B_t under tension, but adapt the parameters A_c and B_c under compression.

However, there is no obligation to do so, and the two thresholds can be calibrated independently.

NOTE 3.3.– As seen in section 3.2.1, the two models operate differently in tension–compression alternating loading:

– with the original model, if damage $D = D_t$ under tension has been created, due to equation [3.1], this becomes intrinsic to the material, and D is preserved not only for unloading, but also for passing into compression. In both sectors (tension and compression), the damage remains the same;

– with the μ model, the pilot variable being $Y = rY_t + (1-r)Y_c$, if traction damage is created ($r = 1$ and $Y = Y_t$), this is retained for any unloading; however, the change to compression leads to $r = 0$, and the pilot variable becomes $Y = Y_c$; this leads to a sudden change from d_t to d_c and therefore from $E(1-d_t)$ to $E(1-d_c)$, making it possible to describe the unilateral nature of the material linked to the closure of the cracks during the tension–compression change.

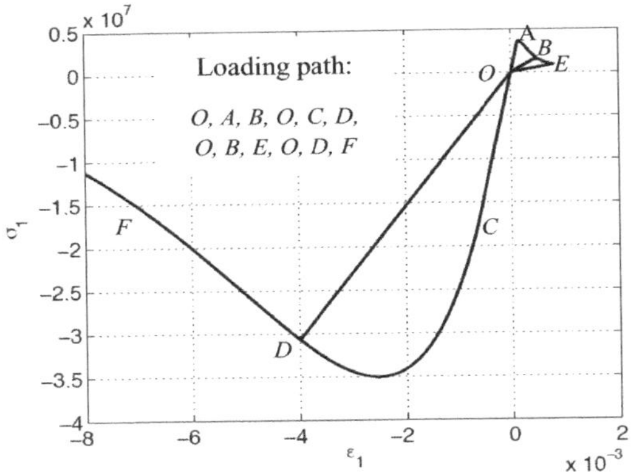

Figure 3.7. *Tension–compression loading highlighting the unilateral effect*

SUMMARY.– *With the adaptation of parameters, both models describe monotonous loads in the same way, but the Mazars model must be strictly reserved for cases of monotonous loads, whereas the μ model also allows for the description of responses to alternating loads (see Figure 3.7). This gives the answer for the case of a specific loading path, from OABE under tension to OCDF under compression, demonstrating this famous unilateral effect.*

3.2.3.2. *Biaxial loading*

To check the efficiency of the model, the biaxial response is interesting because quality experimental results exist for this type of load.

From equation [A.3], the behavior in the principal coordinate system is written as:

$$\varepsilon_i = \frac{1+\vartheta}{E(1-d)}\sigma_i - \frac{\vartheta}{E(1-d)}(\sigma_1 + \sigma_2 + \sigma_3) \qquad [3.41]$$

Considering a biaxial situation, with loading in the plane X_1, X_2, $(\sigma_1, \sigma_2, 0)$, we then obtain:

$$\varepsilon_1 = \frac{1}{E(1-d)}(\sigma_1 - \vartheta\sigma_2),\ \varepsilon_2 = \frac{1}{E(1-d)}(\sigma_2 - \vartheta\sigma_1),\ \varepsilon_3 = \frac{-\vartheta}{E(1-d)}(\sigma_1 + \sigma_2) \qquad [3.42]$$

Hence, we deduce according to equation [3.23]:

$$I_\varepsilon = \frac{(1-2\vartheta)}{E(1-d)}(\sigma_1 + \sigma_2),\ J_\varepsilon = \frac{(1+\vartheta)^2}{E^2(1-d)^2}(\sigma_1^2 + \sigma_2^2 - \sigma_1\sigma_2) \qquad [3.43]$$

$$\varepsilon_t = \frac{1}{2E(1-d)}[\sigma_1 + \sigma_2 + \sqrt{\sigma_1^2 + \sigma_2^2 - \sigma_1\sigma_2}]$$

$$\text{and } \varepsilon_c = \frac{1}{5E(1-d)}[\sigma_1 + \sigma_2 + 6\sqrt{\sigma_1^2 + \sigma_2^2 - \sigma_1\sigma_2}] \quad [3.44]$$

In the $\sigma_3 = 0$ plane, Figure 3.9 gives the traces of the surface threshold of damage ($d = 0$). Each point of this surface is obtained by determining the damage threshold for each pair (σ_1, σ_2): $Y - Y_0 = 0$, where $Y_0 = r\varepsilon_{t0} + (1-\text{r})\varepsilon_{c0}$. For example:

– for $\sigma_1 > 0$ and $\sigma_2 = 0$ (uniaxial tension), we obtain, as seen in section 3.2.3.1, $r = 1$, $Y = Y_t$, and $Y_0 = \varepsilon_{t0}$, but since $Y = Y_t = max(\varepsilon_1)$, the threshold is reached when $max(\varepsilon_1) = \varepsilon_{t0}$;

– for $\sigma_1 < 0$ and $\sigma_2 = 0$ (uniaxial compression), we obtain, as seen in section 3.2.3.1, $r = 0$, $Y = Y_c$, and $Y_0 = \varepsilon_{c0}$, but since $Y = Y_c = max|\varepsilon_1|$, the threshold is reached when $max\ |\varepsilon_1| = \varepsilon_{c0}$;

– for $\sigma_1 = -\sigma_2 > 0$ (pure shear), from the relations seen in equation [3.29], we deduce $r = 0.5$ and so, $Y_0 = 0.5\ (\varepsilon_{t0} + \varepsilon_{c0})$ and $Y = 0.5\ (Y_t + Y_c)$; moreover:

$$\varepsilon_1 = \frac{\sigma_1(1+\vartheta)}{E},\ \varepsilon_2 = \frac{-\sigma_1(1+\vartheta)}{E} = -\varepsilon_1 \text{ and } \varepsilon_3 = 0 \quad [3.45]$$

At the same time, the relationships in equations [3.21] and [3.22] lead to:

$$Y_t = \frac{\varepsilon_1\sqrt{3}}{2(1+\vartheta)} \text{ and } Y_c = \frac{6\varepsilon_1\sqrt{3}}{5(1+\vartheta)}, \text{ that is, } Y = 0.5\,\frac{1.7\varepsilon_1\sqrt{3}}{(1+\vartheta)} \quad [3.46]$$

at the threshold, $Y = Y_0$; thus, the threshold is reached when:

$$\varepsilon_1 = \frac{(1+\vartheta)}{1{,}7\sqrt{3}}(\varepsilon_{t0} + \varepsilon_{c0}) \text{ and } \sigma_1 = \frac{\varepsilon_1 E}{(1+\vartheta)} = \frac{E}{1.7\sqrt{3}}(\varepsilon_{t0} + \varepsilon_{c0}) \quad [3.47]$$

– for $\sigma_1 = \sigma_2 < 0$ (biaxial compression), in this case, $r = 0$, $Y = Y_c$ and at the threshold $Y_0 = \varepsilon_{c0}$.

After all the calculations, we find that at the threshold:

$$\varepsilon_{c0} = \frac{-4\varepsilon_1}{5(1-\vartheta)}, \text{ that is, } \varepsilon_1 = \frac{-5(1-\vartheta)}{4}\varepsilon_{c0} \text{ and } \sigma_1 = \frac{\varepsilon_{c0}E}{1-\vartheta} \quad [3.48]$$

In section 3.2.3.1, we saw that the uniaxial compression threshold was $f_{c0} = \varepsilon_{c0}E$; here, the threshold is therefore $1/(1-\nu)$ times greater.

With regard to the evolution beyond the threshold, for uniaxial cases of tension and compression, we refer to section 3.2.3.1. In both cases, during loading in strain, the stress passes through a maximum (very close to the damage threshold under tension and several times the value of the threshold under compression), and we call this threshold the "failure threshold".

In shear, the curve shape is the same as under tension, but with a higher damage threshold and failure threshold (#1.5 times). Note that, as indicated in equation [3.29], the asymptotic value at the end of the curve is controlled by the coefficient $k = A_{r=0.5}/A_t$, which is taken to be 0.7 in the example of Figure 3.9.

In biaxial compression loading, we have just found that, in the case of equality of the two stresses, the damage threshold is amplified by $1/(1-\nu)$ and the same applies to the failure threshold. With $\nu = 0.2$, they are therefore 1.25 times greater than those of simple compression.

We presented the biaxial test of Kupfer and Gerstle (1973) in Chapter 1. These tests were carried out on concrete platens (200 mm × 200 mm × 50 mm) loaded until failure according to prescribed ratios of σ_1/σ_2.

Several types of concrete have been tested. For the present simulations, we selected one of them, the characteristics of which are given in Table 3.1. The model parameters were adjusted to obtain the same strengths (f_t, f_c) under uniaxial tensile and uniaxial compression loads.

The results given by the μ model were compared with the experimental results of a tension-compression test and a bicompression test for ratios of $\sigma_1/\sigma_2 = -0.052$ and +0.52, respectively (see Figure 3.8). It should be noted that only the characteristics of Table 3.1 were used to assign the material parameters of the model. The final experimental point corresponds to the failure of the sample. In both cases, the model gives a very good prediction of the maximum strength and fairly good results for the evolution of the strain in various directions, with the exception of extension ε_3 in the vicinity of the maximum load. This weakness is because the Poisson coefficient ν remains constant in an isotropic formulation.

Experimental data		Modeling parameters							
f_c (MPa)	f_t (MPa)	E (GPa)	ν	ε_{t0}	ε_{c0}	A_t	B_t	A_c	B_c
32.7	3.2	30	0.2	1.07×10^{-4}	3.2×10^{-4}	1	$1 \times 10^{+4}$	1.25	459

Table 3.1. *Characteristics of concrete for the Kupfer and Gerstle tests (1973)*

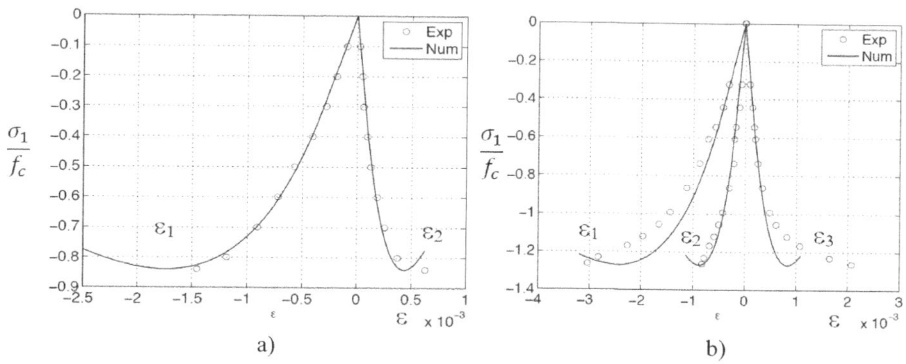

Figure 3.8. *Comparison between numerical curves and experimental points (Kupfer and Gerstle 1973): a) for a tension–compression load with σ_1/σ_2 = –0.052 (σ_2 > 0); b) for a biaxial compressive load with σ_1/σ_2 = 0.52 (σ_2 < 0)*

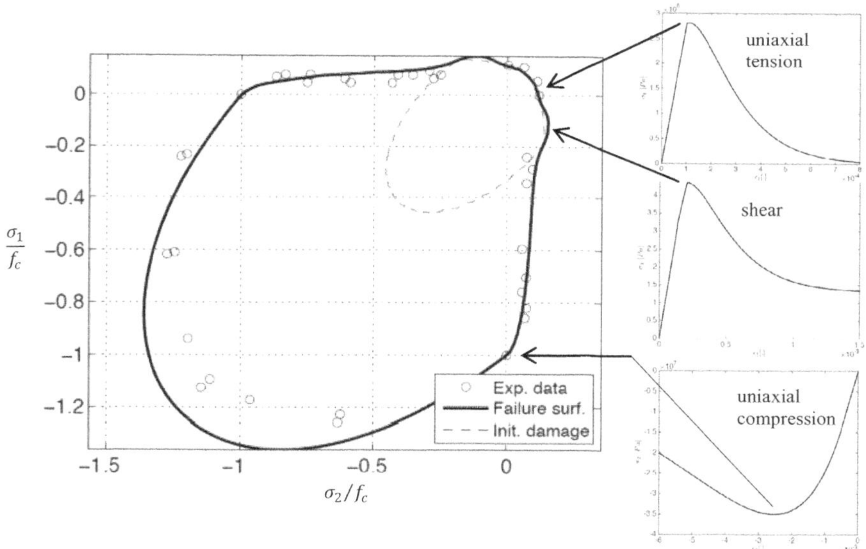

Figure 3.9. *Plane σ_3 = 0: traces of damage initiation (dashed line) and failure surfaces (strength envelope obtained from different loading paths) compared to experimental results obtained by Kupfer and Gerstle (1973) (from Mazars et al. 2015)*

The same authors carried out similar work for many loading paths, making it possible to deduce in particular the trace of the failure surface of the material in the plane $\sigma_3 = 0$. The resulting points (see Figure 3.9) represent the maximum stresses borne by the specimen, which are standardized by the compressive strength f_c. In the same figure, the μ model's forecasts appear with respect to the damage surface threshold (dotted lines) and failure surface threshold (solid lines). Moreover, three specific curves are given in this same figure: tension ($\sigma_2 = 0$), shear ($\sigma_2 = -\sigma_1$), and compression ($\sigma_1 = 0$). The set was plotted from the numerical values given in Table 3.1.

It is found that this model performs well, rather better than the original model. Only a few differences appear close to the bisector of the bicompression zone.

3.2.4. *Adaptation of the μ model to the case of confined loadings*

Confinement corresponds to a tri-compression stress state. The triaxiality coefficient r, introduced in equation [3.26], is in this case equal to 0. As shown in Table 1.2, the modes of damage depend on the presence or absence of extensions. At this stage, two distinct domains can be considered:

– the domain with authorized extensions (AE), in which, for a given load, a positive strain exists in at least one direction; these are the type of stresses referenced as A in Table 1.2;

– a domain without authorized extension (NAE), this is the case where the loading prevents any type of extension; this domain covers the types referenced as B, C, C′, in Table 1.2.

Positive strain in the domain (AE) can generate local mode I damage, and as we have seen in section 3.2.3, the μ model is a good candidate for describing this situation. On the other hand, in the field (NAE), the phenomena are twofold:

– a strong confinemen leads to a gradual collapse of the porosity of the cement matrix (type of loadings referenced C or C′);

– the shear, caused by the triaxiality of the stresses, generates mode II local cracking (type of loading referenced B).

It has been shown that such phenomena can advantageously be modeled by the introduction of consolidation and plasticity effects. The coupling of a damage model with a plasticity model was carried out in the PRM model (Pontiroli et al. 2010), but we have not considered this option here.

While remaining in the spirit of simplicity, to deal with situations of moderate confinement (low and medium), we propose, in sections 3.2.4.1 and 3.2.4.2, some adjustments to the μ model to meet this requirement.

3.2.4.1. *Low confinements*

In the zone (AE), the authorized extension results in $I_{\varepsilon+} = \langle\varepsilon_1\rangle_+ + \langle\varepsilon_2\rangle_+ + \langle\varepsilon_3\rangle_+ \neq 0$. This is integrated into a new definition of the variable Y_c, which is a component of the damage pilot variable:

$$Y_c \text{ becomes } Y_c^* = C\, Y_c$$

where

$$C = 1 \text{ if } r > 0$$

$$C = \frac{\Sigma_i \langle\varepsilon_i\rangle_+}{\Sigma_i \langle\varepsilon_{io}\rangle_+} = \frac{I_{\varepsilon+}}{I_{\varepsilon0+}} \text{ if } r = 0 \qquad [3.49]$$

$I_{\varepsilon0+}$ is the projection of $I_{\varepsilon+}$ on the nearest $\sigma_i = 0$ plane, calculated by setting the compression of lower intensity to 0. For example, if $|\sigma_i| > |\sigma_j| > |\sigma_k|$, then $I_{\varepsilon0+}$ is calculated with $|\sigma_k| = 0$.

Starting from this new definition, the calculation of d is carried out as in equation [3.27], but from the new pilot variable: $Y = rY_t + (1-r)Y_c^*$. Compared to the examples discussed in sections 3.2.3.1 and 3.2.3.2, this changes neither the uniaxial nor the biaxial results because with $\sigma_3 = 0$, we find $I_{\varepsilon0+} = I_{\varepsilon+}$ and therefore $C = 1$.

3.2.4.2. *Medium and high confinements*

The high confinements are characterized by the absence of extension (NAE), and thus $I_{\varepsilon+} = 0$, the coefficient, defined in equation [3.49], is $C = 0$. In this case, $Y_c^* = 0$, *crushing* damage no longer occurs, and therefore, $d = 0$, regardless of the intensity of the load. At the same time, this level of confinement generates compaction of the microporous matrix, which creates the risk of a new form of damage that corresponds to degradation of the cementitious matrix, comparable to consolidation in the sense of soil mechanics. The cause of this consolidation damage is the high level of spherical stress defined in equation [A.10].

In the remainder of this section, for practical reasons, the compressions, contractions and pressures are considered positive. In this context, Figure 3.10 gives the experimental behavior of concrete under hydrostatic loading (spherical stress state). The coordinate system presented gives the volume variation as a function of the confinement, $p = f(\Delta V/V)$ with $p = (\text{trace}\ \boldsymbol{\sigma})/3 := I_\sigma/3$ and $\Delta V/V = \text{trace}\ \boldsymbol{\varepsilon}$.

In the linear elastic zone, the bulk modulus is defined such that $K = p/(\Delta V/V)$, that is, $K = E/3(1-2\nu)$.

The experimental curve shows that the consolidation damage takes place between two thresholds characterized by particular values of (trace $\boldsymbol{\varepsilon}$) = $\Delta V/V = I_\varepsilon$:

– the threshold for the occurrence of this compaction damage is $I_{\varepsilon 0}$;

– the end threshold of this damage is $I_{\varepsilon 1}$; beyond, the behavior is essentially due to the hydrostatic compression of the solid matter (degraded matrix and aggregates) and liquid matter (free water) (see Figure 1.15).

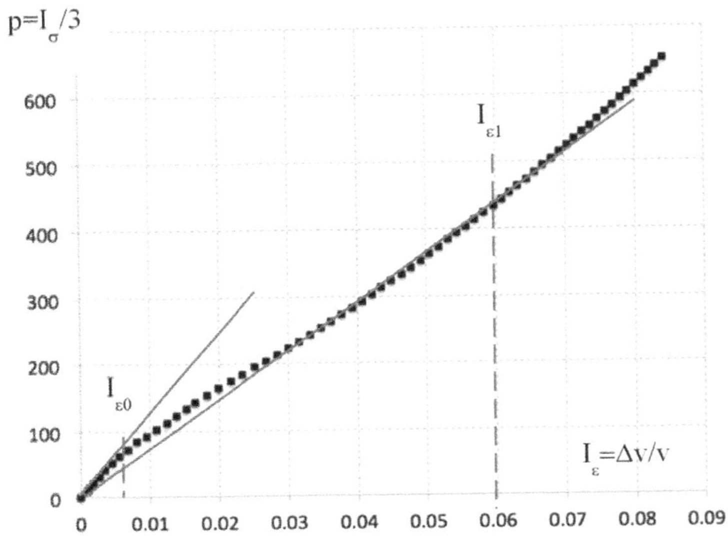

Figure 3.10. *Response of concrete under hydrostatic loading demonstrating the effects of consolidation damage. For a color version of this figure, see www.iste.co.uk/mazars/damage.zip*

Between these two thresholds, the nonlinearity is described by introducing the effects of consolidation damage d_{hc}. This damage is coupled with the damage of *crushing-cracking* d, which leads for the general case to the following constitutive law:

$$\boldsymbol{\sigma} = \Lambda(1-d)(1-d_{hc}):\boldsymbol{\varepsilon} \qquad [3.50]$$

A number of hypotheses are associated with the effective damage d_{hc}:

– it is assumed that the latter is only effective (action on the loss of stiffness of the material) between $I_{\varepsilon 0}$ and $I_{\varepsilon 1}$; below this, the material under hydrostatic pressure remains linear elastic and beyond this, the degraded material compacts, which hardly modifies the stiffness;

– starting from the curve response to hydrostatic pressure, the elastic part is characterized by its slope $p/I_\varepsilon = E/3(1-2\nu) = K$. Moreover, it can be considered that between the thresholds $I_{\varepsilon 0} < I_\varepsilon < I_{\varepsilon 1}$, the hydrostatic response can be approximated by a straight line. Thus, using the usual definitions of damage, we obtain:

$$p = I_\varepsilon(1 - D_{hc})K, \text{ where } D_{hc} = 1 + \mathrm{A} + {}^{\mathrm{B}}/_{I_\varepsilon} \qquad [3.51]$$

– but it was found, in comparison with the experimental results, that the incidence of this damage D_{hc}, taken as in equation [3.51] during a triaxial loading (different stresses on the three axes), causes premature failures. Hence, the idea of using an effective damage $d_{hc} = 1 + a + b/I_\varepsilon$, with $a = \alpha\mathrm{A}$ and $b = \delta\mathrm{B}$; a and b being two constants.

We then obtain:

– if $I_\varepsilon \leq I_{\varepsilon 0}$: $d_{hc} = 0$

– if $I_{\varepsilon 0} \leq I_\varepsilon \leq I_{\varepsilon 1}$: $d_{hc} = 1 + a + {}^{b}/_{I_\varepsilon}$

– if $I_\varepsilon > I_{\varepsilon 1}$: $d_{hc} = d_{hc1}$ [3.52]

$I_{\varepsilon 0}$, $I_{\varepsilon 1}$, a, b and d_{hc1} are material parameters that are to be determined from high confinement triaxial tests coupling the two modes of damage. An example of an axisymmetric triaxial test is discussed in section 9.3.3.

4

Numerical Calculation of Damage

The idea, in Chapter 4, is to show how to move from mathematical expressions, translating the constitutive laws of models, to numerical calculations using finite elements. Two types of calculations are preferred: classical finite element calculations (2D or 3D) and so-called simplified calculations based on the use of Timoshenko beam elements (1945).

4.1. Reminders on concepts governing the use of finite elements

The medium studied is divided into elements adapted to its morphology (bar, beam, plate, shell, massive 3D element). The elements are linked together by nodes (only one for a bar, two for a triangle with three nodes or a quadrangular element with four nodes, etc.). In all cases, the calculation is intended to determine the displacement of the nodes under the effect of an external load. An interpolation function, depending on the type of element, makes it possible to find the displacement at any point inside the element as a function of the displacements at the nodes, $(u)_e : (u) = [N](u)^e$. Thus, it is possible to define the deformations inside an element as a function of the nodal displacements: $(\varepsilon) = [B](u)^e$. Section A.2 describes in more detail the use of these interpolation functions and the calculation of the matrix $[B]$, which contains the derivatives of these functions.

Using the relationship that links the stresses to the strains $(\sigma) = [\Lambda](\varepsilon)$ and translating the equilibrium of the element through the virtual work theorem, we obtain the internal forces to which element e is subjected:

$$(F)^e = \int_{Ve}[B]^t\,[\Lambda][B]dv(\,u)^e = [k]^e(\,u)^e \qquad [4.1]$$

$[k]^e$ is the stiffness matrix of the element.

By assembling the elementary matrices (the technical details of which are given in section A.2), the overall rigidity matrix $[K]$ is obtained. Thus, by extending it to all of the elements, the problem amounts to solving the following matrix equation:

$$\{F\} = [K]\{U\} \tag{4.2}$$

$\{F\}$ is the vector of external forces and $\{U\}$ is the vector of nodal displacements.

From this field of displacements $\{U\}$, the local strains and, more generally, the strain field are deduced:

$$\{\boldsymbol{\varepsilon}\} = [B]\{U\} \tag{4.3}$$

With the constitutive behavioral laws having been integrated, the stress field is then deduced:

$$\{\boldsymbol{\sigma}\} = [\Lambda][B]\{U\} \tag{4.4}$$

$[\Lambda]$ is the matrix of mechanical characteristics, which are likely to include the damage effects. In this case, the resolution becomes nonlinear, and the expression of internal forces [4.1] is replaced by:

$$(F)^e = \int_{Ve} [B]^t \, \sigma dv \tag{4.5}$$

In this expression, σ nonlinearly depends on $\{U\}$ and, consequently, the force vector $(F)^e$ as well.

The assembly of these elementary force vectors $(F)^e$ makes it possible to obtain a global “internal” force vector $\{F\}^{\{int\}}$. Equation [4.2] is, therefore, replaced by the nonlinear equilibrium equation [4.6]:

$$\{R\} = -\{F\}^{\{int\}}(\{U\}) + \{F\} = \{0\} \tag{4.6}$$

In this equation [4.6], the displacement vector $\{U\}$, such as the residue $\{R\} = \{0\}$, is sought to ensure equilibrium. To do this, a Newton–Raphson method is used (see section A.2.3.2), which requires $\{R\}$ to be linearized by searching for a tangent operator $[K] = \frac{\partial \{F\}^{\{(int)\}}}{\partial \{U\}}$. By using an iterative process, the objective $\{R\} = \{0\}$ is reached, and the solution vector $\{U\}$ then makes it possible to return to $\{\boldsymbol{\varepsilon}\}$ and $\{\boldsymbol{\sigma}\}$, as well as to any other variable of interest, such as that of damage.

These points are discussed in more detail in section A.2 for four specific categories of elements that will be used below: bar elements, Timoshenko beam elements, Tri3 surface elements (triangles with 3 nodes) and Quad4 surface elements (quadrilaterals with 4 nodes). In addition, the literature offers numerous books on the subject (see Gallagher 1976; Zienkiewicz et al. 2013).

In the remainder of this chapter, we will discuss numerical procedures associated with the models presented in Chapter 3. On the basis of specific applications, we will see, particularly in Chapters 6 and 7, how to enrich the modeling to deal with a certain number of loading situations that cause damage, particularly in dynamics or at an early age.

4.2. Principle flowcharts

4.2.1. *Flowchart for the Mazars model*

1) Data:

- geometric characteristics;
- boundary conditions;
- material data (E, ν, ε_0, A_t, B_t, A_c, B_c).

2) Load increment (force or displacement).

3) Construction of the stiffness matrix.

4) Calculation of displacements.

5) Calculation of strains and principal strains:

- calculation of the equivalent local strain:

$\varepsilon_{eq} = \sqrt{\sum_i \langle \varepsilon_i \rangle_+^2}$ (see equation [3.2]);

- partition of the strains into ε_{ti} and ε_{ci};
- calculation of: $\alpha_t = \sum_i \mathrm{H_i} \frac{\varepsilon_{ti}(\varepsilon_{ti}+\varepsilon_{ci})}{\varepsilon_{eq}^2}$, $\alpha_c = \sum_i \mathrm{H_i} \frac{\varepsilon_{ci}(\varepsilon_{ti}+\varepsilon_{ci})}{\varepsilon_{eq}^2}$ (see equations [3.6]–[3.8]).

6) Damage criterion:

$\varepsilon_{eq} < \varepsilon_0$ $\mathrm{D} = 0$

$\varepsilon_{eq} < k_D = \varepsilon_{eq}$ (max. reached) $D = \text{const}$ (see equations [3.4], [3.9] and [3.10])

$\varepsilon_{eq} > k_D = \varepsilon_{eq}$ (max. reached)

so: $D_t = f(\varepsilon_{eq}, A_t, B_t)$, $D_c = f(\varepsilon_{eq}, A_c, B_c)$ and $D = \alpha_t D_t + \alpha_c D_c$.

If the damage increases, then a convergence loop is activated between 6 and 3 to ensure equilibrium by minimizing the residues calculated on the evolution of the strains between two consecutive loops.

7) Calculation of stresses:

$\boldsymbol{\sigma} = (1-D)\boldsymbol{\Lambda}: \boldsymbol{\varepsilon}$ and of external forces (see equation [A.1]).

8) Graphical outputs.

4.2.2. Flowchart for the μ model

1) Data:

- geometric characteristics;

- boundary conditions;

- material data (E, ν, ε_{t0}, ε_{c0}, A_t, B_t, A_c, B_c, k).

2) Load increment (force or displacement).

3) Construction of the stiffness matrix.

4) Calculation of displacements.

5) Calculation of strains:

- determination of the principal strains and principal effective stresses (based on the characteristics of the undamaged material):

$\boldsymbol{\sigma} = \lambda Tr(\boldsymbol{\varepsilon})\mathbf{1} + 2\mu\boldsymbol{\varepsilon}$ (see equations [A.2] and [A.3]);

- calculation of the equivalent strains ε_t, ε_c (see equations [3.21] and [3.22]) and the triaxiality factor r (see equation [3.26]);

- determination of the pilot variable:

$Y = rY_t + (1-r)Y_c$ ($Y_{t,c} = \max(\varepsilon_{t0,c0}, \max \varepsilon_{t,c})$);

- calculation of $A = f(A_t, A_c, r)$ and $B = g(B_t, B_c, r)$ (see equation [3.28]).

6) Damage criterion:

$Y < Y_0 = r\varepsilon_{t0} + (1-r)\varepsilon_{c0}$, so $d = 0$

$Y < r\, max\varepsilon_t + (1-r)max\varepsilon_c$, so $d = \text{const}$

$Y > r\, max\varepsilon_t + (1-r)max\varepsilon_c$, so $d = f(Y, A, B)$

If the damage increases, then a convergence loop is activated between 6 and 3 to ensure equilibrium by minimizing the residues calculated on the evolution of the strains between two consecutive loops.

7) Calculation of stresses:

$\boldsymbol{\sigma} = (1 - D)\boldsymbol{\Lambda}: \boldsymbol{\varepsilon}$ and of external forces (see equation [A.1]).

8) Graphical outputs.

NOTE 4.1.– For the μ model, the search for the loading-unloading criterion requires some precautions, mainly related to the value of r. Details of this problem can be found in section 9.4.1.

NOTE 4.2.– Section 4.5 discusses non-local damage. This concept can be easily integrated into the above flowcharts. For this purpose, reference is made to section 9.2.2, which describes the procedure.

4.3. Data preparation

Data preparation is an essential aspect for the success of a numerical calculation. The time spent on this subject is greatly improved compared to the one which consists of groping with data values by performing calculation upon calculation with the hope of converging toward a good result. This preparation mainly concerns three points:

– the fineness of the finite element mesh, which must be adjusted to sensitivity to damage; an initial "rough" calculation may make it possible to estimate its location and extent;

– boundary conditions, when dealing with materials with low adaptability, such as concrete, have a major impact on the results; a sensitivity study may in some cases be necessary to adjust a particular component;

– material data, generally provided on the basis of experiments on laboratory specimens (otherwise they will be derived from catalog or regulatory data).

This last point, which is specific to the use of the models, is developed in section 4.3.1.

4.3.1. *Identification of modeling parameters*

There are seven parameters for the Mazars model (E, ν, ε_0, A_t, B_t, A_c, B_c) or eight for the μ model (E, ν, ε_{t0}, ε_{c0}, A_t, B_t, A_c, B_c). They must be identified from simple tests, some of which are recommended in sections 4.3.1.1 and 4.3.1.2.

4.3.1.1. *E*, *ν*, ε_{c0}, A_c, B_c *identifiable from compression tests*

This test, generally carried out on a cylinder (preferred to testing on a cube) and whose dimensions in France are 16 cm in diameter and 32 cm in height, allows for generating, in the central zone of the specimen, a state of stress that is practically uniform. It is necessary to measure the loading force and, with instrumentation based on displacement sensors and/or strain gauges, the vertical and transverse strains in the central zone. With a controlled displacement test, curves $\sigma_l = f(\varepsilon_l)$ and $\sigma_l = g(\varepsilon_2)$ are easily accessed. Several phases appear:

– a linear-elastic phase from which the following is deduced between two points: $E = \frac{\Delta\sigma_1}{\Delta\varepsilon_1}$ and $\vartheta = -\frac{\Delta\varepsilon_2}{\Delta\varepsilon_1}$;

– a nonlinear phase with increasing stress up to a maximum, which gives $f_C = \frac{F_r}{S}$, where F_r is the maximum force of the curve and *S* is the cross-section of the cylinder;

– a nonlinear phase with decreasing stress.

From these three phases, the best values are identified for *E*, *ν*, A_c and B_c, which makes it possible to reproduce these curves. However, it is first necessary to set the damage threshold ε_{l0}; for this, the strategies differ depending on whether the Mazars model or the μ model is used.

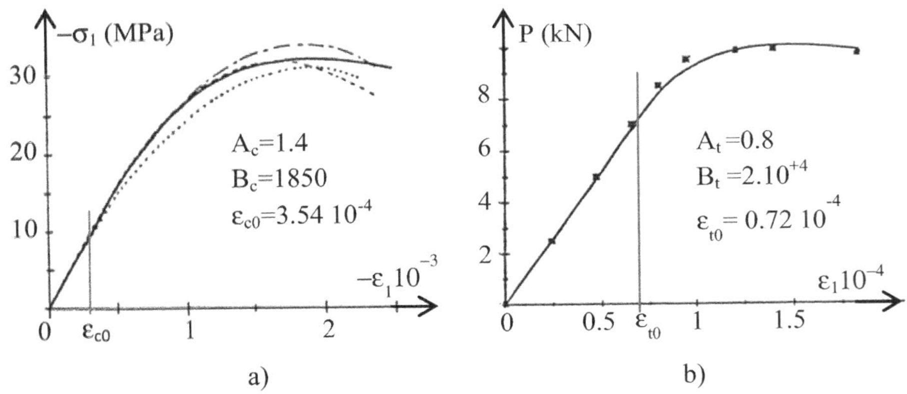

Figure 4.1. *μ model identification (dashed lines or points are experiments, solid lines are calculations using identified parameters)*

COMMENTS ON FIGURE 4.1.– *(a) direct identification of the compression characteristics from three compression tests on a cylinder; (b) indirect identification of the tensile characteristics from a 4-point bending beam test (length: 1.4 m, width: 0.22 m, height: 0.15 m): curve showing the evolution of the strain at the location of the crack on the tensioned side of the beam as a function of the overall load P.*

For the Mazars model, we observed in Chapter 3 that under compression, the threshold was such that $\varepsilon_{10} = -\frac{\varepsilon_0}{\vartheta\sqrt{2}}$, whereas under direct tension, $\varepsilon_{10} = \varepsilon_0$. ε_0 is the value of ε_{eq} at the damage threshold. It is therefore preferable to deduce the threshold ε_0 of a direct tension test and to use this value to set ε_{10} under compression and then identify the values of A_c and B_c on the curve $\sigma_1 = f(\varepsilon_1)$ (see Figure 4.1(a)). Furthermore, it goes without saying that several tests are necessary to improve the reliability of the result.

For the μ model, we have seen that under tension, ε_{t0} and under compression, ε_{c0}, the thresholds differ. Consequently, curves $\sigma_1 = f(\varepsilon_1)$ and $\sigma_1 = g(\varepsilon_2)$ will be used to determine, at the end of the linear part, the best value that leads to $\varepsilon_{10} = -\varepsilon_{c0}$ (see Figure 4.1(a)).

4.3.1.2. *E, ε_{t0} (ε_0) identifiable from tensile tests (direct or indirect)*

The direct test is shown in Figure 4.2(c). For common concretes, we can assume that the damage threshold corresponds to the peak of the σ–ε curve. Thus, for a specimen whose section S is known, the measurement of the resistive force F_r gives $f_t = F_r / S = E\varepsilon_{t0}$, which gives ε_{t0} if E is known. In this section, only ε_{t0} will be considered, but it goes without saying that all the relationships found are the same for ε_0 (see Mazars model threshold).

If the tensile test piece is instrumented (displacement sensor or strain gauge system at its center), then the complete curve σ–ε can be plotted, and E and ε_{t0} can be deduced. If this is not the case, E may be deduced beforehand from a compression test (see section 4.3.1.1).

Since the direct tensile test is not easy to carry out because it requires the specimen heads to be glued to the platens of the testing machine, it is sometimes preferred to carry out a tensile test by bending (see Figure 4.2(a)), carried out on a prismatic test piece. The tensile strength of the material is deduced from the force P_r at failure of the specimen. It is assumed that linear behavior up to failure appears on the tensioned side in the zone of maximum bending moment between the two loads. Thus, we have:

$$f_t = \frac{3P_r}{a^2} = E\varepsilon_{t0} \qquad [4.7]$$

which gives ε_{t0} if E is known.

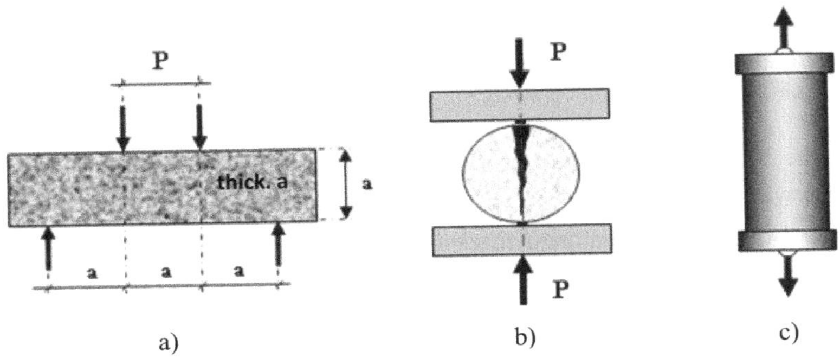

Figure 4.2. *Tests on specimens to identify the characteristics of concrete under tension: a) indirect tensile test by bending; b) indirect splitting test; c) direct tensile test*

An instrumentation by gauge on the tensioned side of the beam makes it possible to trace the curve $P = f(\varepsilon)$ from which it is possible to identify, by inverse analysis, the various parameters such as E, ε_{t0}, A_t and B_t (see Figure 4.1(b)). The writing of the problem that links P and ε in the framework of the theory of beams is analytical. The reasoning takes place in the central part of the beam (the bending moment M is constant), and it is written in a straight section of this zone as follows:

– the resultant of the torsor of the generalized stresses is zero; the stress is then expressed as a function of the local damage, which makes it possible, for a given loading stage, to locate the neutral fiber (that is, where $\sigma = 0$);

– the moment resulting from the generalized stresses is equal to the bending moment ($M = Pa/2$), which makes it possible to relate the location of the neutral fiber to the applied load.

The analytical expressions thus constructed are a function of ε_{t0}, A_t and B_t, and a resolution by successive approaches makes it possible to find the best trio to describe the experimental curve (see Figure 4.1(b)). This method is more reliable than that which consists simply of using the load at failure to deduce f_t, since it allows for better definition of the nonlinearity threshold which always appears before reaching P_r.

Often, the so-called "splitting" test is the preferred of these two tests (see Figure 4.2(b)). The results are evaluated by considering a linear behavior up to failure (P_r), which appears under tension on a vertical plane and gives:

$$f_t = P_r/\pi Rh \qquad [4.8]$$

R and h are the radius and length of the test cylinder, respectively.

This being the case, the experimenters note that, for the same concrete, the f_t values found by the three types of tests differ: f_t direct tension < f_t tension by bending < f_t. tension by splitting.

The explanation of this fact is mainly linked to the scale effect, which is of two types:

– the volume effect, which leads to the observation that the strength decreases when the volume of the specimen increases;

– the structural effect, which leads to the observation that the presence of non-uniform stresses (bending) or combined stresses (splitting) increases the strength.

The first of these effects can be treated by the Weibull distribution (see Mazars 1984), which leads, for two volumes (V_1 and V_2) from direct tension specimens, to:

$$\frac{\varepsilon_{t0}^1}{\varepsilon_{t0}^2} = \left(\frac{V_2}{V_1}\right)^{\frac{1}{m}} \qquad [4.9]$$

m is the Weibull parameter (of the order of 10 for common concrete).

The second is more difficult to process; it implies taking into account the volume share of the specimen highly loaded under tension:

– For example, for the specimen in Figure 4.2(a); if it is loaded under direct tension by both ends, the volume V_1 to be considered is $3a^3$, whereas under tension by bending, the volume is limited to the central zone below the neutral fiber, that is, V_2 # $a^3/2$, which, from equation [4.9] leads with m = 10 and a = 0.1 m, to:

$$\varepsilon_{t0}^1 = \varepsilon_{t0}^2 \left(\frac{V_2}{V_1}\right)^{\frac{1}{m}} \# \ 0.85\varepsilon_{t0}^2 \qquad [4.10]$$

Thus, the result under tension by bending is of the order of 15% higher than under direct tension for a specimen of the same volume.

– If we consider that specimen 2 is a cylinder tested under tension by splitting, of the same volume as specimen 1 tested under direct tension, it has been shown that the volume highly loaded under tension by splitting is of the order of 5% of the total volume, which leads, using equation [4.9], to the following evaluation:

$$\varepsilon_{t0}^1 = \varepsilon_{t0}^2 \left(\frac{0.05\, V_1}{V_1}\right)^{\frac{1}{m}} \# \ 0.74\varepsilon_{t0}^2 \qquad [4.11]$$

The result under tension by splitting is of the order of 25% higher than that under direct tension for a specimen of the same volume.

These considerations lead to a subsidiary but fundamental question: what should be considered as the value for ε_{t0}, for a structure whose dimensions of elements are clearly different from those of the test specimens?

A proposal to consider the V_{eq} volume for a structural element, particularly if it is massive, can be found in (CEOS.fr 2017):

$$V_{eq} = \text{r. } (\min (1.25, \text{L}). \min (1.25, \text{b}). \min (1.25, \text{h}_{\text{ef}}) \quad [4.12]$$

where L, b and h_{ef} are the length, width, and effective height, respectively. The effective height is equal to the total height for a tie rod and is of the order of half the height in bending. r is 1 under pure tension and $1/(m + 1)$ under pure bending (m, Weibull coefficient).

Thus, for a test tie rod (the one proposed in the applications of section 6.2), the dimensions of which are L = 5.10 m, b = 0.5 m, and h = h_{ef} = 0.8 m, we find with r = 1, V_{eq} = 0.5 m^3. Assuming that laboratory tests are carried out under direct tension on conventional cylindrical specimens of 0.16 m in diameter and 0.32 m in length (V_{td} = 0.0064 m^3), we get from this tie rod:

$$\varepsilon_{t0}^{td} = \varepsilon_{t0}^{T} \left(\frac{V_{eq}}{V_{td}}\right)^{\frac{1}{m}} = 1.55\ \varepsilon_{t0}^{T} \quad [4.13]$$

However, if, as in the present case, the laboratory tests were carried out by splitting, for which we have seen in equation [4.11] $\varepsilon_0^{td} \# 0.74\varepsilon_0^f$ (the indices used are *td*, *f*, *T*, for direct tension, tension by splitting and for the tie rod, respectively), this leads to:

$$\varepsilon_{t0}^{f} = \frac{1.55}{0.74}\varepsilon_{t0}^{T} \# 2.1\varepsilon_{t0}^{T} \quad [4.14]$$

In conclusion, the results of the splitting test are divided by 2.1 to address the calculations on the proposed tie rod.

4.4. Concrete fracturing energy

This energy is denoted as G_F; it characterizes the density of energy consumed to fracture a piece of concrete. For a specimen loaded under tension, as shown in Figure 4.3(a), G_F may be determined by considering the area under the $\sigma - \omega$ curve (w overall elongation of the reference length and w_r elongation at failure):

$$G_F = \int_0^{w_r} \sigma\, w dw \quad [4.15]$$

Figure 4.3(b) presents the principle for determining this energy recommended by RILEM (1990). The test is carried out on a pre-notched bending beam, the loading is carried out with imposed displacement, to have access to the post-peak curve, and the area under the curve (P-δ) gives the energy consumed to fracture the specimen. The fracturing surface A is that of the ligament remaining above the pre-notch, and so we obtain:

$$G_F = \frac{\int P\delta d\delta}{A}, \text{ expressed in N/m} \qquad [4.16]$$

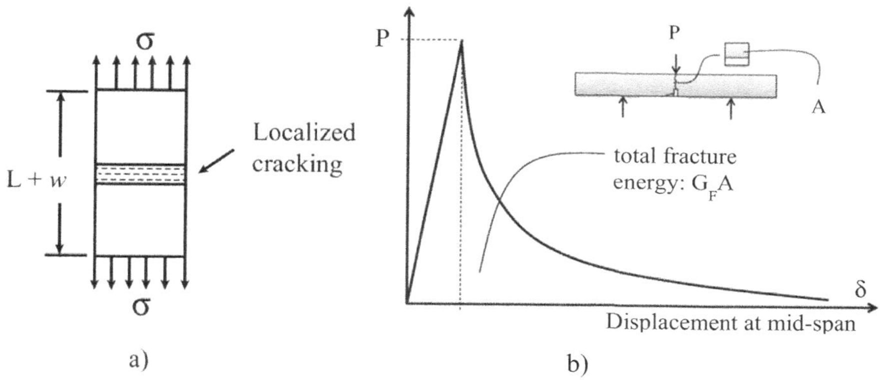

Figure 4.3. *Fracturing energy: a) test of fracturing under tension; b) G_F determination procedure (RILEM 1990 test)*

4.4.1. *Intrinsic and extrinsic energy*

Based on work done by Ritchie (2013) on bone fracturing, we propose an interpretation of concrete fracturing that leads to distinguishing two types of energy. As a pre-crack is created, the progression of this crack takes place in two stages (see Figure 4.4(a)):

– phase of damage, beyond the tip of the visible crack, which will gradually lead to the development of this same crack;

– an opening phase of the crack necessary for the progression of the crack front to take place.

Both phenomena consume energy. We call "intrinsic energy" G_f, which is consumed to damage until the crack front progresses, and "extrinsic energy", which allows the crack to open. Indeed, material bridges oppose this opening, especially when created by the embedding of the aggregates in the cement matrix, which yield by consuming a part of the fracturing energy: $G_F - G_f$.

This presentation is similar to that of Planas et al. (1992) and Bazant (2002), who proposed the distinction of two types of energy in the curve in Figure 4.4(b) and the use of a trilinear tension behavior curve, as shown in Figure 4.4(c). According to some authors, the ratio $G_F/G_f \approx$ 2 to 3 (see Muralidhara Rao et al. 2009).

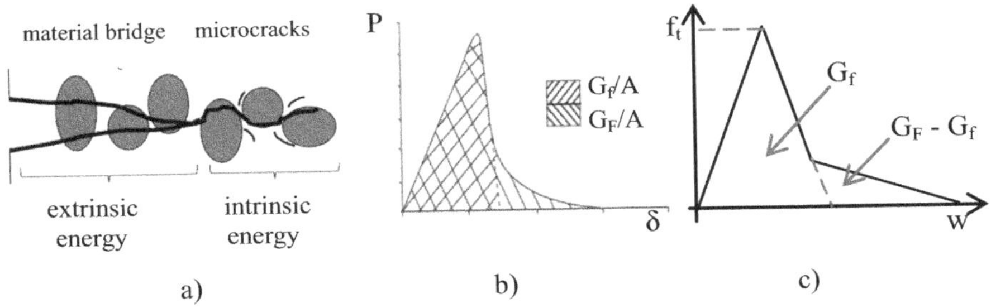

Figure 4.4. *Intrinsic energy G_f and extrinsic energy (G_F-G_f). For a color version of this figure, see www.iste.co.uk/mazars/damage.zip*

COMMENTS ON FIGURE 4.4.– *(a) distribution diagram of the zones of intrinsic and extrinsic energy; (b) typical force–displacement response showing the distribution of these energies on a behavior curve (from Muralidhara Rao et al. (2009)), (A: fracture area); (c) post-peak part of a stress–crack opening curve modeled by two straight lines.*

4.4.1.1. *Use of fracturing energy for identification of A_t and B_t*

When G_F, and therefore Gf, is known, (on the basis of bending tests on pre-notched specimens), and on the basis of the above results, the recommended identification procedure is as follows:

The Mazars and μ models use the same behavior curve:

$$\sigma_1 = E(1-d)\sigma_1 \text{ and } d = d_t = 1 - \frac{(1-A_t)\varepsilon_{t0}}{\varepsilon_t} - A_t \exp\left(-B_t(\varepsilon_t - \varepsilon_{t0}\right) \quad [4.17]$$

with, for the Mazars model, the variables D_t, ε_{eq} and ε_0.

This is reflected in the curve proposed in Figure 4.5. We again find the pre-peak straight line with directional coefficient E, the post-peak curve of exponential and asymptotic nature whose shape is determined by the values of A_t and B_t.

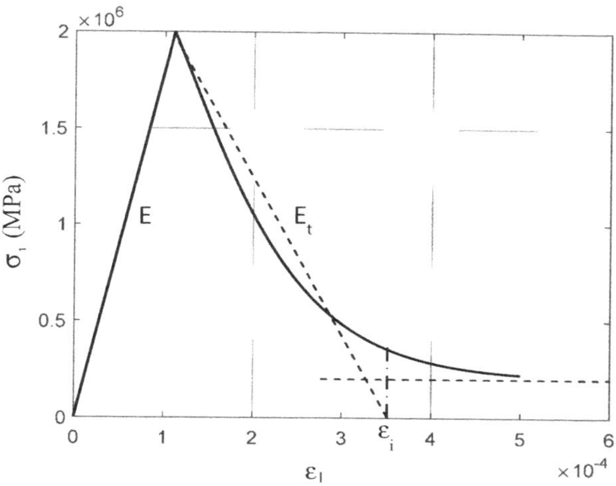

Figure 4.5. *Tension curve of the Mazars model and the μ model*

The asymptote is conditioned by the value of A_t: for $A_t = 1$, it merges with the X axis, and for $A_t = 0$, the post-peak curve becomes a horizontal linear plateau from the peak. The experimental results call for an A_t close to 1, but as we will see in the applications, the numerical stability of damage calculations is improved if A_t is slightly lower than 1. Thus, in the context presented in the introduction of section 4.4.1, we share the area of the curve in the intrinsic part and in the extrinsic part, and assume that the curve is chosen to better adhere to the bilinear post-peak plot proposed in Figure 4.4(c), with:

– the horizontal straight line (asymptote) defined by A_t set in advance to a value of the order of 0.9, which ensures the extrinsic part of the fracturing energy $(G_F - G_f)$;

– the straight line leading coefficient E_t, which ensures the intrinsic part G_f.

B_t will be determined in a way that the exponential post-peak is divided into two. Until ε_i, we obtain the intrinsic part:

$$G_f = h \int_0^{\varepsilon_i} \sigma_1 \varepsilon_1 \, d\varepsilon_1 \quad [4.18]$$

h is the dimension on which the analysis operates, for example, the size of a finite element of unit thickness that undergoes fracture.

The extrinsic part appears beyond the strain ε_i, which is the strain characteristic of the opening of a visible crack; it is defined by:

$$\varepsilon_i = \frac{2G_f}{hf_t} \quad [4.19]$$

4.4.2. *Crack band concept, Hillerborg regularization*

The crack band concept, defined by Bazant and Oh (1983), is based on the fact that the fracturing of a specimen loaded under tension leads to a localization of strains that foreshadow the failure, as shown in Figure 4.3(a). In a classic 2D finite element calculation that uses a post-peak behavioral law, as presented in Figure 4.5, we find that this phenomenon is translated by a localization of strains (and thus of damage) on a band of elements. In section 9.3.2, there is an exercise detailing this point.

If this band of elements (considered to be of unit thickness) has a width h, then, as seen in equation [4.18], the intrinsic energy that will be consumed during the fracturing will be:

$$G_f = h \int_0^{\varepsilon_i} \sigma_1 \varepsilon_1 \, d\varepsilon_1 \quad [4.20]$$

Being an intrinsic characteristic of the material, G_f has a constant value during the process. Thus, this relationship creates a dependency between the behavior curve σ–ε and the dimension of the band of elements h. Based on the post-peak bilinear representation, we obtain from equation [4.12]:

$h = \frac{2G_f}{f_t \varepsilon_i}$, or with the notations of Figure 4.5

$$h = 2\frac{G_f}{f_t^2}\left(\frac{1}{E} - \frac{1}{E_t}\right)^{-1} \quad [4.21]$$

If h is fixed by the mesh, then ε_i and thus E_t will be determined by these relationships (or inversely, if the curve is fixed, we get h from it). This approach is called the Hillerborg method (see Hillerborg et al. 1976).

4.5. Non-local damage concept

The problems that the localization of strains present led researchers to find procedures that allow us to avoid this dependence between behavior and meshing, as seen in section 4.4.2. For this purpose, Pijaudier-Cabot and Bazant (1987) proposed the concept of non-local damage. There are two major points to consider:

– the definition of a characteristic length on which damage will occur around the cracking zone. This length l_c is linked to the size of the heterogeneities (l_c # 2 to 3d_a, d_a dimension of the largest aggregates);

– the choice of distribution of these damages around the principal crack: the farther the distance, the less marked the damage; this distribution is often chosen Gaussian.

Thus, by taking the example of the Mazars model in a 2D finite element calculation, the concept is used in the following manner:

– in the zones susceptible to damage, we determine around each point with coordinates ($\boldsymbol{x}$) of the calculated object, a vicinity V_r: circle centered on the point and with the radius l_c;

– to each Gauss point of the mesh with coordinates ($\boldsymbol{s}$) which are in this vicinity, we attribute a weight $g(\boldsymbol{x} - \boldsymbol{s})$;

– to a given stage of calculation, we classically determine the equivalent strain ε_{eq} in each point, but before determining the damage, we will "force" the distribution of damage using the concept thus presented, and this will be done irrespective of the size of elements:

- for the point considered ($\boldsymbol{x}$), we determine an equivalent strain called non-local strain, which integrates the effects of the vicinity V_r:

$$\bar{\varepsilon}_{eq} = \frac{\int_{V_r} g(\boldsymbol{x}-\boldsymbol{s})\varepsilon_{eq} dV}{\int_{V_r} g(\boldsymbol{x}-\boldsymbol{s}) dV} \qquad [4.22]$$

where:

$$g(x-s) = \exp\left[-\left(\frac{2(x-s)}{l_c}\right)^2\right] \text{ if } |x\text{-}s| < l_c \text{ and } g(x-s) = 0 \text{ if } |x\text{-}s| > l_c$$

- damage is calculated for each Gauss point using the non-local equivalent strain $\bar{\varepsilon}_{eq}$ instead of ε_{eq}.

This procedure spreads out the damage and eliminates the effects of localization of strains. It is shown (see application in section 9.3) that, provided that there is a sufficiently fine mesh in the damage zones ($h < l_c/3$), the dependence on the mesh disappears, and the stability of the calculation is found to have improved.

For example, we propose in Figure 4.6 the result of the effects of non-localization on two examples of application (see Saouridis and Mazars 1989, 1992).

1) A three-point bending concrete pre-notched beam (see Figure 4.6(a)) for which we find, as shown above, that, contrary to classic calculation, the damage spreads in the vicinity of the crack, which appears in the extension of the pre-notch. The calculation carried out with several meshes showed the independence of the result with the fineness of the mesh.

2) The splitting test: the particularity of this loading does not lead to good results in the local approach. The reason is a localization that concentrates under the applied load and leads to a collapse of some elements connected to this zone (mainly in high compression) and obtaining a very premature ultimate load. In contrast, the non-local approach leads to a diffusion of the damage in the central zone, which is under horizontal tension, allowing for a very correct simulation of the vertical cracking and the failure load (see Figure 4.6(b)).

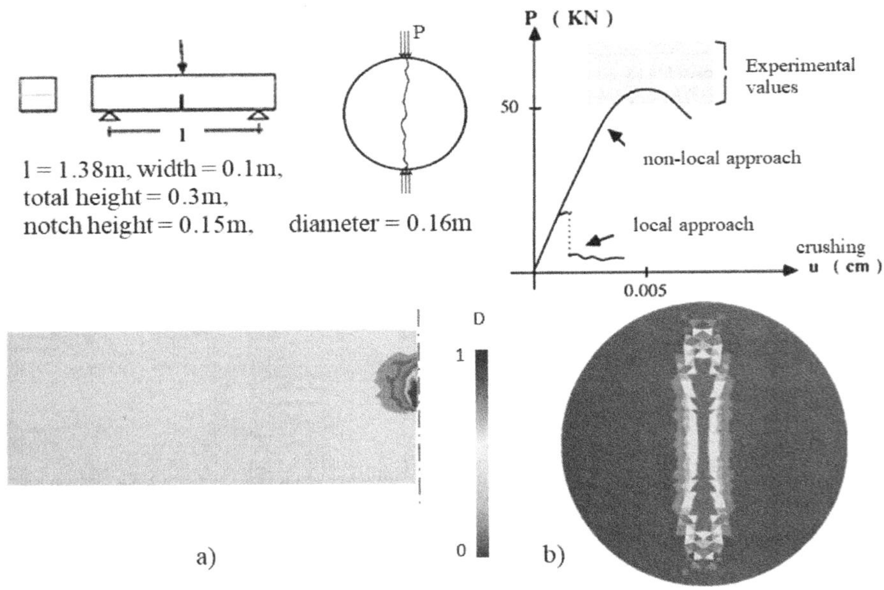

Figure 4.6. *Non-local damage: a) pre-notched concrete beam; b) splitting test (from Saouridis and Mazars 1989, 1992). For a color version of this figure, see www.iste.co.uk/mazars/damage.zip*

These two applications show that, in the case of simulations associated with the cracking of concrete, the non-local procedure is valuable for, thanks to the diffusion of damage, regularizing local instabilities. When concrete is reinforced, the problem is less crucial because the rebars have this regulatory power. This being the case, this diffusion has its setback, that of partially eliminating the characteristics of the crack that opens (path and opening). Recent works have revealed more refined procedures for recovering these characteristics (see Giry et al. 2011).

NOTE 4.3.– In the case of a classic finite element calculation, the last two paragraphs have shed light on the following:

– localized damage which must be handled by connecting the behavior to the size of elements (Hillerborg method);

– diffused damage, which results in the integration of the concept of non-local damage.

In Chapter 5, we will see that localized damage is handled by the crack band integrating the Hillerborg regularization works correctly for reinforced concrete calculations, which are at the heart of this book. This approach is particularly useful if we want to characterize the cracking. This being said, the non-local damage favors the stability of calculation, and we will see that the use of Timoshenko's beam elements generates, in most situations of loading of slender reinforced concrete structures, an automatic diffusion of damage (see section 5.2), which presents the advantages of non-local damage procedures without having to implement them.

Finally, it is good to signal that non-local procedures are also able to deal with the scale effect, which was discussed in section 4.3.1.2, for its structure effect component; we find some of the results on this subject in Mazars et al. (1991).

5

Applications to Common Reinforced Concrete Structural Elements Cases

The applications proposed in this chapter are mainly linked to severe loading situations on reinforced concrete structures. In all cases, the ultimate limit states are aimed for, whether they have monotonous, cyclic loads or dynamic loads. Among these applications, the effects of earthquakes are particularly representative: they generate nonlinear cyclic loads on structural elements, and the velocities of application are sufficiently low to not be taken into consideration. In this context, cyclic loadings will be of particular concern. The case where the velocity of loads has an actual incidence on behavior will be seen in the rest of the book (see section 6.1).

In this chapter, we will discuss the factors of numerical calculation under two aspects: the use of conventional 2D finite elements (FE) (see section 5.1) and the use of Timoshenko multifiber beam elements (see section 5.2), and both of these in the context of monotonous or cyclic loads without the effect of velocity. The μ model, adapted to cyclic loads, will serve as the basis for the study.

5.1. 2D FE calculation

The LMT laboratory of ENS Paris-Saclay (France) conducted an experimental campaign on reinforced concrete elements to study the phenomena that play a major role in the response of structures during an earthquake (Ragueneau et al. 2010; Crambuer et al. 2013). Phenomena such as the evolution of damage during an increase in load, unilateral effects, and the dissipation of energy due to cyclic loads have all been analyzed. These results will serve as a reference for the applications below.

5.1.1. *Details of the experimental program*

Reinforced concrete beams were tested under bending, as indicated in Figure 5.1. The specimens measured 1.65 m in length, 0.22 m in height and 0.15 m in width. The entire campaign considered different longitudinal rebar ratios; we chose analyses that followed only specimens reinforced with four rebars of 12 mm (Figure 5.1(a)). The concrete is a regular C30/37, the characteristics of which are listed in Table 5.1.

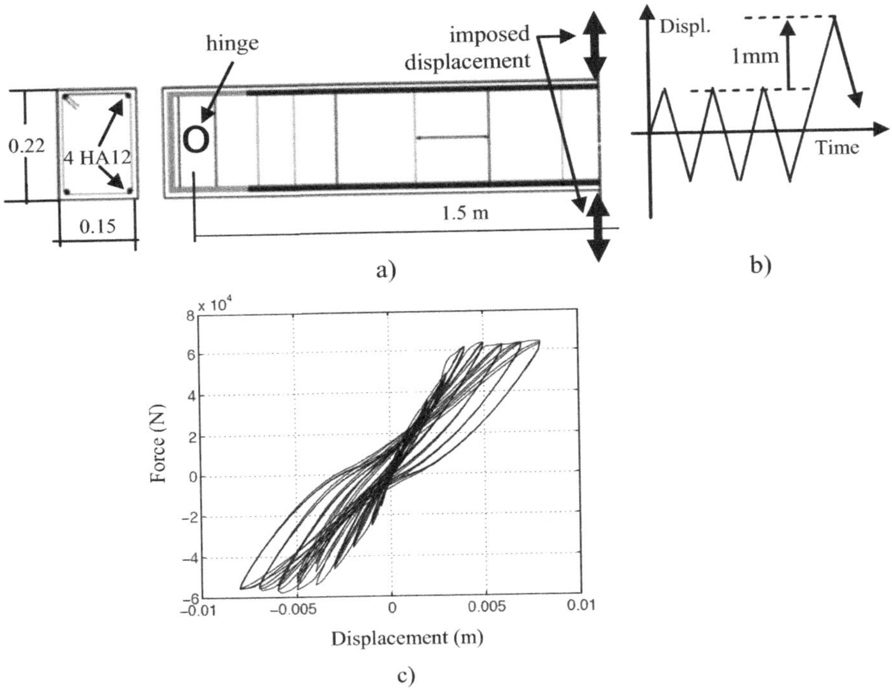

Figure 5.1. *Three-point bending beam tests on RC: a) geometry of the specimen and boundary conditions; b) loading program: series of three cycles from ±1 mm to ±8 mm; c) force–displacement response for the entire loading program (case of beam no. 2)*

It is a three-point bending loading, with vertical displacement when ascending and descending. A device for a specific articulation ensures free rotation to the ends of the beam while preventing vertical displacement. The loading is displacement controlled and consists of sets of three cycles with progressively increasing intensity (±1 to ±8 mm; see Figure 5.1(b)).

Experimental data							
Steel			Concrete				
E (GPa)		f_e (MPa)	E (GPa)		f_c (MPa)	G_F (N/m)	
205		530	28		35	90	
Parameters for the μ model							
E (GPa)	ν	ε_{t0}	ε_{c0}	A_t	B_t	A_c	B_c
28	0.2	1×10^{-4}	3.6×10^{-4}	0.8	7,000	1.25	395

Table 5.1. *Three-point bending tests on reinforced concrete beams: experimental data and material parameters*

Figure 5.1(c) shows the force-displacement response of beam no. 2 for the entire loading program. We find:

– a progressive decrease in the stiffness due to concrete damage during the first cycle series;

– an appearance of plasticity of rebars (in the central zone) after ±4 mm cycles and its prevalence on behavior beyond this step.

5.1.2. *Numerical processing*

The trial specimen was modeled using four-node quadrilateral FE-type Q4 (see section A.3.2) and bar elements for the rebars (see section A.3.1). The problem is considered in the plan with, taking into consideration the thickness of the beam, the plane strains hypothesis ($\varepsilon_3 = 0$). Moreover, the symmetry of the problem is used, and the mesh for the half-beam (711 nodes, 776 elements) is uniform on the central part of the beam. Boundary conditions are defined in such a way as to correctly represent the experimental test (see Figure 5.2(a)). The imposed displacement Uy is applied on the higher and lower parts of the central part of the beam.

To avoid mesh sensitivity, the crack band approach, based on the Hillerborg regularization, was used for this application (Bazant and Oh 1983). The G_F experimentally estimated at 90 N/m led to the use of G_f = 30 N/m for numerical calculations (see section 4.4.2), which signifies that in the central part of the beam (i.e. where damage and plasticity are concentrated), the size of the elements is consistent with the width of the "crack band" (see section 4.4.2):

$$h = \frac{2G_f}{f_t^2}\left(\frac{1}{E} - \frac{1}{E_t}\right)^{-1} \quad [5.1]$$

E_t is the tangent stiffness after the peak for a triangular form respecting the equivalence in intrinsic cracking energy (see Figure 4.5).

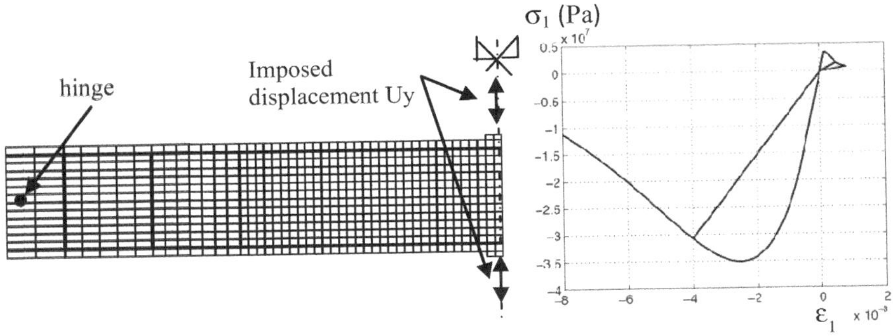

a) b)

Figure 5.2. *Three-point bending tests: a) mesh used for the half-beam; b) behavior under tension-compression of the simulation concrete with the μ model*

The values of the parameters of the selected model were adopted in accordance with the data provided in Table 5.1, and the resulting behavior for concrete under tension–compression is given in Figure 5.2(b), showing the unilateral character of the μ model.

With regard to rebars, bar elements are used with a perfect elastic–plastic behavior model, consistent with the experimental data in Table 5.1.

5.1.3. *Results*

5.1.3.1. *On a global scale*

Various situations have been modeled. Figure 5.3 compares the experimental measurements with the load–displacement curves resulting from the simulations:

– for the total load path, the calculation (see Figure 5.3(a), continuous line), carried out without any cycle, gives a curve in good agreement with the cycle envelope for the set of experimental curves;

– for the loading trajectory for cycles up to ±2 mm, the calculation-experiment comparison shows very good agreement (see Figure 5.3(b));

– for a cyclic load of ±5 mm, including the plasticity of the rebars (see Figure 5.3(c)), the results are also satisfactory.

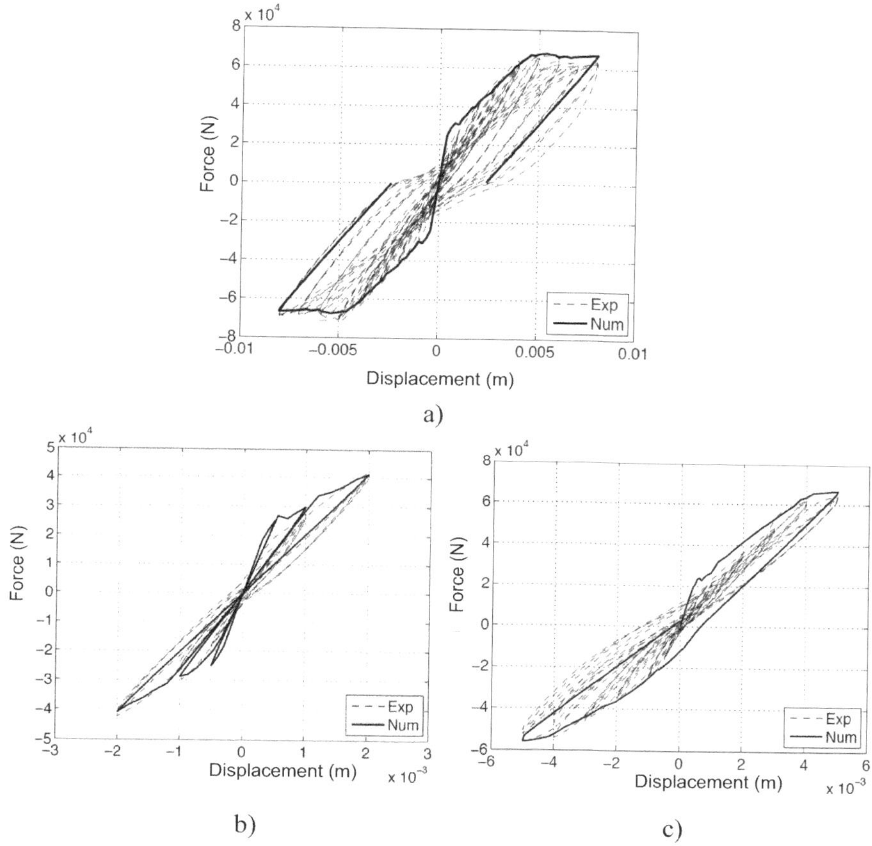

Figure 5.3. *Bending test on RC beams, 2D experiment-calculation comparison. a) Comparison of the total experimental response with an envelope calculation, performed without any cycle; b) comparison for the complete path up to ±2 mm; c) simulated cycle at ±5 mm, compared to the overall experimental response up to ±5 mm*

5.1.3.2. *On the local scale*

The precision of 2D FE modeling allows access to local information relevant to the changes that occur in areas that degrade, both for concrete and rebars. Figure 5.4 shows, at a given stage of the load (±5 mm), the strain field on the lateral surface of the beam. In Figure 5.4(b), the strain field is obtained using the digital image correlation technique (Ragueneau et al. 2010), which in turn allows active cracks to be identified. Figure 5.4(a) shows the strain locations corresponding to the crack band, within which the damage obtained by calculation is also located. The dark blue zone corresponds to a compression strain, which is amplified in the compressed zones due to previous damage. Good agreement can be observed between the experimental and numerical images.

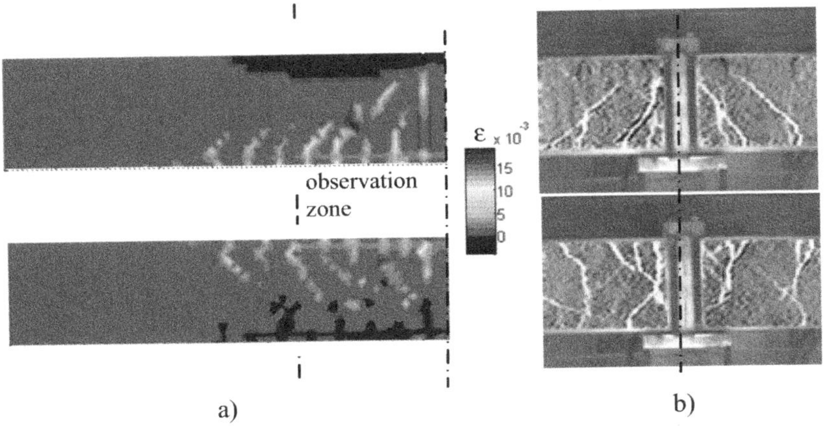

Figure 5.4. *Strain field on the lateral surface of the beam at a given stage (–5 mm at the top and +5 mm at the bottom), giving an image of the crack opening state: a) results of calculations; b) results of tests obtained by digital image correlations (DIC). For a color version of this figure, see www.iste.co.uk/mazars/damage.zip*

During cyclic loading, the effective damage d changes locally until a maximum value of the load applied in one direction is reached. When the load is reversed, the local stress is also reversed. As seen in Figure 3.6, due to the changes, first the triaxiality factor r, then the damage pilot variable Y (see equation [3.26]), the value of d is set to zero. This specific property can be used as an indicator of crack opening. In this context, Figure 5.5 shows the location of the effective damage d for two different steps after a series of cycles up to ±2 mm. The effective damage area is considered to be representative of a crack opening sequence, which, as shown in Figure 5.5, is consistent with the image correlation analysis.

Furthermore, if necessary, it is possible to visualize, following a given loading path, the total area of damage thanks to the thermodynamic variables Y_t and Y_c (see equation [3.25]). Two images can thus be derived: that of Y_t and the other of Y_c, representative of the total damage under tension and under compression, respectively.

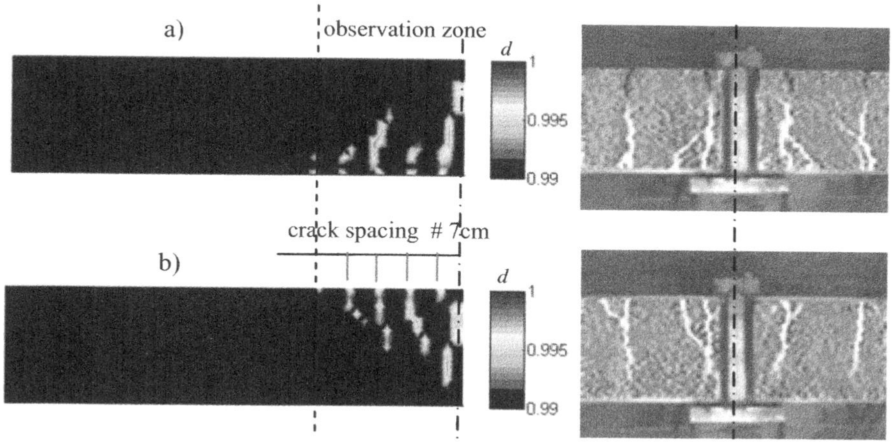

Figure 5.5. *Effective damage d used as an indicator of crack opening (d > 0.99) after cycles up to ±2 mm and cracks observed by digital image correlation for displacements of a) –2 mm and b) +2 mm. For a color version of this figure, see www.iste.co.uk/mazars/damage.zip*

These results show that the μ model is effective in describing the cyclical behavior of reinforced concrete elements at both the global and local levels. It gives, for example, details on strain localization and crack opening. The μ model is therefore a good candidate for dealing with monotonous and cyclical problems. This being the case, on the scale of a complete structure treated in 2D or 3D FE, a description which is necessary if the problems associated with the response to earthquakes are to be approached, it is still heavy to handle, in particular because of the size of the meshes to be processed.

5.2. Calculations by Timoshenko enriched beam elements

The idea here is to show that the problem presented in section 5.1 can be greatly simplified while allowing access to virtually similar information. To this end, two major points must be considered:

– the reduction in the size of the problem requires a simplification of the FE discretization by using, where possible, Timoshenko multifiber beam elements, which not only drastically reduce the number of degrees of freedom, but also lead to the need to manage only uniaxial behaviors;

– the price to be paid is the degradation of the fineness of the results to be expected; this being the case, we will show that a low-cost enrichment of the uniaxial behavior models (concrete and steel) makes it possible to regain a good relevance of the results.

5.2.1. *Strengths and weaknesses of the multifiber beam description*

While "conventional" FE models are powerful tools for simulating the nonlinear behavior of complex parts of structures (joints, connections, etc.), their application to the whole structure may prove to be impractical because of a prohibitive calculation time or the memory size necessary for carrying out this calculation, particularly in dynamics.

On the other hand, a multifiber beam (MF)-type model has the advantage of simplifying the hypotheses of Euler-Bernoulli beam-type kinematics (the straight sections remain straight), Timoshenko kinematics (the straight sections remain flat but not necessarily perpendicular to the neutral axis; Kotronis and Mazars 2005), or enriched kinematics, which makes it possible to describe the warping of the sections under the effect of the shearing force (Capdevielle et al. 2016).

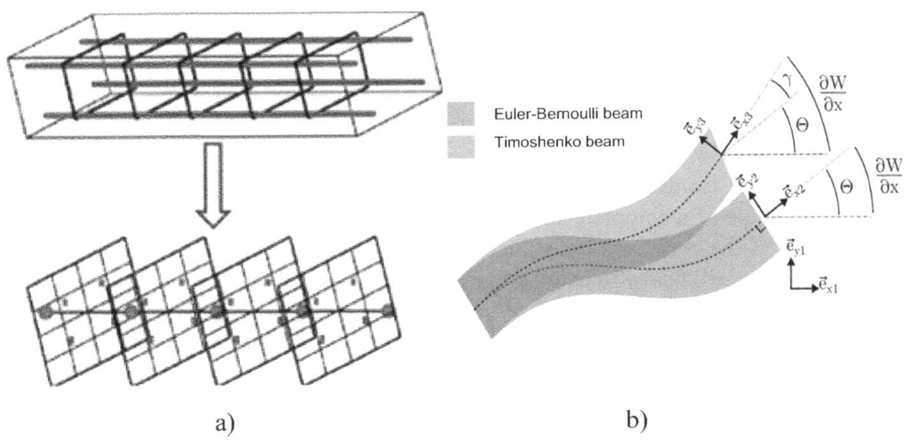

Figure 5.6. *a) Discretization of reinforced concrete by multifiber beam elements; b) nuances between Euler-Bernouilli kinematics and Timoshenko kinematics, which add to the section rotation a component due to shear (from Wassereau (2016)). For a color version of this figure, see www.iste.co.uk/mazars/damage.zip*

As described in section A.3.3.2, we use the Timoshenko kinematics here. This option offers a practical and effective solution for a complex nonlinear 3D analysis of slender elements of composite structures, such as those that may be encountered, for example, in reinforced concrete, in particular by the fact that the behavior of the fibers is a 1D behavior.

This being the case, we shall see in section 5.2.4 that the quality of the calculation results obtained by this approach requires particular consideration with regard to the identification of the parameters used and the enrichment of these models. This presentation builds on the work presented in the publication by Mazars and Grange (2015).

5.2.2. *Bias of results from the choice of material parameters*

Starting from the application discussed in section 5.1, we performed a calculation based on the use of Timoshenko multifiber beam elements. The nonlinearities are described by the μ model (version 1D) for the description of concrete and, as in the FE calculation, by a perfect elastoplastic model for the rebars. The concrete section of the multifiber beam is a matrix containing 5×7 fibers, and in the central zone (the one that is damaged), the beam elements are of sizes comparable to that of the crack bands of the 2D FE calculation. The boundary conditions are identical to those in Figure 5.2.

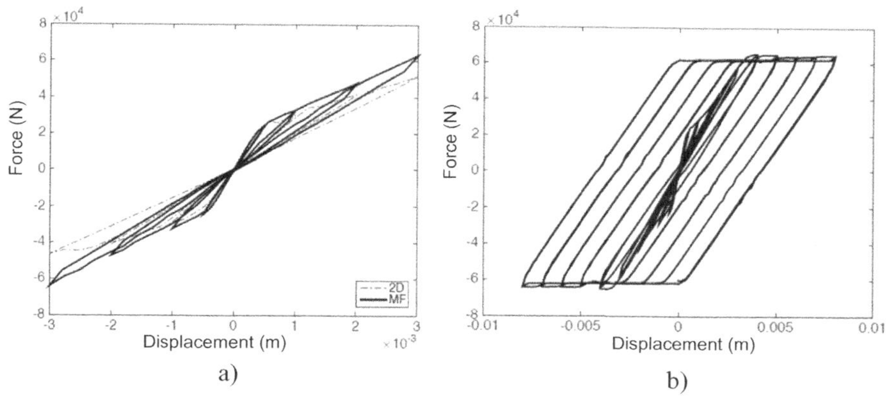

Figure 5.7. *Three-point bending RC beam: bias in the response of the multifiber beam (MF) description using the same parameters as those identified for the 2D FE description. a) Overestimation of the overall behavior; b) high strain cyclic response, showing a consequent deviation from the experimental results shown in Figure 5.1(c)*

Based on the same parameters used for the FE calculation (see Table 5.1), it can be seen that:

– when, in the phase of concrete damage, the rebars remain elastic, the simulation leads to an overestimation of the force (see Figure 5.7(a));

– when the plasticity of the rebars is activated, the simulation leads to an overestimation of the plastic strain (see Figure 5.7(b)) in comparison with the experimental results presented in Figure 5.1(c).

5.2.3. *Multifiber beams and strain localization*

As seen in section 5.1.3.2, the post-peak behavior of concrete results in localization of the strains (see Figure 5.4) and, as seen in Chapter 4, the risk of results depending on the size of the mesh elements.

This problem is well known in conventional 2D-3D FE calculations, for which two main treatment options are available:

– the solution presented in section 4.4.2 with material parameters adjusted to suit the size of the elements (Hillerborg et al. 1976);

– the implementation of regularization methods such as non-local methods (Pijaudier-Cabot and Bazant 1987) presented in section 4.5.

This problem, never studied before in the descriptions by multifiber beams, led to propositions (Mazars and Grange 2015), which we reproduce in this section.

To illustrate the scope of this problem, we refer to the 2D FE reinforced concrete beam in section 5.1 (see Figure 5.1).

On this basis, we performed MF calculations in relation to three types of loading:

– tension;

– three-point bending;

– four-point bending.

The results obtained are summarized in Table 5.2. They suggest that localization appears only in the case of loading under tension and not at all for bending cases.

More generally, it is shown that the description by multifiber beams gives rise to localizations if the overall behavior of the element described has a post-peak behavior.

Thus, there will be localization for:

– an unreinforced concrete element, regardless of its load (tension or bending);

– a reinforced concrete element (RC) if the load is uniaxial and under tension (localization appears when concrete is damaged);

– a reinforced concrete element in bending when the rebar ratio is less than a minimum value, called the fragility percentage (Eurocode 2 2004).

For the other values of the rebar ratio, if the bending is dominant or if the bending and the tension are combined, no localization is present.

Type of loading	Overall behavior (force–displacement)	Damage (at a given loading) and type of response
Tension (tie rod) 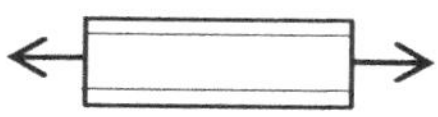	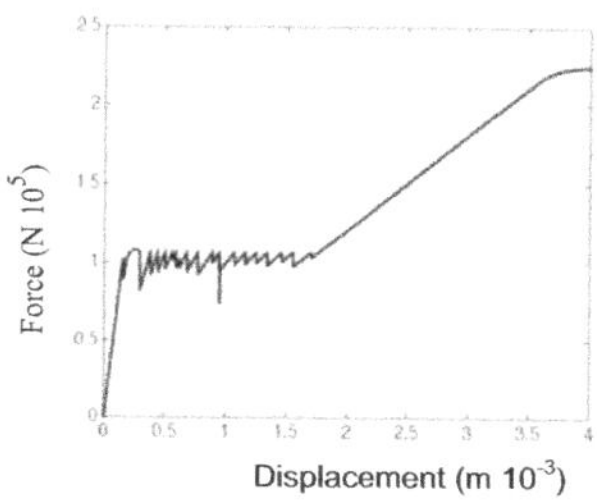	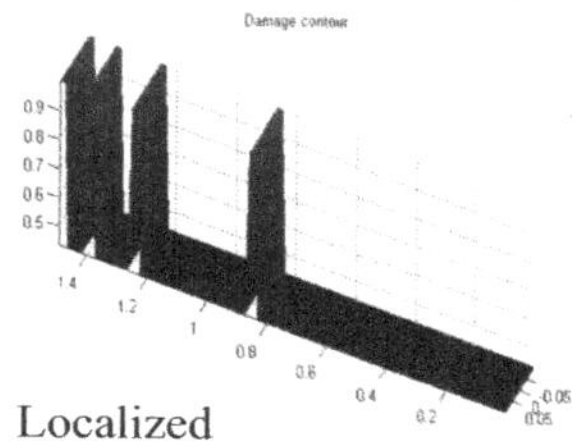 Localized damage
Four-point bending 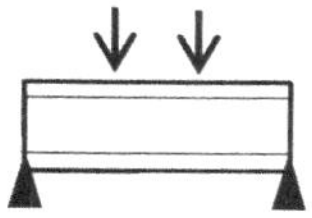	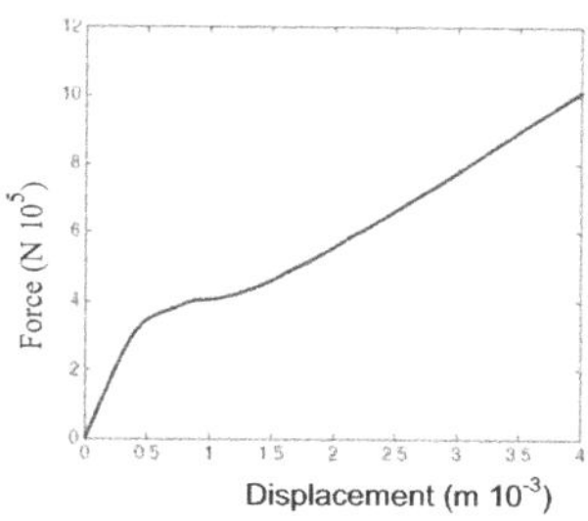	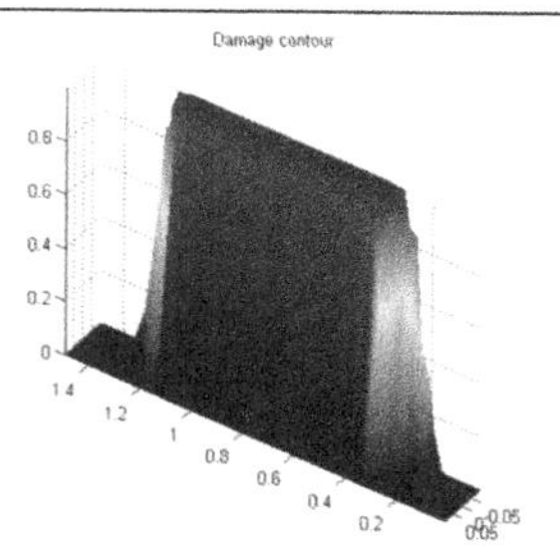 Diffuse damage
Three-point bending 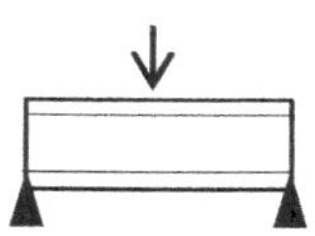	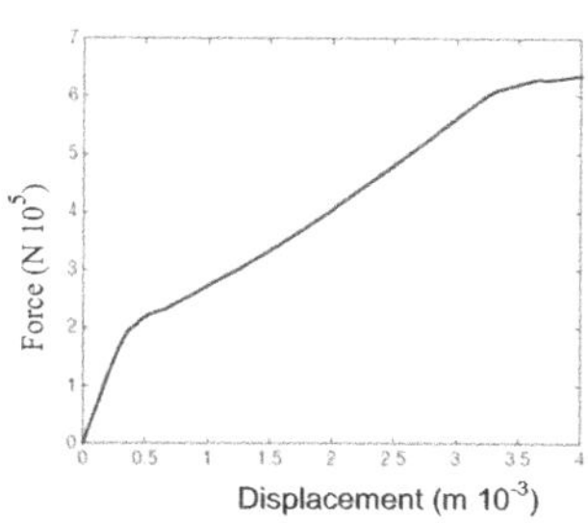	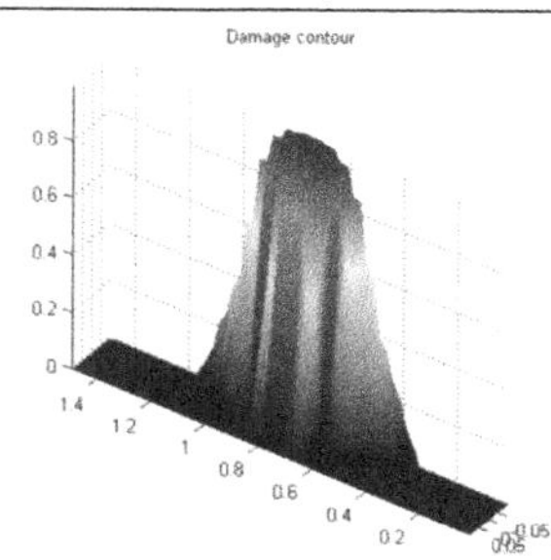 Diffuse damage

Table 5.2. *Multifiber description (MF): various types of loads applied on the same reinforced concrete element and corresponding damage fields. For a color version of this table, see www.iste.co.uk/mazars/damage.zip*

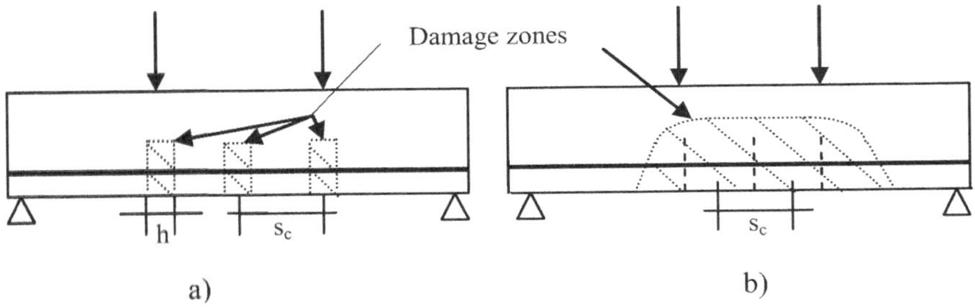

Figure 5.8. *Process of localization on reinforced concrete beams. a) 2D by finite elements (FE) description: damage is localized on bands of elements (size h); b) 1D by multifiber beams (MF) description: damage is distributed along the crack zone (s_c is the spacing between cracks)*

Hillerborg stipulated that the energy dissipated at failure in a unit volume of concrete must be equal to the fracturing energy. In the presence of localization, the control of the objectivity of the results is then the same as for conventional 2D-3D FE calculations. The localization takes place in a band of elements, and the parameters of the material must be calibrated with the size h of these elements (see Figure 5.8(a)). In the absence of localization, crack damage processes are distributed on both sides of the crack over a volume defined by the distance s_c between two cracks (see Figure 5.8(b)). Therefore, the concept of crack bands cannot be applied here. The calculation must be calibrated so that the fracture energy is consumed in a volume of width s_c. Since G_f is an intrinsic parameter of the material, whose value is identical in both types of descriptions, this leads us to write as follows:

$$G_f = s_c \int \sigma \varepsilon d\varepsilon \qquad [5.2]$$

where:

$$s_c \int_{MF} \sigma \varepsilon d\varepsilon = h \int_{EF} \sigma \varepsilon d\varepsilon \qquad [5.3]$$

5.2.4. *Enriched multifiber description and use of suitable parameters*

5.2.4.1. *Cracking process*

As described by equations [5.2] and [5.3], the application of the Hillerborg method can no longer be based on the length of the beam elements, as was the case in the description by FEs, but on the crack spacing s_c. Therefore, s_c must be known in advance. In the present case, s_c was chosen on the basis of the results of the previous FE calculation carried out on the same beams (s_c = 0.07 m) (see Figure 5.5(b)). The tensile behavior has been calibrated according to equation [5.2]. This step is carried out on the basis of the μ model by adjusting the parameters A_t and B_t of the post-peak curve, as proposed in Figure 4.5. The choice for the beam considered here is $A_t = 1$ and $B_t = 8{,}000$, instead of the corresponding values given in Table 5.1 for the 2D FE calculations.

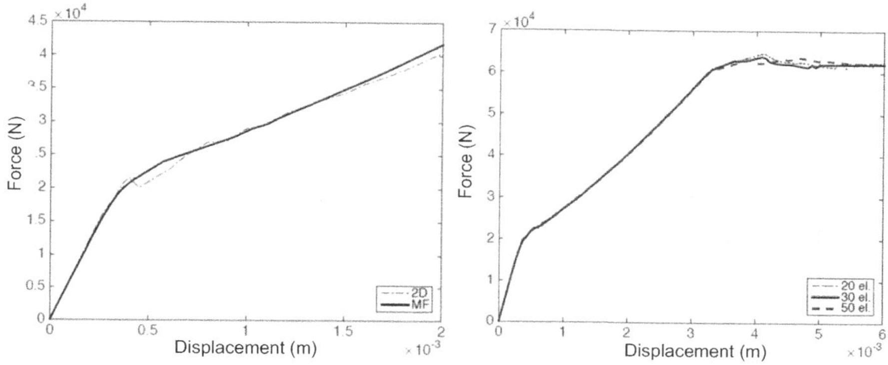

Figure 5.9. *Calibration of multifiber description (MF) model parameters: a) comparison of the results of the previous 2D FE calculation with those obtained with the MF model using parameters derived from crack spacing considerations; b) independence of the MF result from the mesh*

On the basis of these results obtained with this new parameterization (see Figure 5.9), it can be concluded that:

– the overestimation described in Figure 5.7 during the cracking phase disappears (see Figure 5.9(a) for monotonous loading);

– this result is clearly insensitive to a change in the mesh size (see Figure 5.9(b)), which demonstrates that this procedure leads to an objective calculation.

5.2.4.2. *Enriched cyclic behavior of concrete*

The fact that, in a Timoshenko-type multifiber beam, the behavior of each fiber is uniaxial allows, at low cost, the enrichment of the model.

With the idea of improving the reproduction of the cyclic behavior of concrete, we will show that it is possible, without great difficulty, to complete the 1D version of the model presented in Chapter 3, by introducing the simulation of the hysteretic loops, as well as the permanent strains generated by the friction between the lips of the cracks in the load–discharge cycles and the release of the initial stresses resulting in particular from the maturation of concrete.

5.2.4.2.1. Introduction of hysteresis loops

These are described on the basis of a partition of the uniaxial stress: $\sigma_h = \sigma + \sigma_a$, with $\sigma = E(1-d)\varepsilon$ and, as proposed in Pontiroli et al. (2010), the damping stress takes the following form:

$$\sigma_a = (\beta_1 + \beta_2 d_i)\varepsilon . f_i(\varepsilon). sign(\dot{\varepsilon}) \text{ (i = t, c)} \quad [5.4]$$

β_1 and β_2 are used to describe damping, without or with damage. The – sign is for unloading ($\dot{\varepsilon} < 0$) and the + sign is for loading ($\dot{\varepsilon} > 0$). d_i is used to distinguish between tension d_t and compression d_c; $f_i(\varepsilon)$ is a function associated with d_i driven by the variable Y_i that is used to determine the shape and size of the hysteresis loop:

$$f_i(\varepsilon) = 4\frac{\varepsilon^2}{Y_i^2}(\varepsilon - Y_i) \text{ (i = t, c)} \quad [5.5]$$

5.2.4.2.2. Introduction of permanent strains

The principle consists of introducing a slip $(\sigma_{ft}, \varepsilon_{ft})$ of the behavioral axes σ–ε, such as:

$$(\sigma - \sigma_{ft}) = E(1-d_i)(\varepsilon - \varepsilon_{ft}) \text{ (i = t, c)} \quad [5.6]$$

This is done, adhering to the same point of convergence $(\sigma_{fc}, \varepsilon_{fc})$, for the elastic compression unloading (see Figure 5.10(a)), which makes it possible to write as follows:

$$\sigma_{ft} = E(1-d_c)(\varepsilon_{ft} - \varepsilon_{fc}) + \sigma_{fc} \quad [5.7]$$

ε_{ft} depends on the damage obtained under compression; it is equal to ε_{ft0} if $d_c = 0$. If, moreover, it is assumed that the stress remains constant at the time of the closure of cracks, whatever d_c may be, we thus obtain $\sigma_{ft} = E\varepsilon_{ft0}$; by noting in Figure 5.10(a) that $\sigma_{fc} = E\varepsilon_{fc}$, equation [5.7] leads to:

$$\varepsilon_{ft} = \frac{\varepsilon_{ft0} - \varepsilon_{fc} d_c}{1 - d_c} \quad [5.8]$$

NOTE 5.1.– Due to the "slip" of the axes for the creation of the permanent strains, a correction must be made to the parameter A_t for the plotting of the enriched behavior curves. Indeed, the elastic unlodaings of the material damaged under tension converge at the point $(\varepsilon_{ft0}, \sigma_{ft})$ instead of (0,0). In the curve without permanent strain, the asymptotic value of the tensile stress, obtained by tending ε to infinity from the relationship $\sigma = E(1 - d)\varepsilon$, is $E\varepsilon_{t0}$ $(1 - A_t)$. Taking permanent strains into account leads to a change in the variable $\varepsilon \rightarrow \varepsilon - \varepsilon_{ft0}$; thus, $Y_{t0} = \varepsilon_{t0} - \varepsilon_{ft0}$, and the new asymptotic value is written as $EY_{t0}(1 - A_t^*)$, which, with respect to the new axes, must be offset upward by $-|\sigma_{ft}|$. By integrating the value found for σ_{ft} equation [5.7], we obtain $E\varepsilon_{t0}(1 - A_t) - E\varepsilon_{ft0} = EY_{t0}(1 - A_t^*)$; based on this relationship, we determine $A_t^* = A_t(1 + \varepsilon_{ft0}/(\varepsilon_{t0} - \varepsilon_{ft0}))$. This value of A_t^*, deduced from A_t, is to be considered to parameterize the μ model in its 1D version when it is used enriched with permanent strains.

5.2.4.2.3. Introduction of rebar–concrete slip effects

It is well known that, in a reinforced concrete structure, when cracks are formed and open, there then occurs a degradation of the bond between the concrete and rebar. This phenomenon is particularly sensitive (appearance of a slip) when the local strains increase and the steel plasticizes. In the description of a multifiber beam, since there is no interaction between the fibers except at their ends and the processes of damage and failure are distributed over a large area, this slip cannot be reproduced. This leads to an overestimation of the plastic strain (see Figure 5.7(b)).

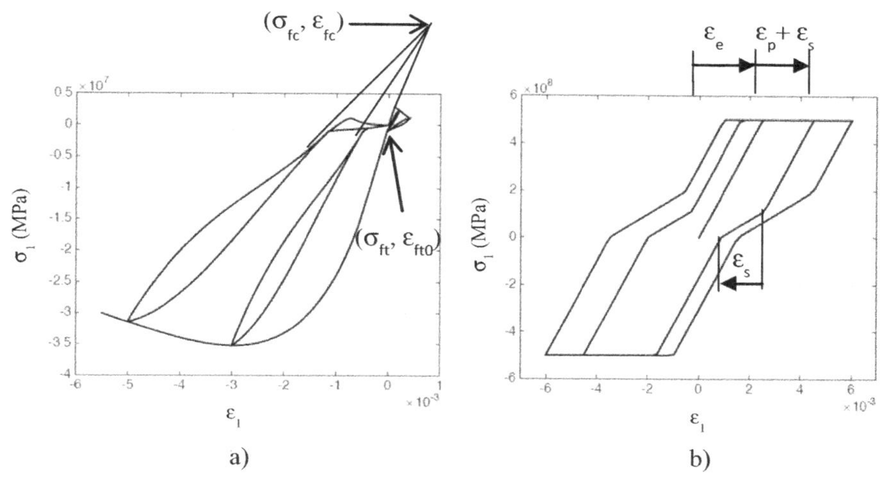

Figure 5.10. *Enriched MF description: a) behavior curve of concrete including hysteresis loops and permanent strains; b) behavior curve of the rebar and its connection to concrete, obtained by introducing a slip phase into the overall behavior of the rebars*

One way to introduce degradation of adhesion and the relative rebar–concrete slip into a multifiber beam description is to combine, in the behavior of bar elements representing the rebars (Wang et al. 2004; Richard et al. 2011), the inherent strain of the steel ($\varepsilon_e + \varepsilon_p$), and the strain ε_s resulting from the slip occurring at the rebar–concrete interface:

$$\varepsilon = \varepsilon_e + \varepsilon_p + \varepsilon_s \quad [5.9]$$

ε_e is the elastic strain ($\varepsilon_e = \sigma/E$), and ε_p is the plastic strain; it is such that, in the case of a perfect elastic–plastic behavior of plasticity threshold f_e:

$$\sigma = \varepsilon_e E \text{ if } \varepsilon_p = 0 \text{ and } \sigma = f_e \text{ if } \varepsilon_p > 0 \quad [5.10]$$

μ model parameters				**Concrete hysteresis**	
E (GPa)	σ_{t0} (MPa)	σ_{c0} (MPa)	G_f (N/m)	ε_{ft0}	ε_{fc}
28	2.8	–10	30	-0.35×10^{-4}	4×10^{-4}
A_t	B_t	A_c	B_c	β_1	β_2
1	8,000	1.25	395	0.05	0.20

Table 5.3. *Concrete parameters used for enriched MF modeling*

Braga et al. (2012) proposed a modified behavioral model for rebar to account for bond slips, reconsidering a nonlinear monotonic relationship for slip. In the same spirit, to minimize the computational effort, this proposal is based on the following hypotheses:

– the slip strain evolves from a threshold and, assuming that the principal slip evolution occurs when plasticity is activated, this threshold is superimposed to be the plasticization threshold of the rebars, when $\sigma = f_e$;

– there is a linear link between plastic strain and slip strain.

Based on these hypotheses, the slip strain used here is:

$$\varepsilon_s = \kappa\varepsilon_p \text{ if } \varepsilon_p > 0 \text{ and } \varepsilon_s = 0 \text{ otherwise } (\kappa \leq 1) \quad [5.11]$$

As seen in Figure 5.10(b) for cyclic loading, when the tensile load decreases (unloading), two steps appear:

– partial recovery of elastic strains, ε_e;

– from a given stage to a zero stress, the progressive disappearance of ε_e and ε_s, both reaching 0 when the sign of the stress changes.

In the rest of the cycle, as long as the stress is negative, the crack is closed; it is then assumed that no slip is possible, $\varepsilon_s = 0$ (see Figure 5.10(b)). The slip reappears progressively when the tensile load resumes up to its previous value and then increases at the same time as ε_p when the loading continues.

5.2.5. *Simulations based on the enriched multifiber description*

With this enriched modeling and the parameters adapted to the use of the MF elements (see Table 5.3), the application to the reinforced concrete beam under cyclic loading, already modeled in section 5.1, leads to the results presented in Figure 5.11.

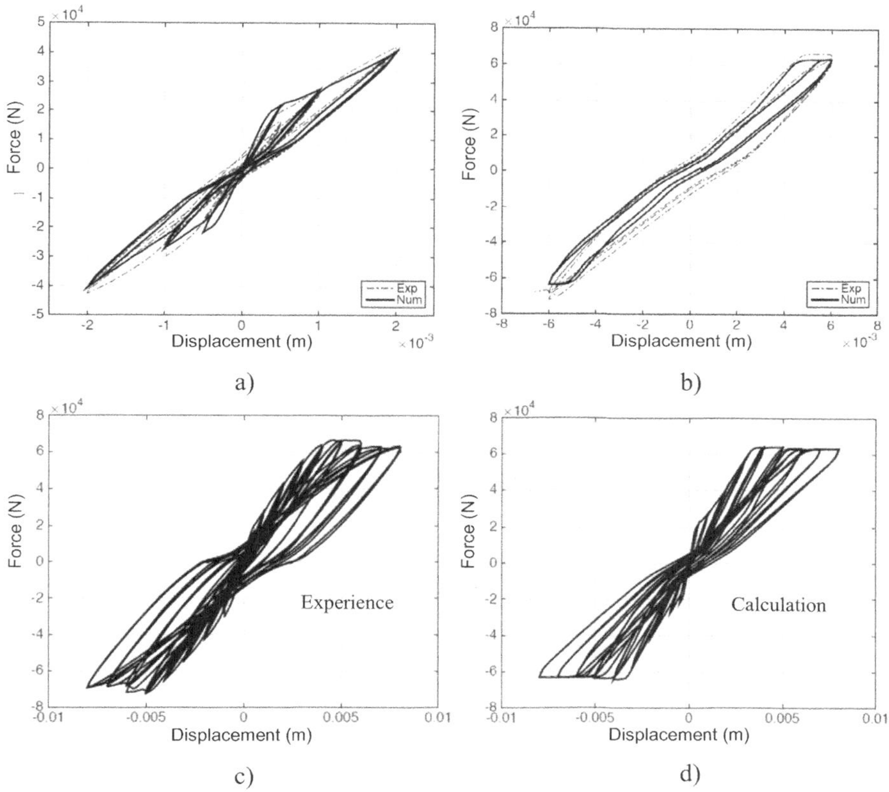

Figure 5.11. *Reinforced concrete beam response under cyclic loads in an enriched MF calculation: a) experiment-calculation comparison up to ±2 mm; b) comparison for a set of three cycles at ±6 mm; bottom of the figure, complete cyclic path within c) experimental results and d) calculation results*

This figure includes a comparative experimental calculation for three different loading path levels:

– a cyclic path during the cracking phase up to ±2 mm (see Figure 5.11(a));

– a cyclic path of three cycles at ±5 mm, for which the plasticity of the steel and the relative concrete–rebar slip are activated (see Figure 5.11(b));

– the total path to ±8 mm (see Figures 5.11(c) and (d)).

It is clear that this modeling strategy leads to quite satisfactory results. Moreover, it should be noted that this type of calculation carried out on the ATL4S platform (Grange 2017) is inexpensive and has perfect stability in nonlinear phases, which, for such paths, is much less so for conventional 2D FE calculations.

5.3. Multifiber calculations and access to cracking indicators

After noting the good quality of the overall results obtained by the enriched MF model, a question arises as to the capacity of this model to provide local information relating in particular to local degradation and indicators related to cracking.

5.3.1. *Damage fields*

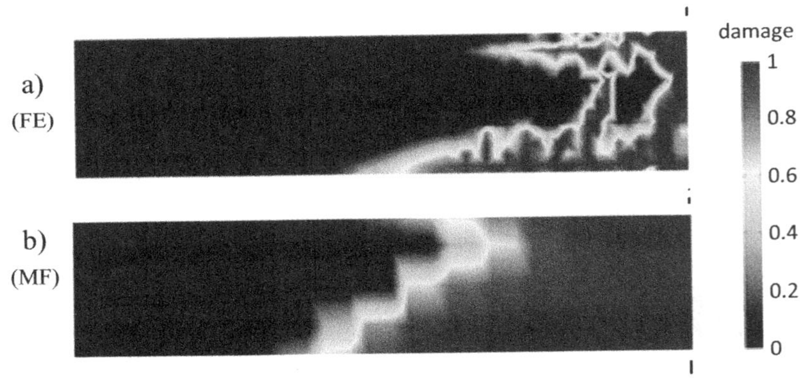

Figure 5.12. *Damage fields in the RC beam at –3 mm after a loading path up to ±3 mm: a) damage field according to a 2D FE calculation; b) damage field according to an enriched multifiber beam calculation. For a color version of this figure, see www.iste.co.uk/mazars/damage.zip*

Figure 5.12 shows, for a given loading stage (–3 mm after an overall loading path up to ±3 mm), a comparison between the field of damage obtained by the 2D

FE description, which highlights the locations of cracks, and the field of damage obtained by the enriched MF description presented in Figure 5.10. As announced in section 5.2.2 (see Table 5.2), the damage described by the MF calculation is diffuse in nature, but it shows a good level of consistency in intensity and extent compared to that described by the FE calculation (see Figure 5.12(b)).

5.3.2. *Opening of cracks*

The crack opening indicators can be obtained by post-processing. In section 5.2.4.1, we saw that a realistic description of the cracking energies leads to the introduction, as given in the enriched calculation, of the spacing of the cracks s_c. However, prior knowledge of these data may pose a problem. Experimental values (physical or numerical) can be used as done here, but we can also rely on the recommendations or rules proposed in construction codes (Eurocode 2 2004; Model Code MC2010 2013). Given this knowledge, the calculation provides an average crack opening value, the relevance of which will be demonstrated in this section.

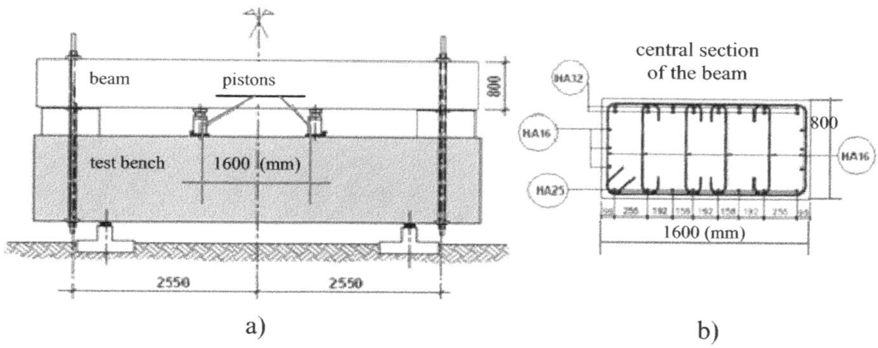

Figure 5.13. *CEOS.fr (2017) program: experimentation on large RC four-point bending beams: a) test system and devices; b) reinforcement drawing for the RL6 beam*

In this type of calculation, it is generally considered that the crack opening appears beyond the stress peak f_t (see Figure 5.14(b)). The diagram in Figure 5.8(b) shows that each crack opens in the middle of a segment of length s_c (equal to the distance between the cracks). Thus, its opening w_c can be calculated as follows:

$$w_c = s_c(\varepsilon - \varepsilon_{t0}) \qquad [5.12]$$

ε is the total strain over the segment of length $s_{c,}$ and ε_{t0} is the strain at the tension peak; $(\varepsilon - \varepsilon_{t0})$ is therefore the part of the strain that contributes to the opening of the crack.

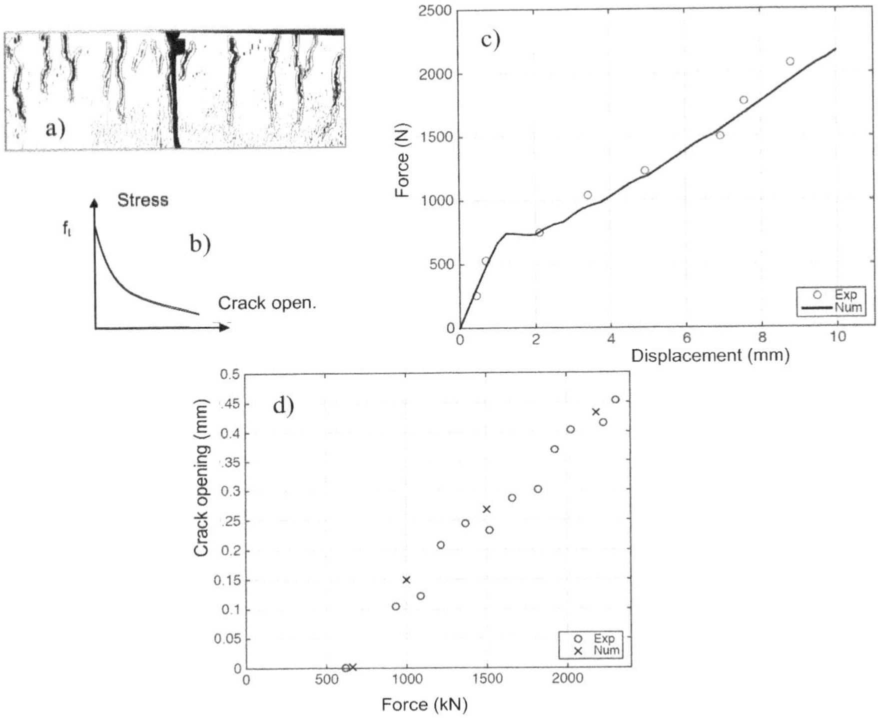

Figure 5.14. *Four-point bending RC beam tests: a) cracking measured from the image correlation analysis in the central area of the beam (mean spacing = 0.206 m); b) stress–crack opening modeled curve; c) force–displacement response; d) crack opening (mean value at the upper layer of rebars) as a function of the applied load*

As part of the French national research program CEOS.fr (Mazars and Coste 2014; Barre et al. 2016; CEOS.fr 2017), tests were carried out on large four-point bending beams (length = 6.1 m, width = 1.6 m, height = 0.8 m) (see Figure 5.13(a)). During these tests, the cracking was analyzed by digital image correlation, which allowed access to both the spacing and opening of the cracks in the pure bending zone (Rospars and Chauvel 2014).

For the beam considered above, this calculation was carried out using a description by the Timoshenko multifiber beam and the μ model. The cross-section of the beam is a matrix composed of 8 × 16 fibers. In addition, the pure bending zone (1.6 m) was divided into 10 elements. The Hillerborg method was implemented on the basis of the crack spacing obtained on the RL6 test beam (mean value: 0.206 m), and this calculation was carried out using the model parameters estimated from the material properties given in Table 5.4.

Experimental data			**μ model parameters**			
E steel (GPa)	E concrete (GPa)	f_t (MPa)	E (GPa)	ε_{t0} (MPa)	ε_{c0} (MPa)	A_t
195	40	4.67	40	1.16×10^{-4}	3.75×10^{-4}	1
–	G_f (N/m)	f_c (MPa)	–	A_c	B_c	B_t
–	75	63.75	–	1.5	355	11,500

Table 5.4. *Four-point bending tests: experimental data and modeling parameters*

Figure 5.14(c) shows a good correlation for the overall behavior (load–displacement). From equation [5.12], the mean crack width is determined at the upper layer of rebars, and the quality of the results obtained attests to the effectiveness of the modeling carried out (see Figure 5.14(d)).

6

Modeling of Situations Related to Specific Loadings

6.1. Simulation of velocity effects

The dependence of the strength of concrete on the strain velocity is a known phenomenon, which has long been a problem for engineers because the experimental investigation on the subject is not simple. The advent of high-velocity press and dynamic failure techniques using Hopkinson bars has allowed great advances in knowledge of the phenomenon. As we will see in section 6.1.1, for concrete, this dependence is important for loading under tension (the strength can be multiplied by 10 in rapid dynamics) and significant for loading under compression (strength at high velocity multiplied by two). The presentation developed in section 6.1 addresses the modeling of the phenomenon; it is the result of the work reported in particular in the paper by Mazars and Grange (2017).

6.1.1. *Analysis resulting from the experiment*

Experimentation at high velocities requires very specific equipment and special know-how. A few years ago, an experimental campaign was conducted by Forquin and his team, first at the LEM3 laboratory in Metz and then at the 3SR laboratory in Grenoble. It focused on quasi-static tests and dynamic tests performed on dry and wet concrete samples (Erzar and Forquin 2011, 2014). A high-velocity hydraulic press was used for the medium velocities, and an experimental device based on the use of a so-called Hopkinson bar (1914) for carrying out Spalling tests (dynamic tensile tests) was used for the higher strain velocities. From these tests, original identification techniques could be developed to deduce the tensile strength of the material. For ordinary concrete, the complete results are given in Figure 6.1(a).

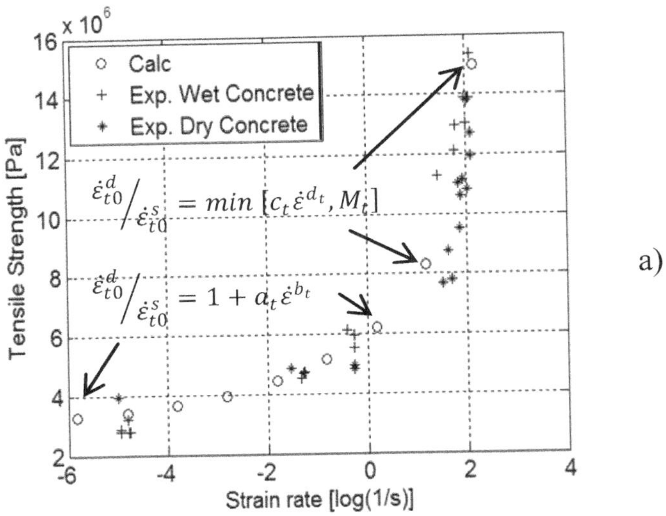

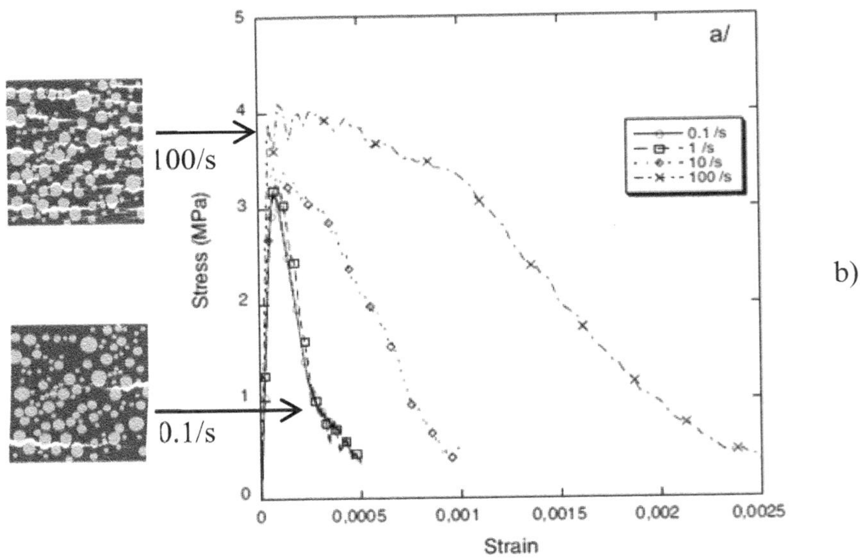

Figure 6.1. *a) Tensile strength of concrete as a function of strain rate: experiments (cross-stars; Erzar and Forquin 2011) and calculations incorporating the concept of retarded damage threshold (points); b) mesoscopic calculation model for concrete showing the effects of strain rate on tensile behavior (Gatuingt et al. 2013)*

As presented in Pontiroli et al. (2010), this effect is taken into account using the concept of retarded damage threshold. A dynamic threshold (ε_{t0}^{d}) is used instead of a static threshold (ε_{t0}^{s}) from a dynamic amplification factor depending on the strain rate ($\dot{\varepsilon} = \frac{d\varepsilon}{dt}$):

$$R_t = \varepsilon_{t0}^{d}/\varepsilon_{t0}^{s} = min[\max(1{,}0 + a_t\dot{\varepsilon}^{b_t}, c_t\dot{\varepsilon}^{d_t}), M_t] \qquad [6.1]$$

– Here, a_t, b_t, c_t, d_t and M_t are material parameters whose values are derived from experimental results.

– In the μ model, used as before in its 1D version, the behavioral law in the field of tension, seen in equation [3.39], is as follows:

$$\sigma_1 = E(1-d)\sigma_1, \text{ where } d = d_t = 1 - \frac{(1-A_t)\varepsilon_{t0}}{Y_t} - A_t\exp\left(-B_t(Y_t - \varepsilon_{t0})\right) \qquad [6.2]$$

To take into account the velocity effect, it suffices to extract the value of $\varepsilon_{t0}^{d} = f(\dot{\varepsilon})$ from equation [6.1] and introduce it into equation [6.2], which gives the damage. As we have already seen in general under tension, the stress peak corresponds to the damage threshold. Thus, the dynamic tensile strength is equal to:

$$f_t^{d} = E\varepsilon_{t0}^{d} = E\, f(\varepsilon_{t0}^{s}, \dot{\varepsilon}) \qquad [6.3]$$

For the concrete corresponding to the Erzar and Forquin (2011) tests, the results have been reported in Figure 6.1(a), thus confirming that the calculations allow this phenomenon to be properly taken into account. It is simple to extend the concept of retarded damage to 3D situations by introducing ε_{t0}^{d} into equations [3.24] to define the initial threshold of the pilot variable Y of damage.

Modeling the effects of strain rate is closely linked to the description scale. Figure 6.1(b) shows that, on the basis of a modeling of the material on a mesoscopic scale (Gatuingt et al. 2011) describing aggregates and cementitious matrix, the degradation within the considered sample changes from monofracturing at low velocities to multifracturing when the intensity of the strain rate increases. This results in a double effect, visible on the overall curves of the behavior of the test sample:

– the increase in tensile strength;

– the increase in the energy consumed to crack the element (area under the σ–ε curve).

In a macroscopic scale description, as is the case here, the second point can be treated by increasing the fracture energy with the velocity. In the absence of experimental data on the subject, the problem was solved as follows:

– the change in tensile properties respects both the increase in tensile strength given by the experiments and the increase in the fracture energy estimated by inverse analysis on the basis of the results of Spalling tests;

– the use of the Hillerborg method is the one presented in Chapter 4, assuming that the equation linking h (or s_c), G_f and the material parameters (as in equations [5.2] and [5.3]) remains the same, regardless of the value of velocity.

On this basis, the proposal for a retarded damage threshold (equation [6.1]), which correctly manages the effects on strength (stress peak in equation [6.3]), is supplemented by acting on the evolution of the fracture energy by a particular management of the post-peak behavior of the σ–ε curve.

In this context, we propose the following approach:

– For low and medium velocities, either when:

$$R_t = \varepsilon_{t0}^{d}/\varepsilon_t^{s} = 1{,}0 + a_t\dot{\varepsilon}^{b_t} \qquad [6.4]$$

– The evolution beyond the threshold is given by the conventional equations of damage (see equation [6.2]). The parameters A_t and B_t are calibrated as a function of the fracture energy G_f known in a static state (see Figure 4.5), with ε_{t0}^{s} as the damage threshold; the evolution of this energy will then take place naturally by the mere evolution of ε_{t0}^{d} with the strain rate according to the relationship in equation [6.4].

– For high velocities, when:

$$R_t = \varepsilon_{t0}^{d}/\varepsilon_t^{s} = min[c_t\dot{\varepsilon}^{d_t}, M_t] \qquad [6.5]$$

The threshold is always a function of the strain rate, but when it reaches the peak, a plateau is introduced followed by a linear evolution up to the intersection with the asymptote defined by A_t. To do this, new parameters are introduced:

– strain segment managing the plateau: $[\varepsilon_{t0}^{d}, \varepsilon_{t1}]$;

– post-peak strain segment:$[\ \varepsilon_{t1},\ \varepsilon_{t2}]$, where $\varepsilon = \varepsilon_{ta}$ at zero stress;

– residual stress segment: $\varepsilon \geq \varepsilon_{t2}$.

The boundary between the two domains is approximately $\dot{\varepsilon} = 10/s$. Figure 6.2 shows the results obtained for uniaxial tension curves at different velocities with the parameter values given in Table 6.1 and, for high velocities, ε_{t1}, ε_{t2}, and ε_{ta} are equal to 3, 7.5 and 8 times ε_{t0}, respectively.

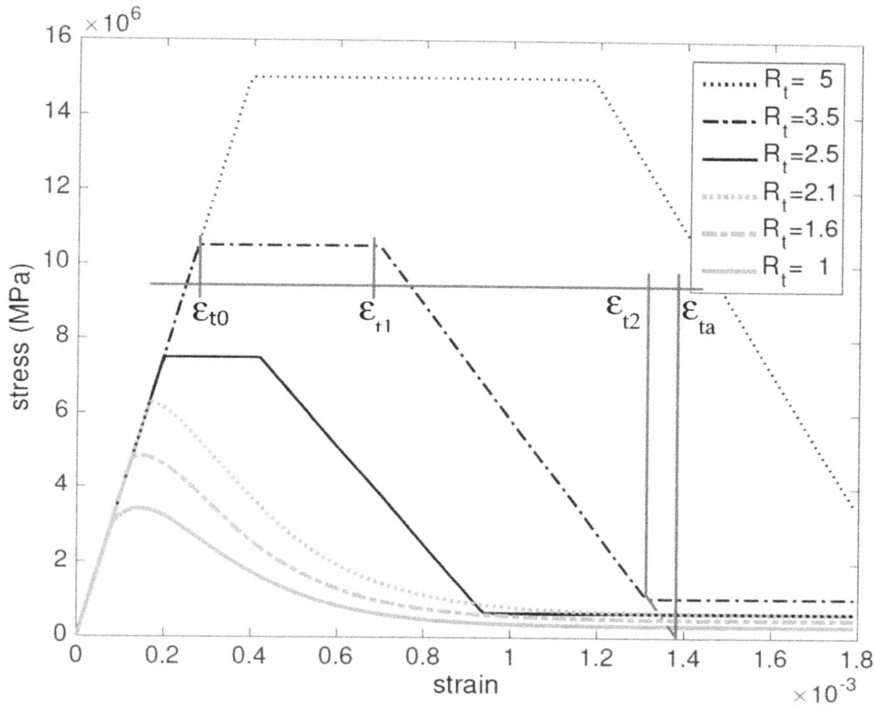

Figure 6.2. *Behavior under tension at different velocities.* $R_t = \varepsilon_{t0}^d / \varepsilon_{t0}^s$ *is the dynamic amplification factor of the damage threshold (gray curves: low and medium velocities,* $R_t < 2.3$*; black curves: high velocities,* $R_t > 2.5$*). For a color version of this figure, see www.iste.co.uk/mazars/damage.zip*

NOTE 6.1.– Under compression, the effects of velocity are much less pronounced than under tension. For example, at $\dot{\varepsilon}\#1/s$, whereas for tension the increase is greter than 100%, for compression it is less than 30%.

To represent this evolution, the same type of law as that seen in equation [6.4] is used for the low and medium velocities: $\varepsilon_{0c}^d / \varepsilon_{0c}^s = 1.0 + a_c \dot{\varepsilon}^{b_c}$. Given the specificity of the applications presented in sections 6.1.2 and 6.1.3, this effect will not be taken into account.

6.1.2. *High-velocity loading: application to Spalling tests*

To demonstrate the model's ability to describe high-velocity problems, one of the Spalling tests performed at the LEM3 Metz (see Figure 6.1(a)) was simulated.

The test specimen was a cylinder (L = 140 mm, Φ = 46 mm) with the left edge in contact with the Hopkinson bar (see Figure 6.3(a)). During loading, a compression wave from the bar was transmitted to the specimen (see Figure 6.3(b)) and propagated along the specimen to its free end, where it was reflected as a tensile wave moving in the opposite direction. When the amplitude of the reflected pulse exceeded that of the incident pulse, a dynamic tensile load propagated through the sample, creating the conditions for high-velocity cracking. Figure 6.3(a) shows the experimental result obtained for an incoming compression pulse whose maximum value $\sigma_{cmax} = 50$ MPa (see Figure 6.3(b)). The result was multifracturing under dynamic tension of the specimen.

Parameters of the μ model				**Velocity effects parameters**		
E (GPa)	σ_{t0} (MPa)	σ_{0c} (MPa)	G_f (N/m)	a_t	b_t	c_t
38	3	– 48	30	1	0.18	0.25
A_t	B_t	A_c	B_c	–	M_t	d_t
0.9	7,000	1.25	400	–	5	1.39

Table 6.1. *"Material" parameters used for the Spalling test*

Figures 6.3(c)–(f) show the results of the calculation carried out on a model of 23 single-fiber beam elements (the problem was considered to be perfectly uniaxial), with the μ model enriched for velocity effects and using the material parameters derived from Table 6.1.

Comparison with the measurements taken during the test showed good agreement for the velocity inside the test piece (left: see Figure 6.3(e); right: see Figure 6.3(f)), for the strain at the crack location, see Figure 6.3(d), and for the resulting damage, see Figure 6.3(c). Three main cracks appeared in the end; this representation is a particular feature of the so-called erosion calculation technique, which, by making the elements whose strain has reached a high value ($\varepsilon > 10^{-2}$), disappear, makes it possible to artificially create a discontinuity. Overall, the results are good for velocities, as well as for strains and damage. The simplified procedure used, discretization in a single-fiber beam and 1D behavior model, leads to a very stable and very fast calculation, only a few seconds on the ATL4S platform (Grange 2017).

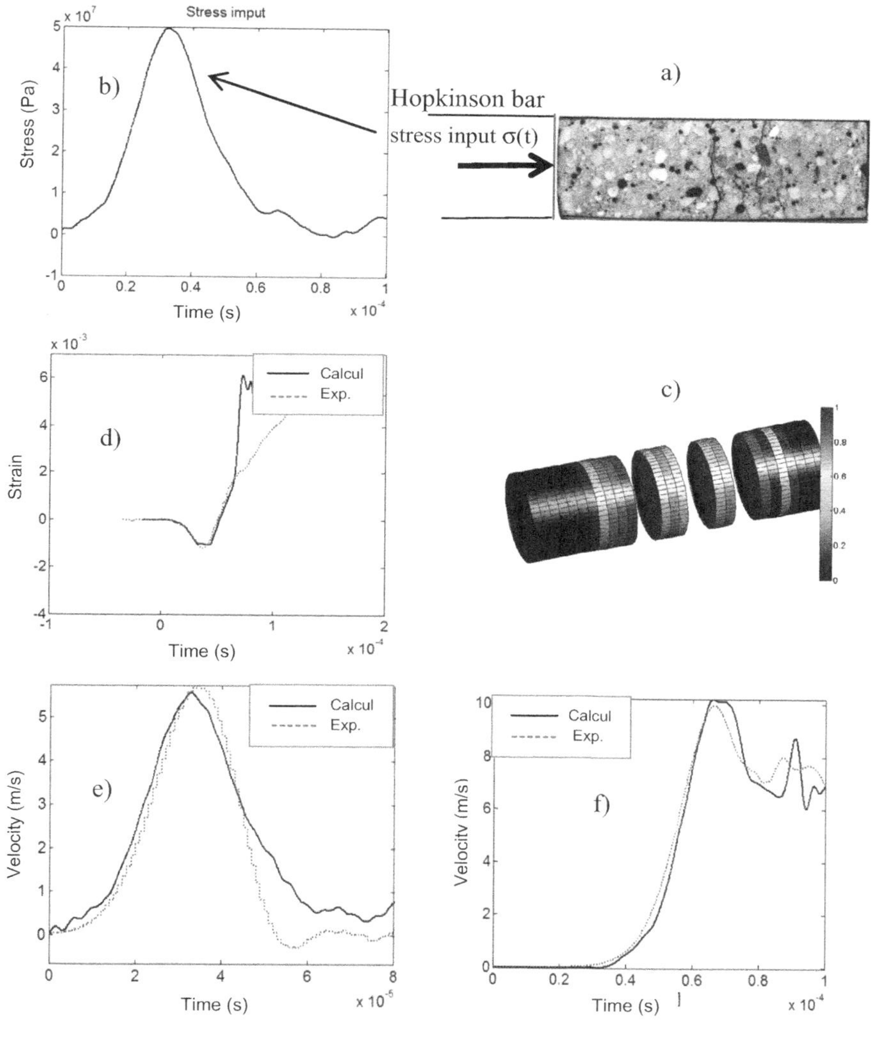

Figure 6.3. *Spalling test: a) final cracking of the concrete test piece (Erzar and Forquin 2011); b) stress wave applied to the left edge of the test piece; c) calculated damage with three main cracks; d) ε(t) response in the cracking zone; e) velocity on the left edge of the test piece resulting from the stress wave; f) output velocity on the right edge of the test piece. For a color version of this figure, see www.iste.co.uk/mazars/damage.zip*

6.1.3. *Loading at medium velocity: impact on a reinforced concrete beam*

An experiment on the analysis of the effects of impacts on reinforced concrete beams was carried out at the FOI laboratory (Sweden). Five beams were tested, each having dimensions according to Figure 6.4 (Agardh et al. 1999).

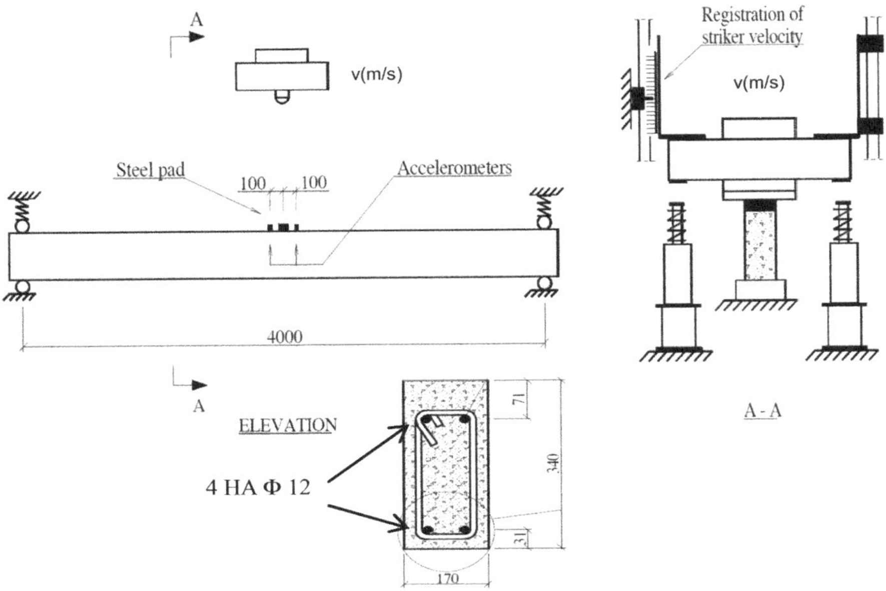

Figure 6.4. *Experimental system of impact applicator on reinforced concrete beams and details of beam reinforcement (Agardh et al. 1999)*

The impact consists of the fall of an object striking the beam at mid-span. The striker, with a mass of 718 kg, is dropped from a height of 2.68 m. Its head strikes a steel pad (50 × 30 mm) located on the beam, with a velocity of approximately 6.7 m/s. The length of the beam between supports is 4 m, and these are considered rigid. To prevent the ends of the beam from moving upward during the impact (rebound effect), retaining devices are positioned at the level of the supports (see Figure 6.4). Two accelerometers are placed on the beam, and an accelerometer is also placed on the head of the striker. In addition, two strain gauges are located on the two lateral surfaces of the beam at the same level as the four arranged in the central zone on the rebars. In addition, an impulse transducer located on the impactor records its velocity, which is valuable information, especially at the moment of impact.

The modeling principles used are as follows:

– the beam mesh comprises 30 Timoshenko multifiber beam elements in length; the multifiber section comprises 10 fibers in height and only two fibers in thickness (the problem is considered to be planar); the reinforcement mesh comprises bar elements for the longitudinal rebars in accordance with the diagram in Figure 6.4 (transverse rebars are not taken into account);

– the μ model is used for concrete in its 1D version enriched with velocity effects and a perfect elastoplastic model is used for steel; the fracture energy is, as in the problem seen in section 6.1.2, treated according to the Hillerborg method (see equations [5.2] and [5.3]), assuming that the spacing of the cracks s_c is independent of the strain rate and therefore of G_f, which changes with velocity here.

Table 6.1 gives the modeling parameters of the materials used for this calculation. As indicated above, the velocity under the impactor was measured in the central part of the beam. This signal was used to simulate the load on the beam (see Figure 6.5(a)). The resulting acceleration and displacement histories are given in Figures 6.5(b) and (c); the experiment-calculation comparison presented in all these figures shows very good correspondences at the global level. Locally, changes in the strains calculated at two points close to one of the strain gauges located on the rebars have been reported; the calculation gives completely correct trends.

Due to the impact, the beam exhibits a multimodal response, which justifies the very specific distribution of the damage areas observed (see Figure 6.5(e)). Three principal ones are observed:

– in the central part of the beam, a zone of severe degradation under tension on the lower part (widely extended to the ends) and a zone of compression damage on the upper part, at the point of impact;

– at approximately one meter from each end of the beam, a zone of tensile damage, on the upper part, more or less connected to the previous one at the end of the test.

These results are consistent with the cracking pattern observed after the test (see Figure 6.5(f)). In addition, the permanent displacements observed after the test are correctly reproduced by calculation (see Figure 6.5(g)).

In conclusion, we can say that the model leads to good simulations of the response of the beam under impact and quite correctly describes the resulting degradations.

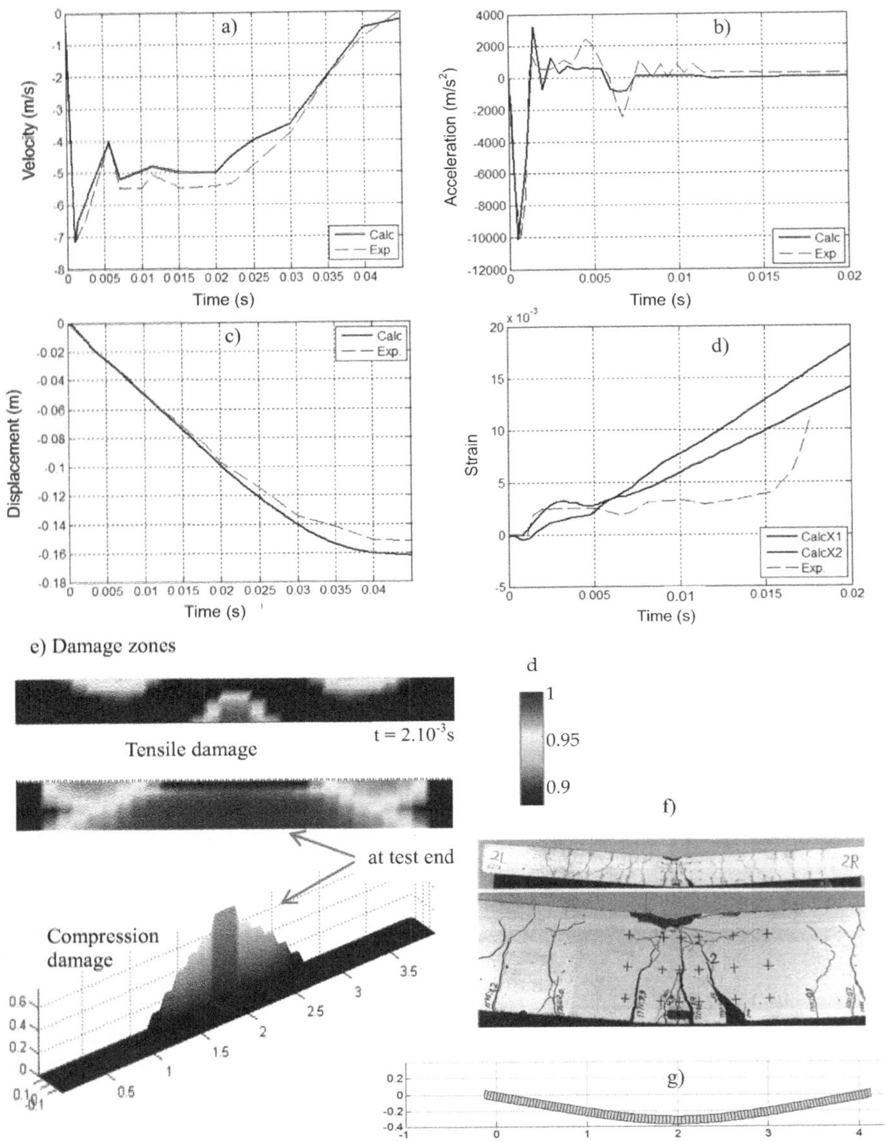

Figure 6.5. *Impact on a reinforced concrete beam. For a color version of this figure, see www.iste.co.uk/mazars/damage.zip*

COMMENTS ON FIGURE 6.5.– *(a) Velocity under the impactor, used as loading; (b) acceleration response at the point of impact; (c) history of the median displacement of the beam; (d) evolution of the strain on a rebar 200 mm from the central section; (e) contours of tensile damage at 2 × 10^{-3} seconds and at the end of the test and under compression at the end of the test; (f) crack pattern (experiment of Agardh et al. (1999)); (g) displacement state calculated at the end of the test.*

Experimental data						
Steel		**Concrete**				
E (GPa)	fe (MPa)	fc (MPa)	E (GPa)	σ_{t0} (MPa)	σ_{0c} (MPa)	G_f (N/m)
200	500	95	45	5	–10	60
Material parameters for concrete (enriched μ model)						
A_t	B_t	ε_{ft0}	σ_{0c} (MPa)	A_c	B_c	ε_{fc}
0.9	2 10^{+4}	-0.33×10^{-4}	–10	1.35	300	5×10^{-4}
β_1	β_2	a_t	b_t	d_t	c_t	M_t
0.05	0.10	1	0.176	1.39	0.255	5

Table 6.2. *Steel and concrete characteristics for testing simulation (Agardh et al. 1999)*

6.2. Simulation of the effects of concrete maturation

In section 1.2.2, we discussed the effects of concrete maturation by showing how this phase of the "life" of the material can permanently affect its behavior over time, due in particular to the risks of cracking that this period of "early age" can generate.

The objective of this part, resulting from published work (Mazars et al. 2018), is to propose an approach for a complete and effective modeling of the effects of concrete maturation, which are all the more important as the structure concerned is massive. This presentation is followed by an application on a large-sized reinforced concrete beam, which served as the basis for an experiment aimed at analyzing the effects of a restrained shrinkage carried out at an early age as part of the CEOS.fr (2017) national research program.

The problem to be solved encompasses many coupled phenomena (increase in temperature, progressive maturation of concrete, creep, effects of restrained shrinkage, etc.), which generates many questions, both theoretically and numerically. Processed by means of a conventional finite element description (Sellier and Buffo-Lacarrière 2011); this implies a very sophisticated numerical simulation. We will show that, provided that the treated structure is sufficiently slender, it is possible to drastically reduce the size of the problem based on a simplified description of this structure using Timoshenko multifiber elements, as used in Chapter 5 or section 6.1 for other loading situations.

6.2.1. *Problems posed by the behavior of concrete at an early age*

As already mentioned, early age phenomena are closely related (Ulm and Coussy 1998). Overall, the following phenomena can be distinguished:

– the maturation of concrete, which leads to an increase in mechanical performance over time;

– the thermal strain of concrete and steel due to the exothermic nature of the cement hydration reaction;

– the autogenous shrinkage of concrete, which is associated with the evolution of maturation that consumes water initially in the fresh concrete;

– the basic creep concrete, which appears with the stress caused by the effects of volume variation as presented in section 1.2.1;

– the elasto-damageable behavior of concrete, particularly activated at low strength;

– the elasto-plastic behavior of rebars, whose strains follow those imposed by concrete.

In addition, the analysis of the structure requires the scale effects on the behavior of concrete to be taken into account, all the more so since this structure is massive.

The reduced complexity of these problems announced in the introduction to section 6.2 requires us, at this stage, to present the case study that we want to deal with.

6.2.2. *Case of a beam in a situation of restrained shrinkage*

6.2.2.1. *Details of the RG8 trial of the CEOS.fr program*

The CEOS.fr research program (Behavior and Evaluation of Special Works with Regard to Cracking and Shrinkage) took place in France from 2008 to 2014. Modeling work has been carried out on the basis of numerous tests, some of them on massive elements in real size, including large beams designed to study the couplings between the effects associated with the maturation of concrete and mechanical loadings.

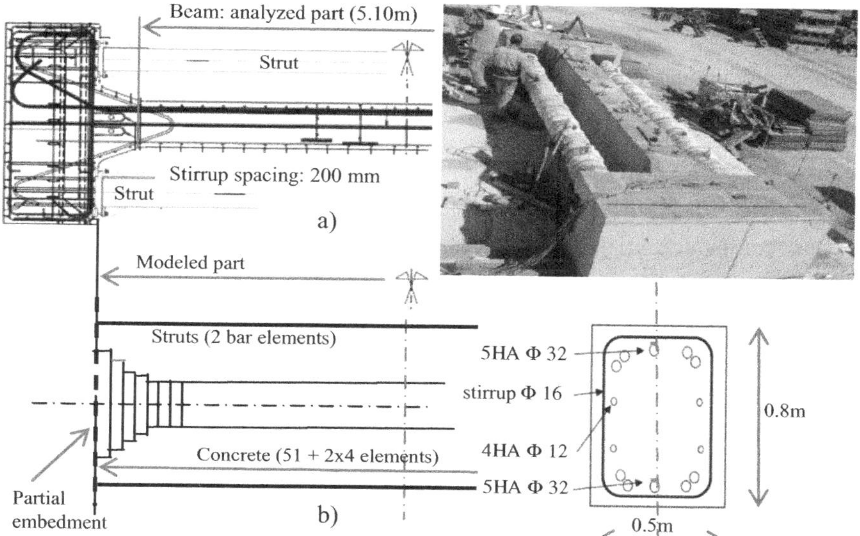

Figure 6.6. *Overview of the RG8 beam and details of the rebars: a) beam reinforcement mapping; b) multifiber beam mesh and boundary conditions. For a color version of this figure, see www.iste.co.uk/mazars/damage.zip*

In this context, the RG8 beam was designed to study the effects of restrained shrinkage within massive concrete structures at an early age. Made with C50/60 concrete, specimen RG8 is composed of three parts (see Figure 6.6):

– a central block (5.10 m × 0.80 m × 0.50 m), plus a 0.4 m transition zone, including a gradual increase in the width between this block and the end heads; the reinforcement mapping is given in the cross-section in Figure 6.6(a); it includes a combination of rebars 32 mm in diameter (× 10) and 12 mm in diameter (× 4); the strirrups have a diameter of 16 mm and are spaced apart by 200 mm;

– two heads (0.9 m × 2.2 m × 0.9 m) are used to support the struts described below;

– two steel struts are fixed on the two heads to limit the evolution of the strain (see the image in Figure 6.6).

Due to the effects of hydration, changes in temperature and shrinkage were substantial in this massive structural element. The two steel struts limit the shrinkage of the central part of the beam and, consequently, cause cracks at an early age. The aim here is to show how to design a model, integrating the different phenomena listed in section 6.2.1, that is capable of reproducing the thermomechanical consequences of such a situation as best as possible.

The test beam was extensively instrumented (nine internal temperature sensors, 24 vibrating ropes, three internal optical fibers, 12 strain gauges on the rebars, etc.). The external temperature and solar exposure of the sample were also measured during the test.

6.2.2.2. *Principles used for the simulation*

In a context where the use of Timoshenko multifiber beam elements is preferred, the details of the discretization are given below:

– Concrete: the useful part (5.1 m) was modeled using 51 elements 0.10 m long. The section was divided into 5 × 8 fibers of 0.10 m. The beam was completed on both sides by a part with a variable section up to the support of the metallic strut (see Figure 6.1(c)). The heads of the beam are not discretized. However, their presence creates a mechanical effect on the discretized part (see Figure 6.1(b)), which results in a moment applied to the two ends of the beam (partial embedment) evaluated on the basis of knowledge of the geometry of these heads, the weight of the beam and the position of the supports during the experiment (placed under the heads).

– Reinforcement: only the longitudinal rebars are represented by bar elements placed in accordance with the reinforcement mapping of the beam (see Figure 6.1(a)).

– Struts: these elements are represented by bar elements respecting the characteristics of the actual struts (tube 0.32 m in diameter with a total measured stiffness of 3.25 GN/m). The struts and rebars are connected to the same partial embedment as the concrete beam.

At this stage, two major hypotheses have been retained:

– the thermal problem is treated independently of the mechanical problem;

– the temperature fields are, at a given moment, assumed to be the same in all sections of the beam studied.

6.2.3. *Thermomechanical model of concrete at an early age*

Cement hydration is a thermally activated chemical reaction. Thus, during the process, the structure heats up under the effect of internal heat sources, in particular because of its massive nature. The exhaustion of the reaction and the exchanges with the outside cause a cooling of the structure to an equilibrium imposed by the ambient temperature. At the same time, the material is subjected to shrinkage, and its mechanical performance improves, which generates changes both in the volume variations and in the local strengths.

6.2.3.1. *Thermal changes*

In this problem, the following basic equations (see (Briffaut et al. 2011)) must be solved. It is solved using a finite element calculation carried out here with the Cast3M (2020) code.

1) The energy balance equation, known as the heat equation, includes a source term:

$$C\dot{T} = \nabla(k\nabla T) + L\dot{\xi}, \text{ with } \dot{T} = \frac{dT}{dt} \text{ and } \nabla T = \left(\frac{\partial T}{\partial x}\frac{\partial T}{\partial y}\frac{\partial T}{\partial z}\right) \qquad [6.6]$$

where T is the temperature (in °K), k is the thermal conductivity ($W \cdot m^{-1} \cdot °K^{-1}$) and C is the volume heat capacity ($J \cdot m^{-3} \cdot °K^{-1}$). These last two terms are taken as constants. L, the latent heat of hydration ($J \cdot m^{-3}$), is obtained directly from the measurement of the temperature rise during the hydration of the cement under adiabatic conditions (without heat exchange with the outside).

2) The equations give ξ, which is the hydration degree of the cement, the kinetics of which are given by the Arrhenius law:

$$\dot{\xi} = A(\xi)exp\left(\frac{E_a}{RT}\right) \qquad [6.7]$$

E_a is the activation energy of the cement ($J \cdot mol^{-1}$), R is the perfect gas constant (R = 8.3145 $J \cdot K^{-1}$ mol^{-1}) and $A(\xi)$ is the chemical affinity (s^{-1}), which is a thermodynamic function calibrating the evolution of the reaction.

ξ changes between ξ_0 (close to 0), the percolation threshold reached when the medium becomes solid, and ξ_∞ the hydration degree at the end of the reaction. To describe the state of the medium at a given instant of the chemical reaction, the maturity M is introduced (Mazars and Bournazel 1998).

$$M = \left(\frac{\xi - \xi_0}{\xi_\infty - \xi_0}\right)_+ \qquad [6.8]$$

M changes between 0 (start of the process) and 1 (end of hydration).

By introducing the concept of maturity into the expression from Sciumé et al. (2012), the chemical affinity, $A(\xi)$, is given by:

$$A(\xi) = \frac{A_i + (A_p - A_i)sin\left[\frac{\pi}{2}\left(1 - \left(\frac{M_p - M}{M_p}\right)_+\right)\right]}{\left[1 + \zeta\left(\frac{M_p - M}{1 - M_p}\right)_+^4\right]} - \left(\frac{A_p}{1+\zeta}\right)\left(\frac{M - M_p}{1 - M_p}\right)_+ \qquad [6.9]$$

A_i is the initial affinity, ζ governs the deceleration phase of the hydration process and A_p is the maximum value of the affinity reached for $M = M_p$. The calibration of these parameters also comes from the curve of evolution of the temperature of the cement hydration under adiabatic conditions, for which we can write:

$$\dot{Q} = \rho C \dot{T} \text{ and } \dot{Q} = Q_\infty A(\xi) \exp\left(-\frac{E_a}{RT_{ad}}\right), \text{ from which we deduce } A(\xi) \quad [6.10]$$

T_{ad} is the temperature at an instant of the adiabatic process, and Q_∞ is the total amount of heat dissipation measured, Q.

3) Relationships dealing with external exchanges:

$$\varphi = h(T_s - T_{ext})n \quad [6.11]$$

φ is the heat flux ($W \cdot m^{-2}$), n is the normal vector at the surface, T_s is the surface temperature (°K) and T_{ext} is the outside temperature (°K). h is the exchange coefficient whose value changes when the mold is removed (from 2, calculated taking into account the insulation provided by the mold at 10 $W \cdot m^{-2} \cdot K^{-1}$ when the latter is removed).

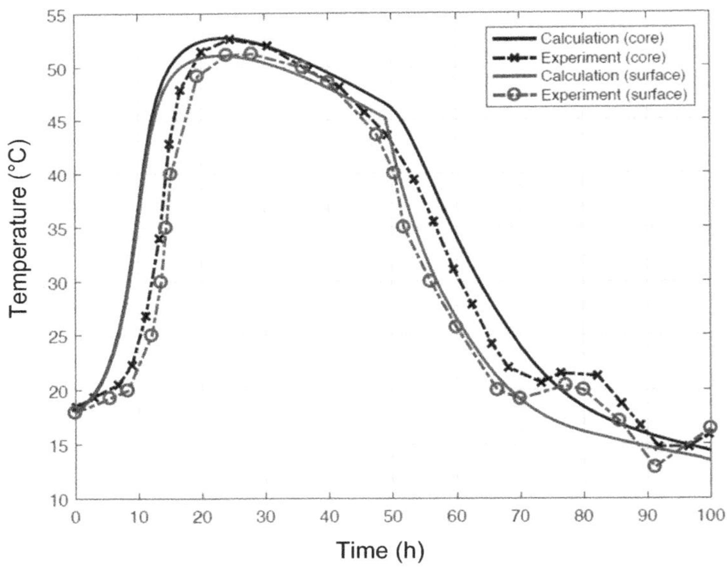

Figure 6.7. *Evolution of the temperature at the center (black curves) and near the surface (red curves) of the beam (experiment calculation comparison). For a color version of this figure, see www.iste.co.uk/mazars/damage.zip*

Figure 6.7 shows the results obtained after the identification of the parameters involved in the temperature change, at the center of the beam and close to the surface (see Table 6.3 for their numerical values). A good level of agreement is obtained with the measurements carried out in situ, both at the surface and at the core of the beam. The fluctuations measured at the end of cooling (beyond 70 h) result from the day-night external temperature cycles, whose curve has been smoothed to carry out the calculations.

6.2.3.2. *Thermo-chemo-mechanical couplings*

On the basis of the modeling of the mechanical behavior, there is the μ model in its 1D version. In the case in question, the mechanical characteristics change as the maturity *M*, which characterizes the state of the medium between the "just solid" state $M = 0$ and the "totally hardened" state $M = 1$ at a given instant.

In this context, *M* will appear, in the same way as *d*, in the equations of behavior. Focusing on simplicity, we discuss below how this introduction is carried out.

– For a given maturity, the elastic modulus is given by:

$$E(M) = M.E \quad [6.12]$$

E is the value of the elastic modulus when $M = 1$, that is, when the concrete is mature.

– In a completely conventional way, the tensile strength of the concrete becomes:

$$f_t(M) = E(M)\varepsilon_{t0} \quad [6.13]$$

A strong hypothesis is that ε_{t0} and ε_{c0} are assumed to be constant regardless of maturity.

– It is therefore also possible to determine the compression damage threshold:

$$\sigma_{c0} = E(M)\varepsilon_{c0} \quad [6.14]$$

These relationships are a specific case of those proposed by de Schutter (1999).

From these results, it is now easy to describe the nonlinear behavior of the material. With i = t under tension and i = c under compression, it is possible to write for $\varepsilon_i > \varepsilon_{i0}$:

$$\sigma_i = M.E(1 - d_i)\varepsilon_i \quad [6.15]$$

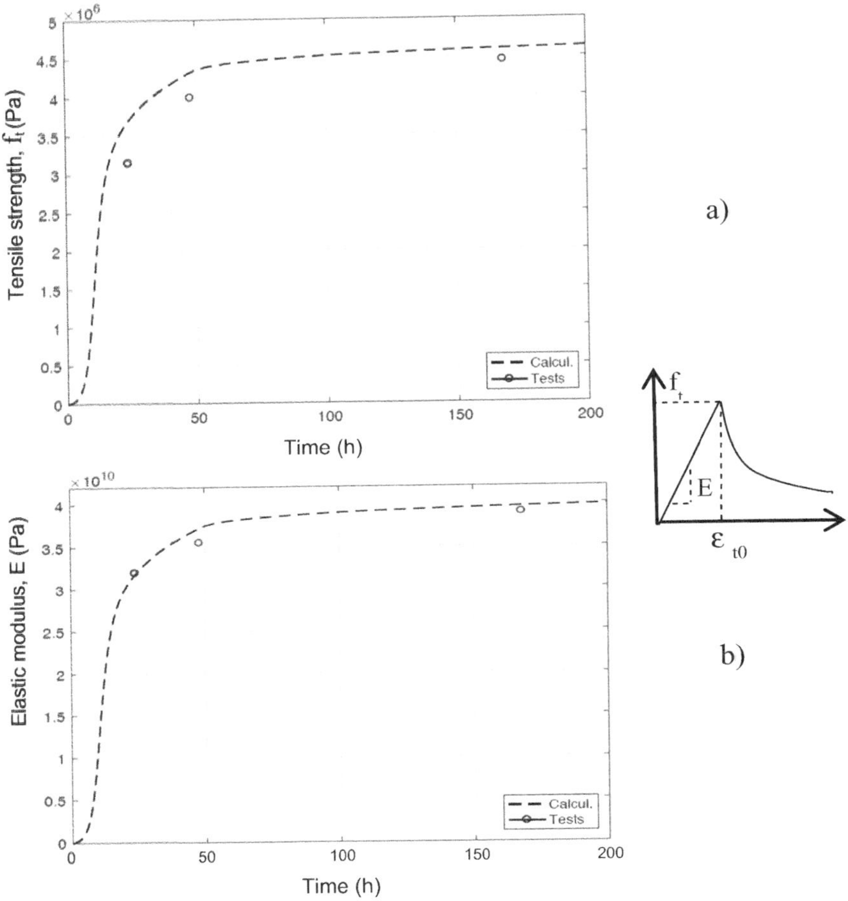

Figure 6.8. *Evolution over time of mechanical properties; test-modeling comparison: a) tensile strength f_t; b) elastic modulus E*

SUMMARY.– *From equations [6.7] and [6.8] and the evolution of the temperature as a function of time, M(t) is calculated at any point. This makes it possible, using equations [6.12]–[6.15], to determine the evolution of the elastic modulus and the strengths as a function of time. For the concrete constituting the RG8 beam (see Figure 6.8), the results obtained for the elastic modulus and the tensile strength are in good agreement with those obtained from tests carried out on specimens at different ages (see Figure 6.8).*

6.2.3.3. *Scale effect*

As indicated in section 6.2.1, the large dimensions of the structure in relation to the size of the specimens that are used to characterize the material require the scale effect to be taken into account. We addressed this problem in section 4.3.1.2, where we have taken as an example the case of data relating to the mechanical characteristics under tension by splitting tests on a cylinder 16 cm in diameter and 32 cm in height used for the modeling of a test tie rod (T) of dimensions: L = 5.10 m, b = 0.5 m and h = h_{ef} = 0.8 m (V_{eq} = 0.5 m^3), which are the dimensions of the RG8 beam loaded under tension in the restrained shrinkage phase. Using Weibull's law, we found (with m = 10) a relationship between the damage threshold (and therefore the tensile strength) between a direct tensile test on a cylinder (Φ = 16 cm, h = 32 cm or V_{td} = 0.0064 m^3) and that which is to be used for the calculation of the tie rod:

$$\varepsilon_{t0}^{td} = \varepsilon_{t0}^{T} \left(\frac{V_{ef}}{V_{td}}\right)^{\frac{1}{m}} = 1.55\ \varepsilon_{t0}^{T} \qquad [6.16]$$

However, the characterization is in fact the result of splitting tests on the same type of specimen. Thus, the relationship between the tension threshold and the splitting tension threshold is (see equations [4.7] and [4.8]):

$$\varepsilon_0^{td} \# 0.74 \varepsilon_0^{f}\text{; thus, we obtain: } \varepsilon_{t0}^{f} = \frac{1.55}{0.74} \varepsilon_{t0}^{T} \ \# \ 2.1 \varepsilon_{t0}^{T} \qquad [6.17]$$

In this approach, it is assumed that the strength of the hardened concrete is such that $f_t = E\varepsilon_{to}$; thus, with a strength estimated at 4.5 MPa per splitting test, a value f_t = 4.5/2.1 = 2.14 MPa will be used in the calculations relating to the beam (see Table 6.3).

6.2.3.4. *Creep expansion-shrinkage: overall description of the problem*

With all this information, we now come to the complete description of the problem, which will be based on the superposition of the following strains:

$$\varepsilon = \varepsilon_e + \varepsilon_{bc} + \varepsilon_{au} + \varepsilon_{th} \qquad [6.18]$$

The overall behavior is written as follows (Briffaut et al. 2011):

$$\dot{\tilde{\sigma}} = E(M)\dot{\varepsilon}_e = E(M)(\dot{\varepsilon} - \dot{\varepsilon}_{bc} - \dot{\varepsilon}_{au} - \dot{\varepsilon}_{th}) \qquad [6.19]$$

$\tilde{\sigma}$ is the effective stress, $\tilde{\sigma} = \sigma/(1-d)$.

Thus, equation [6.19] introduces damage into the overall process, where ε_{th} is thermal strain and ε_{au} is autogenous shrinkage:

$$\varepsilon_{th} = \alpha\theta \text{ and } \varepsilon_{au} = -\kappa_{\infty} M \qquad [6.20]$$

α (°K^{-1}) is the coefficient of thermal expansion, θ is the temperature variation and κ_{∞} (µm/m) is the shrinkage value for $M = 1$.

	C KJ/(°C.m^3)	k W/mK	L KJ/m^3	H W/m^{2}K	κ µm/m	α µm/m°C	$k^1_{bc\_\infty}$ GPa	$k^2_{bc\_\infty}$ GPa
Concrete at an early age	2.400	2.8	157.84	2 and 10	110	12	3×10^{11}	9×10^{10}
	$k^3_{bc\_\infty}$ GPa	A_i (s^{-1})	A_p (s^{-1})	M_p	ζ	ξ_0	ξ_∞	–
	2.5 10^{10}	64.85	1.660	0.172	75	0	0.8	–
Concrete model param.	E GPa	ε_{t0}	G_f N/m	ε_i	ε_{c0}	A_t	B_t	A_c B_c
	40	0.53×10^{-4}	60	3.5×10^{-4}	3.75×10^{-4}	0.815	19.25	1.5 355
Steel character.	E_s GPa	f_e MPa	E_t GPa	α µm/m°C	–	–	–	–
	210	500	5	7.5	–	–	–	–

Table 6.3. *Material properties and parameters used in modeling*

Basic creep is described by a series of three Kelvin–Voigt models (see Figure 6.9), the behavior of which is given by:

$$\tau^i_{bc}\ddot{\varepsilon}^i_{bc} + \left(\tau^i_{bc}\frac{\dot{k}^i_{bc}(M)}{k^i_{bc}(M)} + 1\right)\dot{\varepsilon}^i_{bc} = \frac{\dot{\sigma}}{k^i_{bc}(M)} \qquad [6.21]$$

ε_{bc} is the basic creep, and $\tau_{bc}(M) = \eta_{bc}/k_{bc}(M)$ is the characteristic time for a given Kelvin–Voigt model, where:

$$k^i_{bc} = k^i_{bc\_\infty}\frac{0.473}{2.081-1.608M}M \qquad [6.22]$$

k_{bc_∞} the final stiffness of the corresponding spring, when $M = 1$.

NOTE 6.2.– Principles used for the numerical calculation of couplings.

In the spirit of simplification, it has been accepted that during a time increment *dt*, the temperature *T*, the maturation *M* and the damage *d* remain constant. This hypothesis is possible if *dt* is small enough.

Thus, at a given moment, the creep-damage coupling is managed as follows:

$$\tilde{\sigma} = \frac{\sigma}{1-d} = E(M)\varepsilon_e$$

leading to:

$$\dot{\sigma} = (1-d)E(M)\dot{\varepsilon}_e = (1-d)E(M)(\dot{\varepsilon} - \dot{\varepsilon}_{bc}) \qquad [6.23]$$

This latter expression reflects the rheological behavior of the damage model of Figure 6.9, within which that of the Kelvin–Voigt model no. i is written as:

$$\dot{\sigma} = k^i_{bc}\dot{\varepsilon}^i_{bc} + \eta^i_{bc}\ddot{\varepsilon}^i_{bc} \qquad [6.24]$$

This relationship is identical to equation [6.21] when it is assumed that $k^i_{bc}(M)$ given by equation [6.22] is constant during the increment *dt*.

The stress and its velocity $\dot{\sigma}$ are the same in each of the four elements, and the strain velocity is the sum of the velocities of these elements:

$$\dot{\varepsilon} = \dot{\varepsilon}_e + \textstyle\sum_1^3 \dot{\varepsilon}^i_{bc} \qquad [6.25]$$

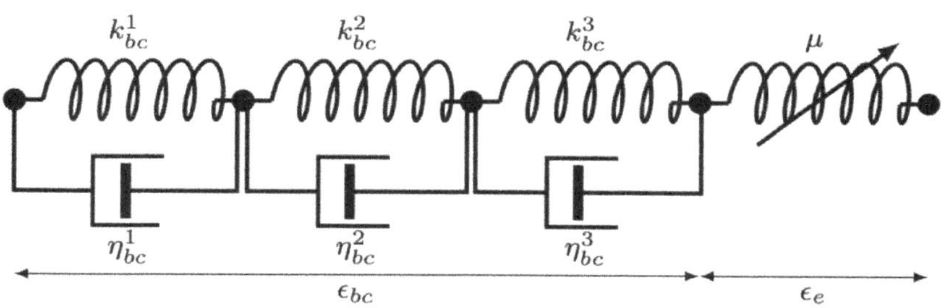

Figure 6.9. *Rheological assembly coupling creep (three Kelvin–Voigt models in series) and damage (spring with variable stiffness managed by the μ model)*

It is on the basis of these three equations that the numerical resolution takes place for each of the fibers of the Timoshenko beam. The resolution of the system is then as follows: at each time step t_{i+1}, the Newton–Raphson "global" resolution algorithm (at the scale of the complete structure) generates for the new iteration $k+1$, a kinematic U_{i+1}^{k+1} assigned to all of the elements of the system, which is retranscribed to each element and behavioral laws in the form of the strain ε_{i+1}^{k+1}. For this global iteration, the stress σ_{i+1} to be returned from the behavioral law is then calculated via another "local" Newton Raphson algorithm within the behavioral law, according to the flow chart that follows:

1) At each instant t_{i+1}, the temperature T and the maturity M are determined for the fiber concerned; from equations [6.20], ε_{th} and ε_{au} are deduced. These strains subtracted from the total strain ε_{i+1}^{k+1}, resulting from the step t_{i+1}, are given according to equation [6.18]:

$$\varepsilon_{i+1}^{k+1} - \varepsilon_{th} - \varepsilon_{au} = \varepsilon_{bc} + \varepsilon_{e} \qquad [6.26]$$

strain which is applied to the assembly of Figure 6.9.

2) The dynamic equilibrium of the chain of models is formulated by assembling the contributions of the four moduli in a matrix form whose strains at each node of the assembly are the unknowns. This is a dynamically formulated equation showing the dampers, springs, and force provided by the nonlinear spring of the damage law.

3) The strains for each of the moduli are calculated at the time step considered by a Newmark temporal discretization and by a Newton–Raphson method (required by the nonlinear character of the μ model). Convergence is obtained when equilibrium is ensured and therefore when the stress σ_{i+1} is the same in each modulus. We then deduce the strain that each of them withstands and the value of d in the damage model.

Next, we pass to the section of the beam, where the stresses σ_{i+1} of the fibers are integrated to calculate the forces at the nodes of the Timoshenko elements to achieve the resolution of the overall structure. This procedure is repeated at the next iteration if the overall equilibrium of the structure has not been reached at time step t_{i+1} or at the first iteration of time step t_{i+2} if the overall convergence has been reached.

NOTE 6.3.– The thermally damaged version of the μ model used here, like the original model, is part of a thermodynamic framework. The thermodynamic free energy potential as seen in equation [2.21] must be completed as follows (Bournazel 1995):

$$\Psi = \frac{1}{2\rho}\{(1-d)\,(\boldsymbol{\varepsilon} - (C(M) + \alpha\theta)\mathbf{1} - \boldsymbol{\varepsilon}_{bc}) : \boldsymbol{\Lambda}(M) : (\boldsymbol{\varepsilon} - ((M) + \alpha\theta)\mathbf{1} - \boldsymbol{\varepsilon}_{bc}) + \Psi_M + \Psi_T\} \quad [6.27]$$

This leads to the following state law:

$$\boldsymbol{\sigma} = \rho\frac{\partial\Psi}{\partial\varepsilon} = (1-d)\boldsymbol{\Lambda}(M) : (\boldsymbol{\varepsilon} - (C(M) + \alpha\theta)\mathbf{1} - \boldsymbol{\varepsilon}_{bc}(M)) \quad [6.28]$$

$C(M)\mathbf{1}$ is the tensor of the autogenous shrinkage strains, $\alpha\theta\mathbf{1}$ is the tensor of thermal expansions and θ is the temperature difference. $\boldsymbol{\varepsilon}$ and $\boldsymbol{\varepsilon}_{bc}$ are the tensors of the total strains and creep strains, respectively. Finally, Ψ_M and Ψ_T represent the terms of maturity and temperature, respectively, in the free energy. Thus, from equation [6.28], the coherence with the strain partition used in equation [6.18] and the expression of the behavior linking the effective stress and the elastic strains to equation [6.19] are found.

6.2.4. *RG8 test: application and results*

From the different equations presented in section 6.2.3 and from the laboratory tests carried out on samples, the different parameters identified are listed in Table 6.3.

6.2.4.1. *Thermal loading, couplings, mechanical loading*

The loading is primarily correlated with the temperature evolution, which concerns both the concrete and rebars. Two principal steps were observed:

– From 00 to 120 h, temperature variation is mainly due to the effects of hydration; thanks to the energy balance equation (equation [6.6]), this evolution was simulated via the Cast3M code; Figure 6.7(a) shows good agreement with the experimental in situ measurements, both in the center and in the vicinity of the surface of the beam section.

– From 120 to 700 h (700 h is the time corresponding to the removal of the metal struts), the temperature variation is mainly due to changes in external temperature (difference between day and night). Due to the massiveness of the structure, this change concerns only the material close to the surface, and for the sake of simplification, the calculation was carried out with an average constant temperature of 12°C. This simplification has not been made for the struts, which are tubular metallic elements and are therefore sensitive to the day–night temperature differential, which results in cyclic expansion/shrinkage. The cycle lasts 24 h, and the range is assumed to be between 9°C and 15°C.

As for the day-night temperature change, due to the massiveness of the structure and the relatively short duration of the experiment (one month), the drying effects are reduced and have not been considered here. This is reflected in the failure to take into account the desiccation shrinkage and hydraulic creep, the importance of which would be major for a long-term simulation.

In this context, laboratory tests have been carried out on cylindrical samples ($\Phi = 12$ cm, h = 24 cm), and the parameters useful for modeling appear in Table 6.3, but for details, reference should be made to the report (Kolani et al. 2012):

– the autogenous shrinkage has been calibrated to obtain κ_∞ for $M = 1$ (equation [6.20]);

– from compression tests loaded at 2 and 7 days, the basic creep model (equations [6.21] and [6.22]) was also calibrated, and to optimize the coverage of the entire duration of the test, the characteristic times of the three Kelvin–Voigt models chosen are $\tau_1 = 0.1$ day, $\tau_2 = 1$ day and $\tau_3 = 10$ days.

From 00 to 700 h, as shown in Figure 6.10 (red curves), the struts find themselves under load: under tension for the first 50 h, due to the thermal expansion of the concrete, then under compression due to the shrinkage of concrete. At the same time, they are subjected to the change in external temperature: cyclical evolution of 24-h periods with an average value of 12°C and an amplitude of 6°C (smoothing of the actual curves). In addition, from the moment of formwork removal (at 50 h, as seen in the temperature evolution in Figure 6.7), it was assumed that the gravity load was active over the entire beam. At 700 h, the struts were removed and, a few days later, the beam was tested on a bending test bench. Since the supports are spaced 5.1 m apart, they are located at the end of the beam portion where the cross-section remains constant. The loading is a four-point bending with 1.6 m spacing between loads (see Figure 6.14).

Due to the complexity of the phenomena concerned, many results have been obtained, but only the principal ones are presented here. The full set of experimental results can be found in previous studies (Kolani et al. 2012; Barré 2017). As indicated in the presentation of the RG8 test, the structure has been extensively instrumented. This device made it possible to accurately follow the evolution of three sections specific to the center and on both sides of the beam, approximately 1 m from the central section. The results that follow relate to these sections, which have been instrumented with strain gauges on the rebars, vibrating cords near the upper and lower surfaces, and three optical fibers distributed on a median plane of the beam covering 2.5 m of its central zone.

6.2.4.1.1. From 00 to 120 h: concrete maturation period

Relative forces that develop in the central section

In a given section, three principal structural elements were considered: concrete, the set of rebars and the struts. From the strain gauges located on the rebars and assuming that the latter remain elastic, it is easy to deduce the stress acting at a given moment on each element ($\sigma = \varepsilon E_s$), from which it is possible to deduce the force F_s supported by the set of rebars in a given section.

Strain gauges were also placed on the struts. By considering the cross-section of these elements, it is also possible to deduce the force F_e that is exerted on them. Moreover, at all times, each section is in equilibrium.

From this equilibrium, the force F_c supported by the concrete section can be deduced:

$$F_c(t) = -[F_s(t) + F_e(t)] = -[\Sigma_i(E_s\varepsilon_i(t)A_i) + 2E_s\varepsilon_e(t)A_e] \qquad [6.29]$$

where E_s is the elastic modulus for steel; A_i and ε_i are the cross-section and the strain of rebar i, respectively; and A_e and ε_e are the cross-section of each strut and the corresponding strain, respectively. The changes in the central section of the beam of these three global forces are compared in Figure 6.10(b).

Attention should be drawn to the following:

– the comparison between calculation (continuous lines) and experience (dotted lines) is generally good, both from a qualitative point of view (change trends) and from a quantitative point of view (extreme values);

– after a period during which the struts are under tension, the concrete and the rebars under compression, the sign of the three forces changes at approximately 65–70 h, while the struts pass under compression, the rebars and the concrete pass under tension and a first crack appears (71 h measured and 76 h calculated). Because of the bending caused by the structure's own weight, the crack is located in the central section of the beam (measurement and calculation area). This phenomenon leads to a peak of force in concrete, followed by a sudden decrease, with a simultaneous increase in the force taken up by the rebars. At 103 h, the calculation provides for a second series of cracks, which, due to the diffuse description of the damage, appears as an extension on both sides of the previous damage area (see Figure 6.10(c)). This second cracking process was observed experimentally at approximately 167 h (see the cracking facies in Figure 6.13(c)).

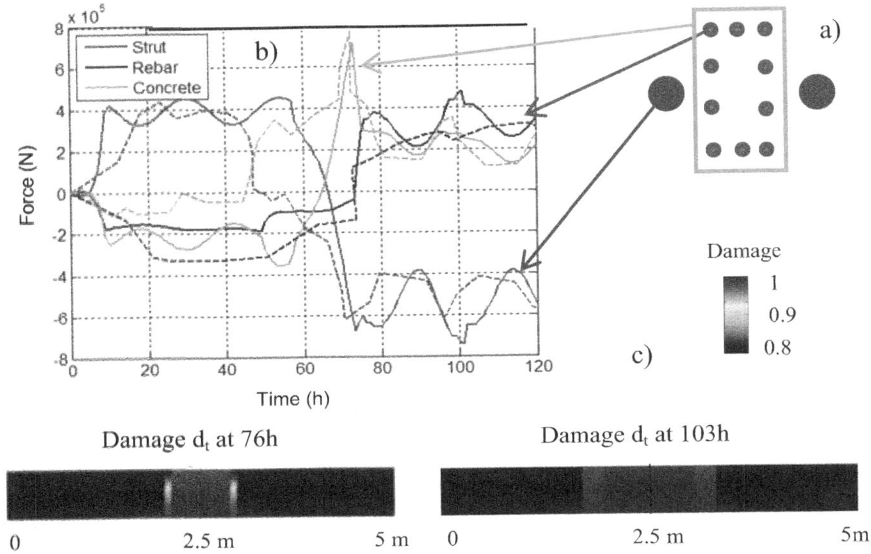

Figure 6.10. *Evolution from 00 to 120 h: a) the three elements of the section: struts (red), rebars (blue), concrete (green); b) overall forces supported by the three systems in the central section of the beam (dotted lines: test; continuous lines: calculations); c) damaged areas at 76 h (date of first damage) and 103 h (date of extension of this damage). For a color version of this figure, see www.iste.co.uk/mazars/damage.zip*

Evolution of local stresses in the rebars

During this restrained shrinkage process, the rebars remain elastic and the strain–stress relationship therefore remains linear. Figure 6.11b shows, for the first 100 h and in the central section, the evolution of the stress over time, corresponding to both the upper rebar (blue line) and the lower rebar (red line).

The principal findings are as follows:

– a jump, clearly visible, caused by the loading of the rebar system due to the cracking of concrete in the central zone; for the lower rebar, this jump has been correctly predicted, and the difference obtained between calculation and measurement for the upper rebar is certainly due to the hypothesis of the remaining planar sections of the Timoshenko approach, which is a little defective here;

– Figure 6.11(a) shows the changes for a section outside the cracking zone (this section is approximately 1 m from the central section); up to approximately 70 h, both for the experiment and for the calculation, there is a similar change in the stresses in the two sections (see Figure 6.11(a) and (b)); beyond 70 h, the results differ due to the passage of the crack in the surroundings of the central section. Note that after this time threshold, the stresses change only slightly in the non-cracking section (see Figure 6.11(b)), and due to the shrinkage of concrete, the latter remains negative (the rebars are slightly compressed).

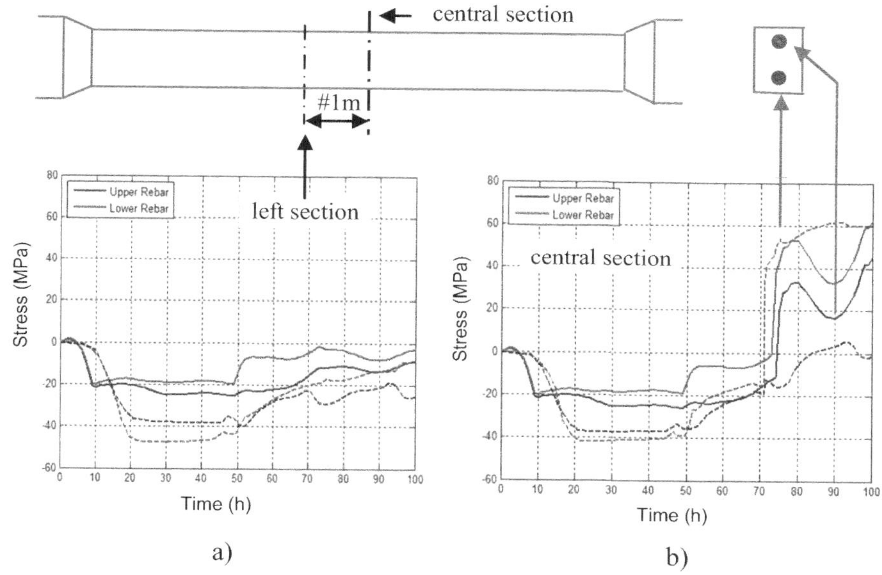

Figure 6.11. *Development of stress in rebars (dotted lines: test; continuous lines: calculations): a) in a section outside the cracking zone (left section); b) in the central section of the beam. For a color version of this figure, see www.iste.co.uk/mazars/damage.zip*

Cracking and opening of cracks

The experimenters placed three optical fibers in the beam. The active zone of the fibers is 2.5 m around the center of the beam. In this area, the changes in displacement were compared. Figure 6.12 shows a discrepancy between the two series of curves (dashed line: experiment; continuous line: calculation). This is due to the difficulty encountered in fixing the reference zero on the optical fibers. In any case, both for measurement and calculation, a displacement jump clearly appears during cracking. This jump corresponds to the crack opening.

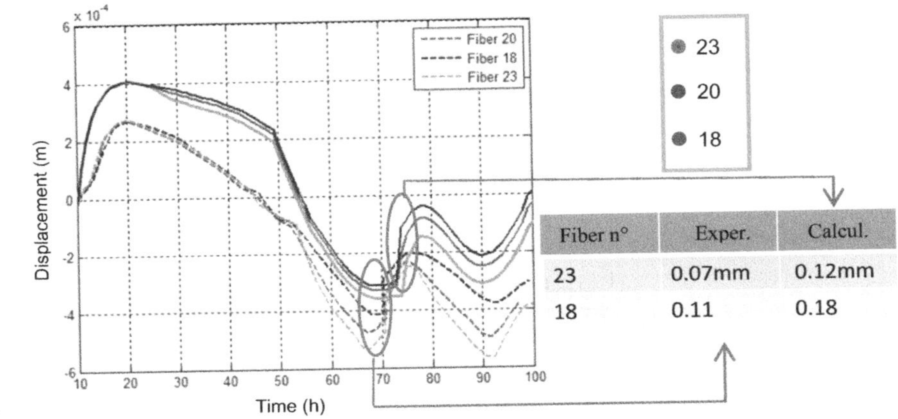

Fiber n°	Exper.	Calcul.
23	0.07mm	0.12mm
18	0.11	0.18

Figure 6.12. *Evolution of displacements in concrete: measurements from optical fibers compared with displacements calculated at the same points (dotted line: test; continuous lines: calculations). The opening of the cracks is directly deduced from the jump of displacement. For a color version of this figure, see www.iste.co.uk/mazars/damage.zip*

From these measurements, it is possible to deduce the crack opening. The values obtained experimentally and by calculation are of the same order of magnitude. Certainly, the calculations overestimate this openness by a few tens of percent, but this remains a good indicator, given the complexity of the phenomena involved and the large number of hypotheses underlying these calculations.

6.2.4.1.2. From 00 to 700 h: effects of restrained shrinkage

The initial period was mainly marked by the heating and cooling of the concrete during hydration. After 120 h, maturation continues and the system suffers the effects of nocturnal and diurnal temperature variations. These variations generate a quasi-cyclical strain in the struts that is transmitted to the beam. Figure 6.13 clearly shows this effect, which results in an evolution of the load in the three systems (concrete, struts, rebars). In addition, the following facts should be highlighted.

1) Three principal cracks are observed during the experiment. The calculation provides for a first crack zone at the center of the beam at 76 h (see Figure 6.10(c)), which corresponds to the first experimental crack appearing at 71 h. At 103 h, the crack zone extends on both sides at the same time, while the experimental cracks appear on the left side at 167 h and on the right side at 239 h (see Figure 6.12(c)). It can therefore be concluded that the calculation does indeed provide for these two cracking phases (to the left and to the right of the central zone), but prematurely in relation to the experiment. The present load is mainly tension combined with

bending due to the weight of the beam. As demonstrated in Chapter 3, in the presentation of simulations by multifiber beam elements, damage is diffuse and there is no sensitivity to the mesh. This diffusion and the symmetry of the mesh and the loading are probably the cause of the simultaneous appearance of the damage zones to the left and to the right of the initial zone.

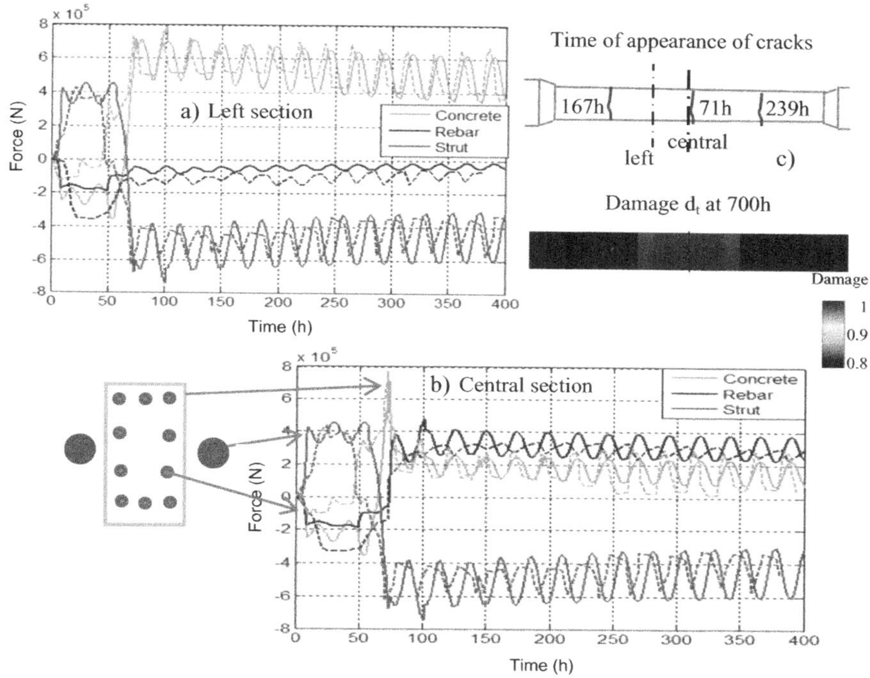

Figure 6.13. *Evolution from 0 to 700 h. For a color version of this figure, see www.iste.co.uk/mazars/damage.zip*

COMMENTS ON FIGURE 6.13.– *(a) Overall forces supported by the three systems in the left-hand section, 1 m from the center of the beam outside the crack zone (dashed line: experiment; continuous line: calculation); (b) overall forces supported by the three systems in the central section (inside the cracking zone); (c) observed cracks and contour of tensile damage at the end of the process (700 h).*

2) Figures 6.13(a) and (b) show the overall responses of the three systems (struts, concrete, rebars). In the central part, once the concrete has cracked, the rebars and the struts are the principal active systems, whereas in the left section, outside the cracking zone, the principal active systems are the struts and the concrete because in the absence of cracking, the load borne by the rebars is close to zero.

3) After 120 h, the mean values of these loads are practically constant, and the small variations that appear are due to the cyclic variations in diurnal temperature, which generate a quasi-cyclic expansion in the struts that is then transmitted to the beam. This limited evolution barely produces changes within the beam, which is why the damage predicted by the calculation is almost the same at 700 h (see Figure 6.13(c)) as that obtained after the second and third cracking stages at 103 h (see Figure 6.10(c)).

6.2.4.1.3. Beyond 700 h: mechanical performance of the beam

After this phase of cracking caused during the maturation process, the experiment was focused on the analysis of the residual mechanical performance of the beam, which is an important issue and has rarely been studied thus far.

To do this, after 700 h (approximately one month), the struts were removed, and a few days later, the beam was placed on a test bench to be tested under a four-point bending monotonous loading. The supports were spaced 5.1 m apart and located at the end of the beam portion, where the section is constant. The loads were spaced 1.6 m apart (see Figure 6.14).

With respect to the overall behavior, Figure 6.14(a) compares three curves:

– the experimental curve (line and black dots); the objective was to analyze the behavior of the beam up to the serviceability limit state (SLS), so the experimental loading was only carried out until this stage;

– the curve calculated for the beam that has undergone the restrained shrinkage load program (blue line);

– the curve calculated for the same beam, but considered to be structurally sound (red curve).

The calculated responses were continued until and beyond the plasticity of the rebars; the behavior of the steel considered here is a perfect elastoplastic model (f_e = 500 MPa). Attention should be drawn to the following:

– the calculated RG8 curve conforms well with the experimental curve;

– the two calculated curves (RG8 and the structurally sound beams) are very close during most of the load evolution;

– a significant difference is observed between the initial stiffness values; the initial stiffness of the RG8 beam is greatly reduced (approximately 50%) due to cracking of concrete produced by the restrained shrinkage.

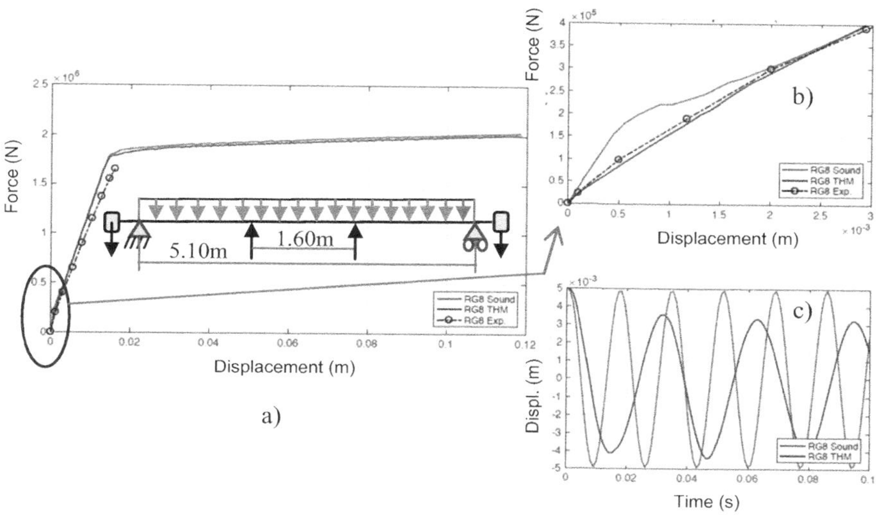

Figure 6.14. *Bending test on the RG8 beam after the cracking phase due to restrained shrinkage. For a color version of this figure, see www.iste.co.uk/mazars/damage.zip*

COMMENTS ON FIGURE 6.14.– *(a) Force-displacement responses (experiment: points; RG8 beam simulation: blue line; structurally sound beam simulation: red line); (b) magnification of the first 3 mm of displacement, showing the great difference between the two original stiffnesses (RG8 beam and sound beam); (c) corresponding responses to free vibrations for these same two beams.*

On this last point, an additional calculation has shown that the first mode of vibration of the same beam is equal to 31.7 Hz instead of 59.2 Hz for the sound beam, which results in a very different dynamic response in free vibrations (see Figure 6.14(c)).

All of these observations lead to the conclusion that the behavior at failure of the beam (stage of the ultimate limit states [ULS]) has not been modified by the restrained shrinkage.

At the same time, its initial stiffness has been greatly reduced, which leads to giving it a very different dynamic response (frequency of the principal mode divided by 2) and therefore the risk of having a reaction to a dynamic loading, of a seismic nature, for example, which is very different from the one imagined assuming that the structure is undamaged.

Taking such an effect into consideration deserves to be integrated into the procedures for designing structures subject to earthquakes.

Evolution of cracking

Figure 6.15 shows the evolution of cracking observed during loading in experimental bending with respect to the evolution of damage obtained by calculation at the same load level.

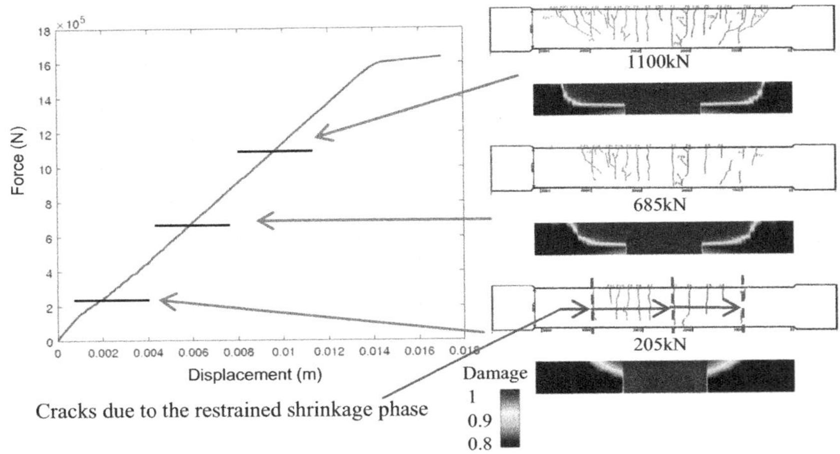

Figure 6.15. *Evolution of cracking observed during loading in experimental bending compared with the evolution of damage obtained by calculation at the same load level. For a color version of this figure, see www.iste.co.uk/mazars/damage.zip*

Attention should be drawn to the following:

– for a low load level (205 kN), cracks appear in the center of the beam between the cracks due to the restrained shrinkage, while the diffuse damage, as developed during the shrinkage phase, is located in the same area and remains relatively unchanged during the first load phase;

– for the following phases (685 kN and 1,100 kN), cracks and damage develop on the upper part of the beam (the load acts from bottom to top); moreover, the contour of damage, which is always diffuse, is entirely consistent with the observed crack map.

6.2.4.2. *Key findings from this study*

6.2.4.2.1. In terms of modeling

Modeling the effects of maturation of concrete is a complex problem; excellent teams of researchers are working on this subject, which is generally approached on a meso- or even microscopic scale and usually leads to models that are admittedly very useful, but require much expertise for their implementation. Furthermore, the resulting calculations lead to cumbersome and complex problems.

We have shown here that, on the basis of a simplified description (multifiber beams) and the use of a 1D model with enrichments adapted to the treated situation, particularly relevant results are obtained. In particular, the fact that these 1D calculations lead to the creation of diffuse damage stabilizes the convergences and drastically reduces the calculation times. All of these calculations were conducted on the basis of the ATL4S platform (Grange 2017) in a few minutes on a completely conventional PC.

6.2.4.2.2. In terms of results achieved

It is observed that when shrinkage is restrained during maturation, which is observed for many actual structural elements, substantial premature cracking can be formed, particularly when the latter are of large dimensions. This involves two major types of consequences:

– on the durability of the structure (Tang et al. 2015): cracks are an entry vector for water and aggressive substances that will cause the corrosion of steel, or in some cases damage due to sulfates, the alkali-aggregate reaction, or the destructive effects of freeze–thaw;

– on the mechanical response: although there are no short-term consequences on the behavior of the ULS, the same is not true for dynamic loads, such as earthquakes (in this case, the first natural frequency has been reduced by almost 50%); moreover, in the long term, the resulting weakening of the durability can have serious consequences on the mechanical performance of the ULS, and sometimes even the SLS. Studies to analyze this problem are underway at the INSA in Lyon (Jaafari 2020).

Finally, it should be noted that, in a structure in a real situation, the hindrance to the establishment of shrinkage can be evaluated a priori by the estimation of a restraint degree. This estimate makes it possible, thanks to specific constructive rules provided by regulations, to act, from the design stage, on the reinforcement provisions. This degree is determined in an exercice presented in section 9.4.

7

Structures Combining Beams and Planar Elements

7.1. Simulation of the behavior of a reinforced concrete wall

Common reinforced concrete structures, particularly buildings, are a combination of slender elements (beams and columns) and planar elements (slabs and walls). Chapters 4–6 have shown that the calculation techniques associated with the use of multifiber elements are a good solution for many types of problems associated with slender structures. The objective here is to see how to approach, using simplified methods, that of planar elements.

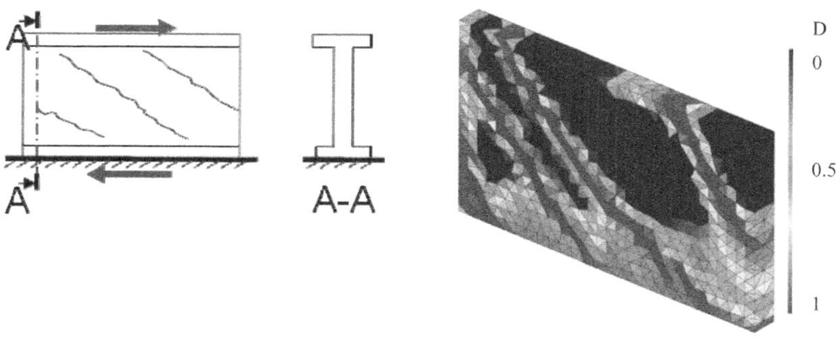

Figure 7.1. *Reinforced concrete wall subjected to monotonous shear loading treated with the non-local Mazars model (LMT – ENS Paris-Saclay calculation). For a color version of this figure, see www.iste.co.uk/mazars/damage.zip*

The simulation of a slab or a wall is often tricky because it requires the use of 2D descriptions, or even 3D descriptions, when there is an out-of-plan component. The most general solution for tackling this problem is to use planar elements (or shells, if the geometry has a curvature), or better still, 3D volume elements. This is what has been done in the case of the wall of Figure 7.1, monotonously loaded in planar shear. The Mazars model was used by integrating a "non-local" calculation procedure to solve the problem of mesh dependency. The results were convincing, but the overall procedure of meshing–calculating–operating the results becomes very cumbersome if it is extended to a complete structure.

Moreover, in the case of alternating loadings, for example, under the effect of an earthquake, this requires using unilateral models such as the μ model of greater susceptibility to convergence to manage the restoration of stiffnesses, particularly in the case of heavy damage and/or with rotation of the principal directions.

For a building, the slabs are considered diaphragms, which mainly have elastic behavior during seismic loading. Under these conditions, using elastic planar elements (available in all calculation codes) does not pose any problem. On the other hand, walls have a structural role blending the recovery of vertical loads and horizontal loads (bracing effect) with, in addition, some out-of-plan effects related to the 3D character of the action. We will see that we can use a 1D-based model, in the same spirit as the processing of beams, to solve all of these situations.

7.1.1. *Model for structural walls: equivalent reinforced concrete*

The problem at hand is that of the representation of a concrete wall of constant thickness reinforced by means of a network of vertical and horizontal rebars placed equidistantly in two layers close to the outer surfaces.

Figure 7.2(a) shows the principle of describing what we call equivalent reinforced concrete (ERC). We start from an elementary representative volume of the wall (E_v) at the intersection of the two rebar networks, vertical and horizontal. The description of this representative volume includes the following:

– a parallelepipedal elementary volume of concrete represented by a rectangular assembly of six bars;

– a rectangle of four bars representing the rebars and coupled to the previous ones at the nodes (for more details, see Kotronis et al. 2003). All of these bars are hinged at their ends and consequently behave uniaxially. Those relating to concrete have elasto-damageable behavior derived from the enriched μ model (see the behavior curve in Figure 7.2(c)), and those relating to steel have perfect elasto-plastic behavior.

On the basis of an energy equivalence between the homogeneous body and the network of bars that is substituted for it, Hrennikoff (1941) determined the section of concrete bars for an elastic problem (see Figure 7.2(b)), with the actual section being taken for the "steel" bars representing the rebars. Note that this discretization imposes an overall Poisson coefficient equal to 1/3 (instead of the current value of 0.2 for concrete), which is not too disadvantageous.

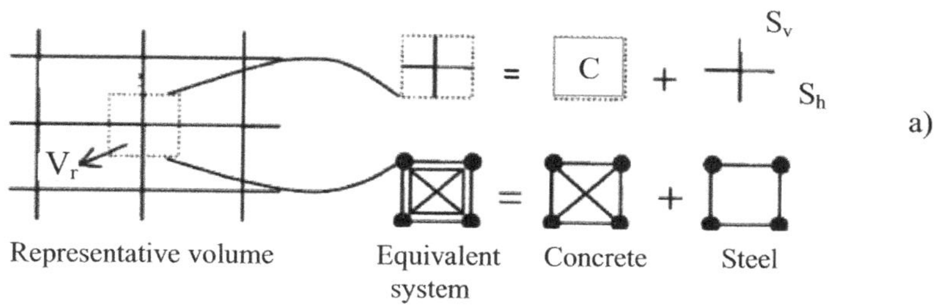

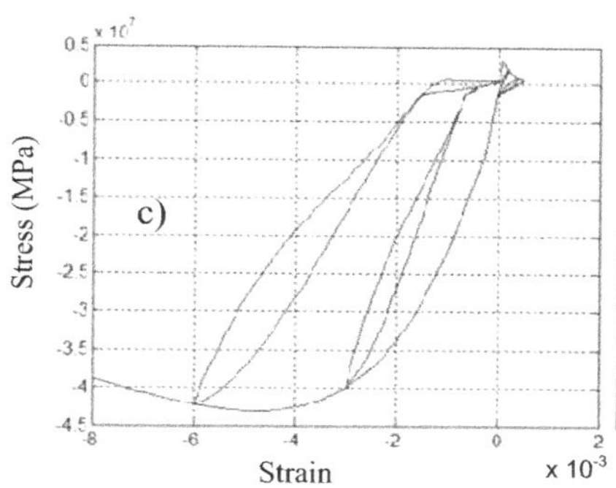

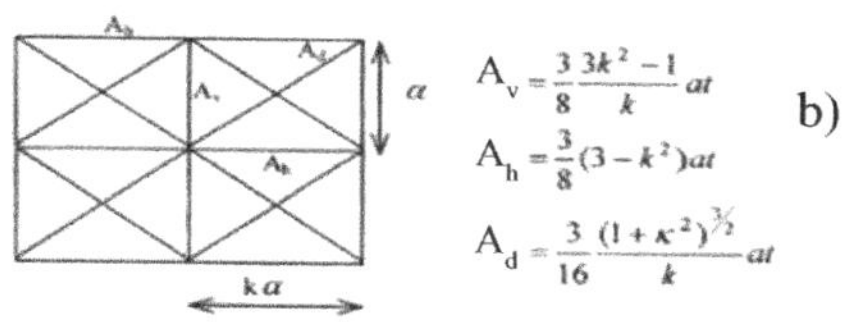

d)

	Concrete		Steel	
E (GPa)	f_c (MPa)	f_t (MPa)	E (GPa)	f_e (MPa)
35	43.5	3.3	200	500

Figure 7.2. *ERC structural model: a) the superposition of concrete-steel bars; b) the section of concrete bars according to Hrennikoff (1941), t is the thickness of the wall; c) the used concrete behavior; d) the characteristics of the materials used for the calculation of the "SAFE" wall*

7.1.2. *Application to the SAFE experiment shear wall case*

The SAFE testing program was carried out at the initiative of EDF and COGEMA on the reaction wall of the Joint Research Centre (JRC) of the European Commission in Ispra (Italy).

The experimental program covered 10 walls (Labbé et al. 2015); Figure 7.3 schematically shows one of them. The choice was to consider only the part of the wall analyzed in a nonlinear lattice model (ERC). The rebars were vertical HA8 and horizontal HA10 with a spacing of 12.5 cm. The assembly was discretized by 10 × 15 base cells (see Figure 7.3(c)). The two reinforcement elements, left and right, are each represented by an elasto-damageable multifiber beam (cross-section:

0.16 × 0.8 m) that has the actual vertical rebar (20 HA12). The assembly was reinforced using an upper beam and a lower beam, which, because of their massiveness, were assumed to be elastic and very rigid (the section of these beams was # 1 m^2). Figures 7.2(c) and (d) show the behavioral law used for concrete and the characteristics of the materials used for modeling.

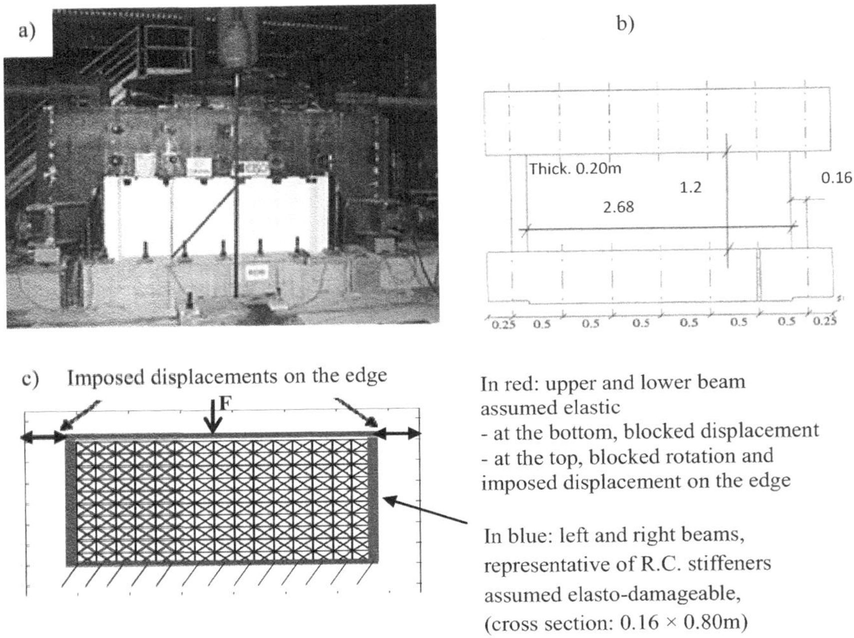

Figure 7.3. *The SAFE wall: a) photo of the experimental assembly showing the metal load application helmet; b) scheme of the wall to be modeled; c) details of the ERC discretization selected for the wall. For a color version of this figure, see www.iste.co.uk/mazars/damage.zip*

The purpose of the experiment was to determine the behavior of these walls under the effect of alternating shear from seismic loading. Tests were carried out on a reaction wall by a so-called "pseudo-dynamic" procedure. This procedure combined measurement results (the current stiffness of the structure) and dynamic calculations to determine, at each instant, the increments of displacements complying with the equation of motion, updated by the measurements and the acceleration increment of the loading program. In addition, to simulate the effects of a gravity load, a constant vertical load was applied per specific actuator (see Figure 7.3).

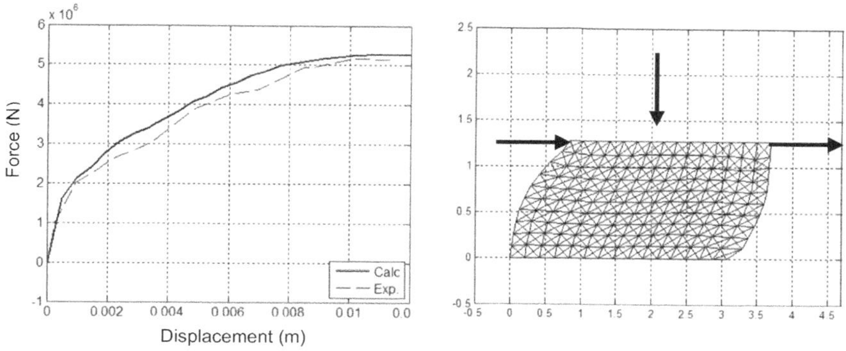

a) b)

Figure 7.4. *Structural response to monotonous shear loading: a) test-calculation comparison of the overall force–displacement behavior; b) deformation of the wall (amplified) at the end of the test*

Resulting from the pseudo-dynamic procedure, we know the succession of displacement increments applied at the top of the wall; these increments have been used for the present simulation. Thus, as time becomes a fictitious time, inertia does not occur, and the load becomes a cyclic load of variable intensity.

Before the seismic tests, monotonous loading of shear until failure was carried out on the first wall. Figure 7.4(a) shows the results obtained. They show an excellent ability of the model to reproduce the overall response of the structure. Figure 7.4(b) gives the amplified deformation of the wall at the end of this test in monotonous loading; it was observed that all of the elements of the structure were mobilized and that the failure began in the right edge at the bottom of the wall, which is consistent with the experimental results.

7.1.2.1. *Overall cyclic behavior*

Figure 7.5 shows, for the most intense loading cycles (cycles 3 and 4), the "force–displacement" comparison between the test results (see Figure 7.5(a)) and the calculation results (see Figure 7.5(b)). It can be seen that the results of the calculations are quite relevant to those given by the tests, both qualitatively and quantitatively, including for cycle 4, which leads to the structural failure of the wall. As part of an organized benchmark on the subject, these results, compared to complete 2D-3D approaches, are among the best.

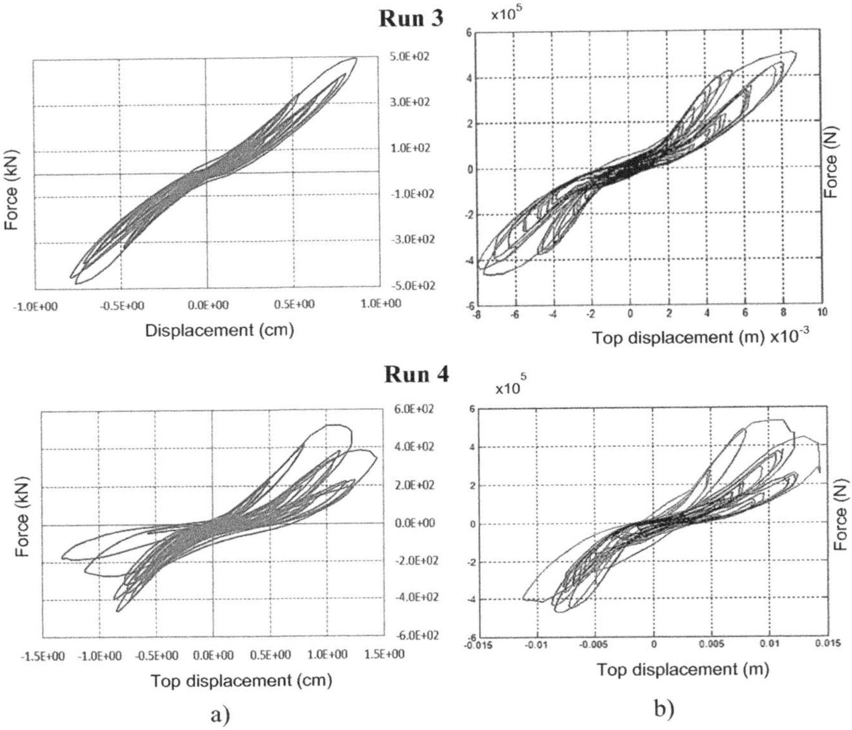

Figure 7.5. *SAFE: Overall cyclic loading force–displacement behavior for Run 3 and Run 4: a) experimental response; b) computational model response*

7.1.2.2. *Local cyclic behavior*

Despite its simplified nature, the calculation gives access to local information such as the strain, damage, and plastic strain fields of steel. Figure 7.6 shows some examples. We see that:

– for rather modest loads (6 mm), a tensile damage field d_t was created, which was relatively homogeneous in the wall, consistent with in situ observations that showed a double network of diagonal cracks distributed over the entire surface of the wall (see Figure 7.6(a));

– compression damage d_c is activated from Run 1 (3 mm), especially at the four edges of the wall (see Figure 7.6(b));

– important localizations were observed, especially in the failure phases (see Figure 7.6(d)) with the creation of an intense strain zone which, from the upper left edge, followed the diagonal line, and joined the lower right corner where the opening of a major crack at the base began, consistent with in situ observations (see the photo at the end of the test at the bottom right of Figure 7.6);

– the plasticization of steel was activated late; Figure 7.6(c) shows the platicization state for the same load level as the strain field in Figure 7.6(d); there was an accentuated platicization in the areas of localized strain.

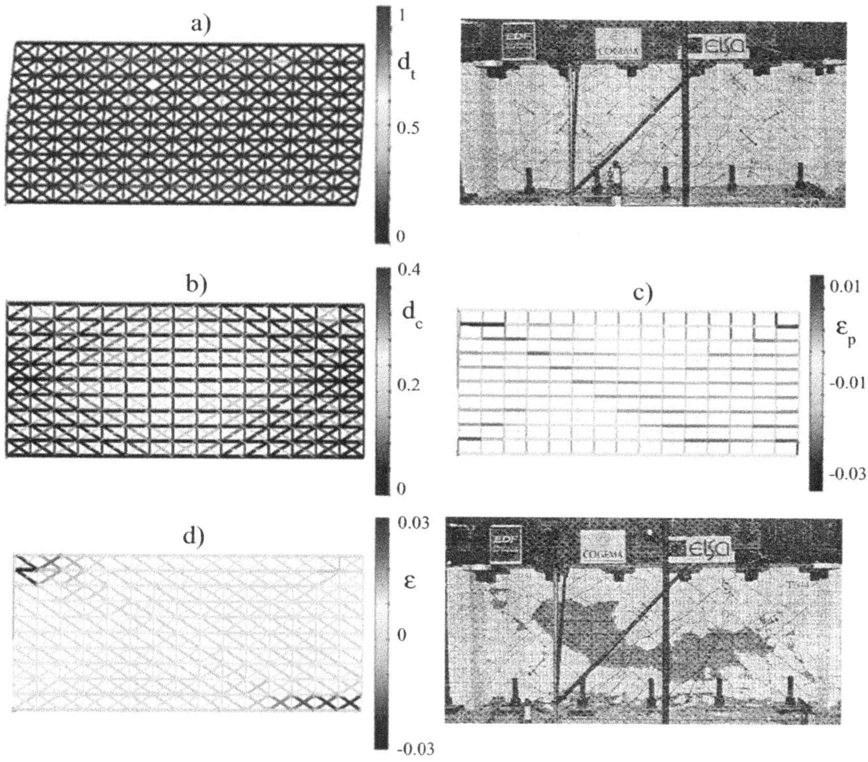

Figure 7.6. *Example of local results (the darker the color, the more intense the local field). For a color version of this figure, see www.iste.co.uk/mazars/damage.zip*

COMMENT ON FIGURE 7.6.– *(a) Deformed mesh and damage field d_t for a displacement of 6 mm (Run 2), to be compared with the crack distribution observed during the test (photo in the top right); (b) compression damage field d_c at 3 mm (Run 1); (c) plastic strain field of steel at 12 mm (Run 4); (d) strain field obtained at 12 mm (Run 4), to be compared with the photo of the experimental ruin in the bottom right.*

NOTE.– This original and inexpensive procedure in computation time, mainly related to the fact that nonlinear behaviors are 1D, requires some precautions to achieve good discretization. The ERC discretization scheme is based on the principle of the so-called "strut and tie" diagrams, which require finding the paths of forces, tension by rebars, and compression by concrete connecting rods within a reinforced concrete beam, the whole schematized by a lattice model.

In the same spirit, the ERC must be constructed so that the hinged bars representing the concrete are oriented as far as possible parallel to the principal directions. This may require preliminary (elastic) calculation to determine these directions.

7.2. Application to a structure combining walls, beams and columns

The aim of the SAFE programme was to analyze the behavior of structural walls under seismic-type loading. In section 7.2, we discuss the behavior of a complete structure subjected to actual seismic loading, the modeling of which requires combining the description of several types of elements (beams, columns, walls, slabs, etc.). We will show that at this stage of the presentation, this has become possible while remaining in the use of the simplified modeling concept. This problem is addressed based on the work carried out as part of the SMART program (EDF-CEA) conducted around tests on a model embedded on the Azalée shaking table (CEA Saclay) and subjected to a series of bidirectional earthquakes.

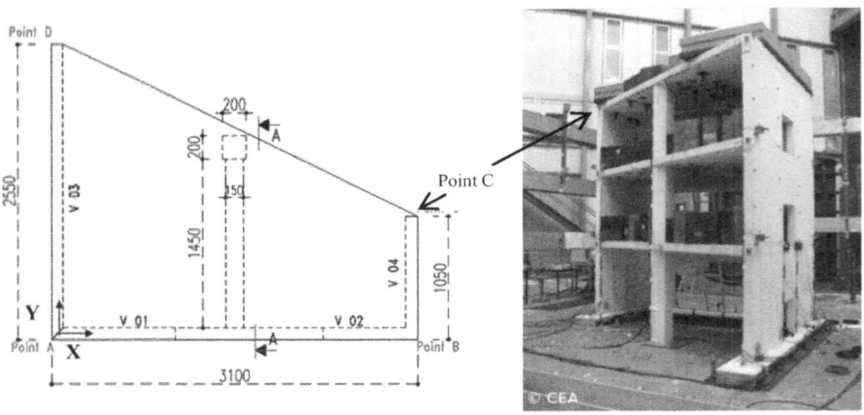

a) b)

Figure 7.7. *SMART mockup. a) Dimensional scheme of the structure; b) general view of the model set on the shaking table (CEA documents). For a color version of this figure, see www.iste.co.uk/mazars/damage.zip*

The SMART program integrated a modeling component, carried out in particular as part of an international benchmark (Richard et al. 2014), which made it possible to confront many numerical approaches with the realities of the experience. In sections 7.2.1 and 7.2.2, we recount the work carried out and the main results obtained by the team of the 3S-R laboratory in Grenoble (De Biasio et al. 2014), which is fully in line with the approach developed in this book. In what follows, the principles and the main results of the study are proposed without going into the details of the calculations (carried out with the Cast3M (2020) code), the aim being to prove the relevance of the approach.

The SMART model (see Figure 7.7) is a three-story trapezoidal structure representative of a real building on a ¼ scale. It is mainly composed of structural walls, with the addition of a column and three beams (under each slab).

The characteristics of the tests carried out are:

– the seismic loading is in two components X and Y (see Figure 7.7(a));

– the structure is completely asymmetrical.

Because of this, the response of the structure is three-dimensional with a displacement component in all three directions. In this context, the modeling choices were as follows:

– columns and beams are modeled using nonlinear multifiber Timoshenko beam elements;

– slabs are modeled by means of elastic shell elements; this part of the structure works as a diaphragm, which either activates nonlinearities or not;

– the ERC approach, is used for modeling walls, but because of the 3D nature of the movement that generates an out-of-plane deformation component, the ERC modeling, as seen in section 7.1.1, must be enriched to cope with this complex situation.

7.2.1. *Enriched ERC modeling*

In general, a structural element can cope with bending and/or torsion loading if it opposes a moment of inertia in the direction of this loading. The wall, as shown in the SAFE test in Figure 7.7, has no inertia in the orthogonal direction to its plane and therefore cannot oppose a transverse action.

Integration of out-of-plane inertia requires the addition of supplementary structural elements:

– leading to this introduction of moments of inertia (two bendings and one torsion);

– not modifying the stiffnesses in the plane, with the risk of changing the behavior of the structure in its plane.

One solution, proposed by De Biasio et al. (2014), is to introduce supplementary "virtual" beams with the mechanical characteristics of concrete (E and ν), located in coupling with each vertical and each horizontal beam, and with an out-of-plane bending inertia and a torsional inertia that will have to be calibrated. The cross-section of these beams and their mass are zero so as not to add additional stiffness in the wall plane and additional grip to the accelerations of the movement.

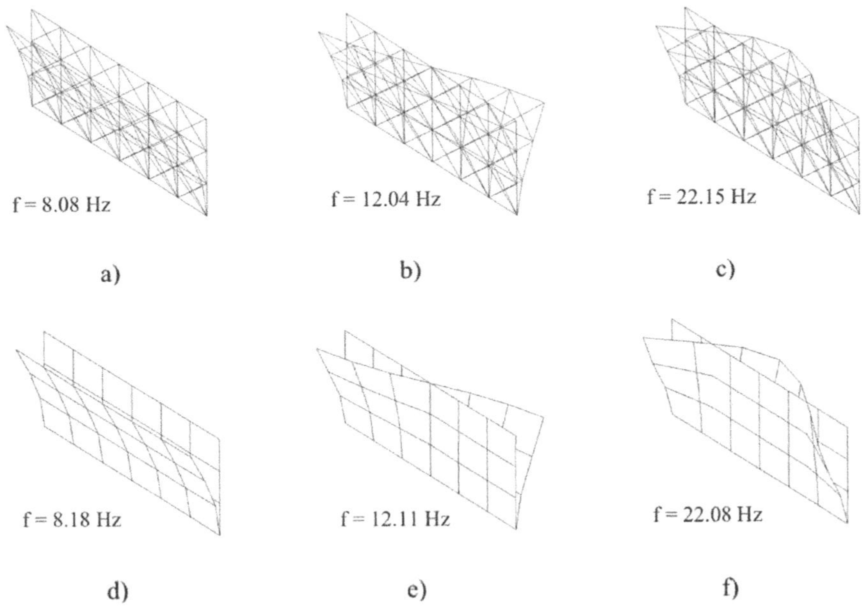

Figure 7.8. *Adjustment of out-of-plane stiffness in enriched ERC modeling*

COMMENT ON FIGURE 7.8.– *The enriched ERC model (a, b, c) has the same dimensions, mass, and meshing as the shell model (d, e, f). The bending moments of inertia (J_{xx}, J_{yy}) and the torsional moment of inertia (J_t) of the added beam layer are matched to reproduce the first three modal frequencies, as well as the shape of the corresponding modes given by the reference shell model (d, e, f) (reproduced from De Biasio et al. 2014).*

This set of bars constitutes a layer of additional elements to the ERC model, of which it is necessary to determine three unknowns: the two bending inertias (J_{xx}, J_{yy}) and the torsional inertia (J_t). The calibration of these inertia values is given in

Figure 7.8 for a wall section representative of those of the SMART model (the size of this section can be chosen arbitrarily). In parallel, this same wall section is modeled using elastic shell elements (see Figure 7.8(b)), respecting all the characteristics (material, geometry, mass). For fixed boundary conditions (an embedment at the base here), the vibration modes (see section A.2.4 for more details) of these two wall sections (the one in FE and the one in enriched ERC) are analyzed. The calibration of the bending and torsion inertias aims to find the same modes for the ERC wall enriched as those of the discretized wall with shell elements. Figure 7.8 shows that we obtain good results from this.

7.2.2. *Modeling the response of the SMART model*

The masses of the structure thus discretized are distributed over the elements (beams, columns, slabs, and walls), while the additional masses necessary to respect the service loads and scale effects are located on each of the upper floors. They are explicitly modeled through massive 3D elements. For bar elements representing rebars, a conventional elastoplastic constitutive relationship has been chosen, whereas the constitutive law of concrete is that of the μ model. The problem will not be further discussed here, but all the necessary information can be found in (SMART 2013).

The model is loaded bidirectionally (X-Y), and the loading program comprises three main phases. In this order:

1) an artificial earthquake complying with the design spectrum with a peak ground acceleration of 0.22 g;

2) natural Northridge earthquake with a peak ground acceleration of 1.1 g;

3) a replica of the Northridge earthquake (1994, near Los Angeles (USA)).

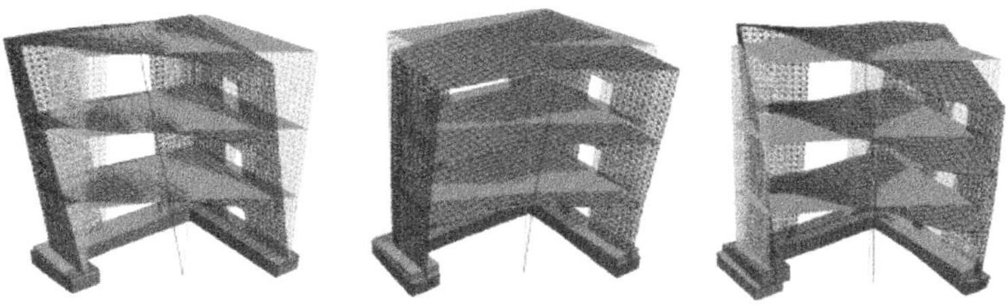

a) b) c)

Figure 7.9. *Vibration modes of the model (for readability, the shaking table and additional masses have not been visualized): (a) bending direction X; (b) bending direction Y; (c) torsion (reproduced from De Biasio et al. 2014). For a color version of this figure, see www.iste.co.uk/mazars/damage.zip*

Mode	Bending X	Bending Y	Torsion
Exp. frequency	6.4 Hz	9.1 Hz	17.8 Hz
Calc. frequency	6.3 Hz	7.9 Hz	16.5 Hz

Table 7.1. *Comparative values (experiment – calculation) of the first three eigenfrequencies of the SMART model*

Figures 7.9 and 7.10 compare the numerical and experimental results in terms of modes and displacements. Figure 7.9 shows the first three vibration modes of the structure from modal analysis; the associated frequency values are shown in Table 7.1. Comparison with the experimental values confirms the relevance of the discretization choices.

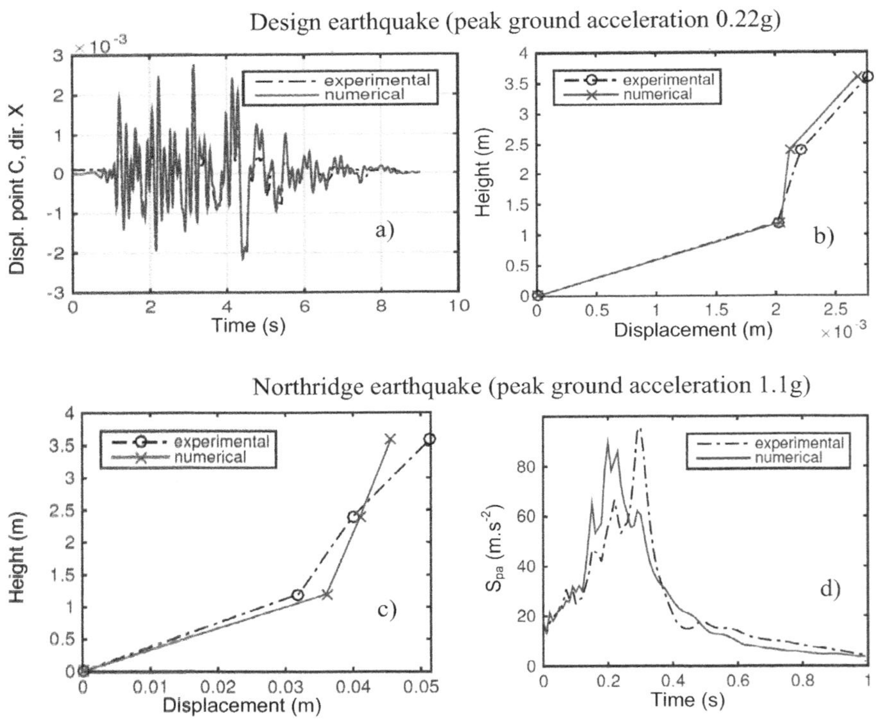

Figure 7.10. *Experiment-calculation comparison for different earthquakes. For a color version of this figure, see www.iste.co.uk/mazars/damage.zip*

COMMENT ON FIGURE 7.10.– *(a) Design earthquake: history of displacements at point C, third floor, direction X; (b) design earthquake: maximum displacement at each floor, point C, direction X; (c) Northridge earthquake: maximum displacement at each floor, point C, direction X; (d) Northridge earthquake: acceleration spectrum at point C, third floor, direction X.*

Numerous results were obtained in this program, and only an overview is given here. Among the various tests carried out, Figure 7.10 presents some displacement results on two specific tests (reference point C is located in Figure 7.7):

– test no. 9 (design level), for which the experimental and calculation displacement history (point C, direction X, third floor) and the maximum displacement at point C in direction X of each floor are compared;

– test no. 19 (Northridge earthquake signal), for which the maximum experimental and calculation values of the displacement in direction X at point C of each floor and the acceleration spectrum at point C of the third floor in direction X are compared.

Overall, these results are good, in any case as good as those obtained during the benchmark (SMART 2013) by much finer discretizations (2D, 3D), resulting in significantly more costly calculations (Richard et al. 2018). It can be recalled that the modeling strategy chosen here leads only to 1D nonlinear calculations, both for the beam-type elements as well as for the wall-type elements, and it goes without saying that this gives rise to little or no problems of calculation stability and that the convergences are very fast.

7.3. Calculation combining 2D finite elements and multifiber beams

The results given in sections 7.1 and 7.2 show the value of a 1D description for dealing with the behavior of slender structural elements. This being the case, real structures, even those based on beams and columns, cannot always be represented solely by the combination of structural elements based on beam elements. Figure 7.11 shows that the connection of two structural elements simply modeled with beam elements (see Figure 7.11(d)) cannot represent the constructive details (see Figure 7.11(b)) and the ultimate states when the critical sections are beam-column connections (see Figure 7.11(c)). We can be satisfied with a simplified representation, such as that shown in Figure 7.11(d), if the critical zones are not nodes, but beams or columns, and if these same connections are not damaged, or are only slightly damaged. Otherwise, when the connections are critical areas, the question that arises is how to numerically deal with the behavior of nodes and, more generally, the whole structure. The characteristics of the arrangement of rebars in these zones and the multiaxiality of the local stress fields generated by the loading guide the description of nodes by a richer description, 2D or 3D, depending on whether the problem to be treated is planar.

In section 7.3, we will show that this description of nodes does not prohibit dealing with the beams and columns with simplified descriptions, as we have seen them thus far using Timoshenko beam elements.

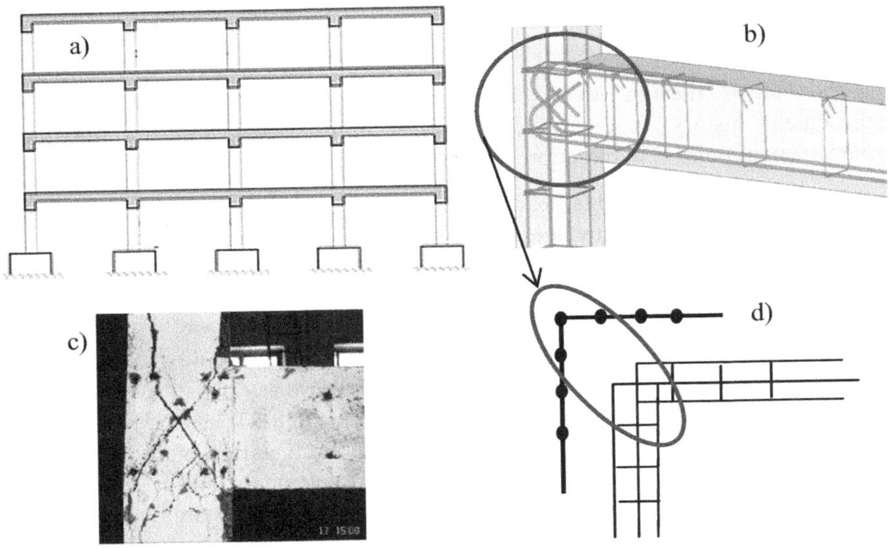

Figure 7.11. *a) Slab-beam-column structure; b) detail of a beam-column connection; c) failure facies of such a connection; d) weakness of the connection in a description with only beam elements. For a color version of this figure, see www.iste.co.uk/mazars/damage.zip*

7.3.1. *Case study: loss of bearing capacity of a column in a structure*

To this end, we propose here an application on a representative structure of a part of a double-span and multistory gantry for which, following an accident (impact, explosion), there appears to be a loss of bearing capacity. The tested element represents a beam of this structure (Chen and Forquin 2019). The primers of the three support columns are integrated, the experimental system replicates the boundary conditions of the presence of the end columns, and the load is applied at the central column (see Figure 7.12(a)). Details of the reinforcement of the structure are given in Figure 7.12(b).

The problem is considered to be planar and symmetrical. Consequently, one beam and two connections are to be discretized. Discretization of the left node and the central half-node are in 2D finite elements, and the beam is discretized by multifiber Timoshenko beam elements (see Figure 7.12(c)).

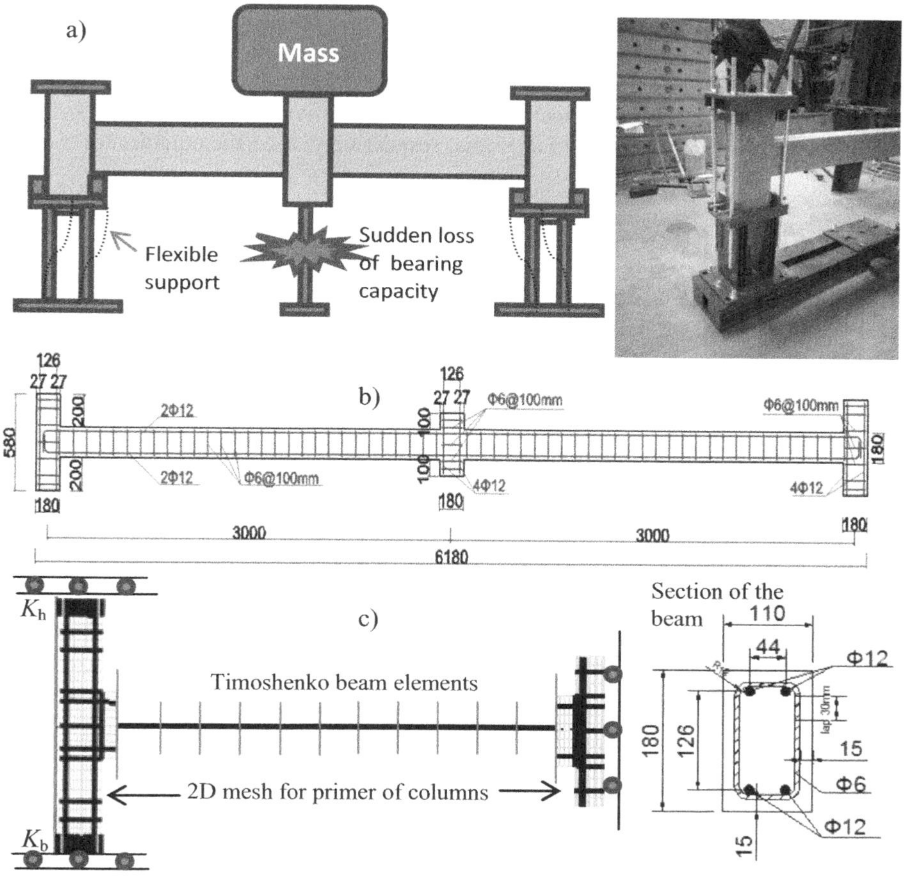

Figure 7.12. *Bearing capacity loss test on a beam with two spans (Chen and Forquin 2019): a) details of supports and loading system; b) characteristics of the beam; c) mesh and boundary conditions used (Kh and Kb are the stiffnesses of semi-rigid connections). For a color version of this figure, see www.iste.co.uk/mazars/damage.zip*

The 2D discretization represents the node itself and a primer of the beam. The multifiber Timoshenko description induces the fact that the sections of the beam remain planar during deformation, so the coupling with the 2D node must take place on a plane.

Webb (1990) showed that, in the case of static problems, this type of kinematic stress can be applied by a so-called "elimination" method. Taking up this work, we will generalize the concept for dynamic problems here. This method consists of finding a projection matrix $[B]$ of the degrees of freedom linked to the 2D domain to condense them on the terminal node of the last Timoshenko beam element. If $\{F\}$ and $\{U\}$ are the force and displacement vectors, respectively, and if $[M]$,$[C]$,$[K]$ are the matrix of mass, damping and stiffness, respectively, then the equation of motion is written as:

$$^{t}[B][M][B]\{\ddot{U}\} + {}^{t}[B][C][B]\{\dot{U}\} + {}^{t}[B][K][B]\{U\} = {}^{t}[B]\{F\} \qquad [7.1]$$

Equivalent to the use of the Lagrange multiplier, this method has the advantage of reducing instead of increasing the system size (for more information, see Grange (2017)).

Details of this assembly are shown in Figure 7.12(c):

– for the beam part, multifiber elements are 10 cm long and have an 8 × 2 fiber section;

– for nodes, the elements are Quad4 of the 1.5 cm side;

– for reinforcement, bar elements shall be used;

– to represent boundary conditions associated with the end node, a horizontal degree of freedom and flexibility at the interface with horizontal supports are introduced. This flexibility is identified, by inverse analysis, from experimental results.

7.3.2. *Calculation-experiment comparison results*

The experimental program involved different beams and many load situations that can be found in (Zhao 2019). As an example, we present here the modeling of one of these tests, which consisted of loading the beam at its center with a mass representative of service loads. Since this loading is less than that which causes the collapse, the sudden loss of the central support leads to obtaining a nonlinear dynamic response of the whole system. We will show that this modeling strategy is relevant to simulate this response.

	E (GPa)	σ_{0t} (MPa)	G_f/x (N/m^2)	x (m)	A_t	B_t	σ_{0c} (MPa)	A_c	B_c
FE calculation	16	2.2	2,000	1.5×10^{-2}	0.9	3,000	–18	1	215
MF calculation	16	2.2	400	7.5×10^{-2}	0.9	14,500	–18	1	215

Table 7.2. *Material parameters and characteristics used for 2D-1D coupling (post-peak tensile parameters are different for FE and MF)*

Table 7.2 shows the different values of the parameters and characteristics used for the presented calculations. It should be noted that the elastic modulus of concrete has been reduced to take into account the initial damage related in particular to the formwork and the storage conditions of the structure, which generated shrinkage cracks leading to an apparent reduction in stiffness (Chen and Forquin 2019). It should also be noted that the cracking characteristics are not the same for the two types of discretization (FE and MF): the fracture energy G_f to be considered is the same, but because of the localization, or not, the size x over which this energy dissipates is different ($x = h$ for the FE calculation and $x = s_c$ for the MF calculation) (see equation [4.15]). In addition, in the absence of dedicated experiments, parameters related to the strain rate effects were chosen to be identical to those presented in Table 6.2.

Loading takes place in a sequential order. The following shall be applied in order:

– gravitational effects;

– bearing capacity loss, by a sudden release of the condition of vertical locking of the central node; this generates an inertia load created by the 675 kg mass and the weight of the beam.

Since this load is less than the load that causes the collapse (>1,600 kg), the system will damage without breaking and vibrates.

Figure 7.13 shows that the calculation leads to a good reproduction of the history of the maximum velocities and displacements reached. The few frequency deviations (# 20%) are related to the difficulty of quantifying the actual boundary conditions, which appeared to be nonlinear during the test.

At the local level, it is observed that damage is created in the vicinity of the interface between the node and the beam, generating crack opening and a high localization of the strains on the longitudinal rebar ($>1.5 \times 10^{-3}$). As a result, damage appears around the corresponding rebar. From the central node, this cracking continues, but in a diffuse manner, on the multifiber Timoshenko beam element. All of these results are in good conformity with experimental observations.

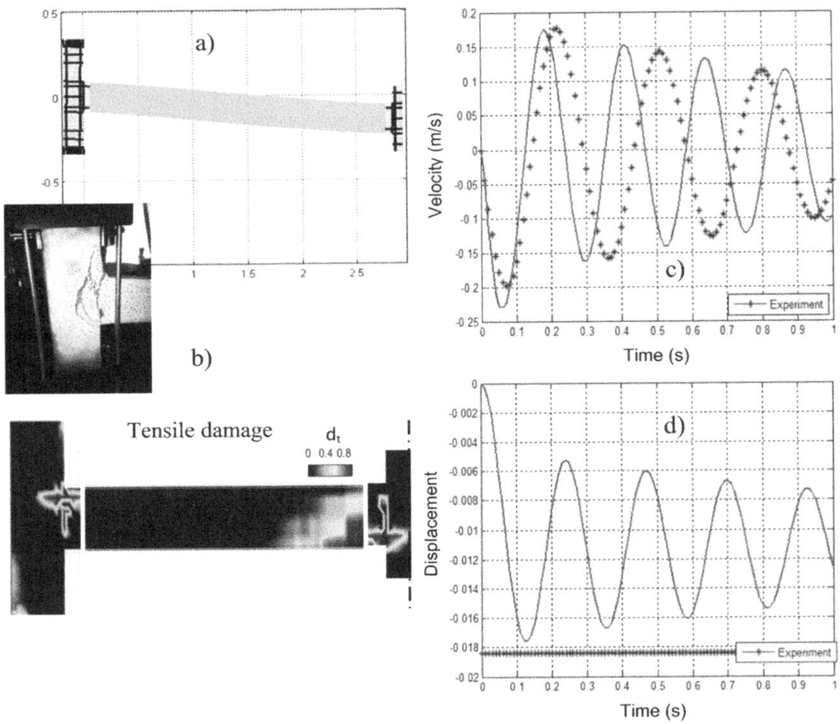

Figure 7.13. *Bearing capacity loss on a two-span beam. For a color version of this figure, see www.iste.co.uk/mazars/damage.zip*

COMMENT ON FIGURE 7.13.– *(a) Deformed mesh after loading with a mass of 675 kg; (b) contour of damage and state of cracking in the left node; (c) time history of the velocity in the central node (solid line is calculation); (d) time history in the displacement in the central node (the straight dotted line gives the maximum value measured during the experiment).*

7.4. Conclusion

The applications proposed in this chapter show that there is no limit to associating and combining several approaches to respond to a given specific problem: the introduction of out-of-plane stiffness to cope with the 3D responses of walls, the introduction of 2D and 3D descriptions for a good representation of complex local fields associated with damage to connections while retaining the simplified representations for slender elements of the structure. It is on the basis of an in-depth analysis of the problem to be addressed that all of an engineer's expertise must be used, in order to make progress on the best choices to be made and this at the stage of preparation of the simulation, which is by far the most critical phase of a calculation.

8

Assessment of Cracking by Limit Analysis

8.1. Characterization of cracking: case of homogeneous fields of tensile elements

The objective of this chapter is to propose a simplified but effective modeling approach to simulate the behavior of situations where the strain field remains homogeneous until the appearance of tensile cracking. This is particularly the case for tie rods or reinforced concrete cylindrical tanks under liquid or gas pressure. In the spirit that has prevailed thus far, the idea is to lead by a simplified calculation to a diffuse description of damage. Next, we will show that post-treatment, based on a dialog between the situation of diffuse damage and localized cracking established on the basis of approaches resulting from limit analysis concepts, we arrive at convincing results relating to cracking indicators.

8.1.1. *Limit analysis and yield design theory*

Yield design is an ancient concept that, in civil engineering and particularly in soil mechanics, has become an actual science. In particular, it makes it possible to approach the estimation of failure conditions of a soil massif with a given geometry and soil type, under the effects of a system of loads, including gravity, in a "simple" way. This is done by imposing a failure surface within this soil massif (Coulomb prism (1773), Fellenius circles (1936), Rendulic spiral (1935), etc.).

This concept is part of the broader theory of limit analysis, which is defined as follows (Salençon 1983):

> The existence of the limit load of a system of perfectly plastic elastic material on a given loading path appears to result solely from the necessary compatibility between the equilibrium of the system and the condition of perfect plasticity of its constituent material.

The objective of limit analysis is to directly determine limit loads from this mathematical definition alone.

The yield design theory focuses on the same problem by replacing the condition of perfect plasticity with the concept of the strength condition. This is defined, at each point of the system, by the data of a convex strength domain assigned to the stress tensor or, more generally, to internal forces. The advantage is that it is a direct method that frees itself from a step-by-step calculation along a loading path to failure, thus avoiding the difficulties inherent in the latter type of approach, which greatly simplifies the problem.

Our purpose here is to adapt the concept not to the failure or – in a more common vocabulary in civil engineering – to the ultimate limit states (ULS), but to the cracking limit state (CLS).

Let us take the example of a reinforced concrete beam element of rectangular cross-section; if it is subjected to a bending moment, a relative rotation between two sections that will generate a strain distribution will follow in the maximum moment zone (ε_t on the lower side, ε_c on the upper side).

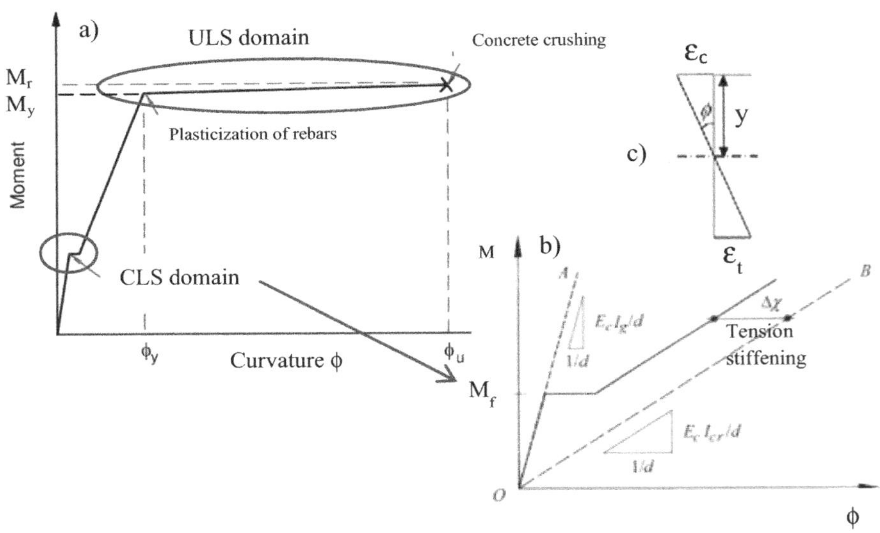

Figure 8.1. *Moment–curvature behavior curve of a reinforced concrete beam showing: a) the domain of the ultimate limit states (ULS); b) cracking limit state (CLS); c) the definition of the curvature Φ*

If y is the location of the upper fiber relative to the neutral fiber of beam (see Figure 8.1(c)), then $\emptyset = -\varepsilon_c/y$ is called the curvature. An increasing loading of the beam section leads to the moment–curvature response of Figure 8.1. This curve consists of three phases:

– $0 < M < M_f$ (M_f cracking moment), the behavior is linear elastic;

– $M_f < M < M_y$ (M_y moment of entry of steel into plasticity), the behavior is nonlinear;

– $M_y < M < M_r$ (M_r failure moment), it is the failure phase, the domain of the ULS.

The cracking moment stage M_f shows a more or less marked plateau according to the characteristics of the beam (see Figure 8.1(b)); it is at this time that the cracking of the element is triggered. This phase will be of particular interest to us here.

8.1.2. *Case of reinforced concrete beams in bending*

For example, consider the case of a reinforced concrete beam subjected to bending (beam on two supports loaded in its center). Assuming that it responds at the failure moment to a perfect plastic behavior, let us study this beam using the limit analysis concepts and the plasticity situation (ULS) caused by the load (see Figure 8.2).

– *"Static" approach*: a necessary condition is written for the "stability" or the holding of the structure to be ensured and for there to be compatibility between the quasi-static balance of the structure and the strength capacities of its constituent material. If the failure moment of the structure is M_r, the static approach leads to:

$M_r = P_r\, l/4$, then the failure load is $P_r = 4M_r\,/l$ [8.1]

– *"Kinematic" approach*: by mathematical dualization, by means of the principle of virtual works (PVW), a "kinematic" approach is highlighted to conveniently determine loads for which the "instability" of the structure under the indicated conditions is certain. In the case of our beam, we imagine a failure pattern compatible with connections (see Figure 8.2(c)), leading here to a mechanism and therefore to the instability of the system, for example, a "plastic hinge" in the center of the beam. PVW stipulates the equality of virtual work during the failure caused by a virtual displacement δ:

$P_r\delta = M_r\, 2\theta$, where $\delta = l\,\theta/2$, we then obtain: $P_r = 4M_r\,/l$ [8.2]

Thus, assuming that all unplasticized elements are assumed to be stiff, the result is the same as with the static approach. In the general case, the static approach is called "from the inside" and the kinematic approach "from the outside"; the two approaches frame the real solution (here, equality shows that the real solution is achieved).

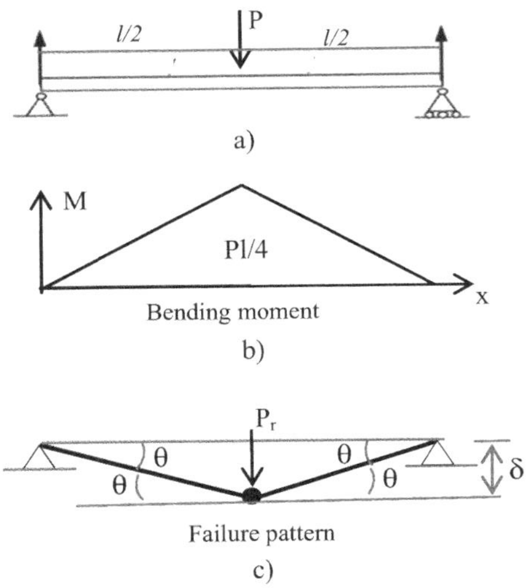

Figure 8.2. *Limit analysis applied to a three-point bending beam: a) characteristics of the beam; b) variation of the bending moment; c) failure pattern with a plastic hinge located at the maximum moment point*

Now consider the same problem, but with a four-point bending beam under the conditions of Figure 8.3(a):

– The bending moment is constant between the two loads, is equal to $Fa/2$ and the static approach leads to:

$$F_r a/2 = M_r, \text{ that is, } F_r = 2M_r/a \quad [8.3]$$

We obtain the same result as previously if $b = 0$ and $2a = l$.

– The kinematic approach involves the choice of a failure pattern; starting on the same principles as in the 3-point bending of Figure 8.2, let us put a "plastic hinge" under each load, which in fact appears simultaneously. In this case, the PVW leads to:

$$2F_r\delta/2 = M_r 2\theta; \text{ as } \delta = a\theta, \text{ we obtain } F_r = 2M_r/a \quad [8.4]$$

The result is the same as with the static approach, and shows that the choice of a single plastic hinge, under a load or between loads, leads to exactly the same result.

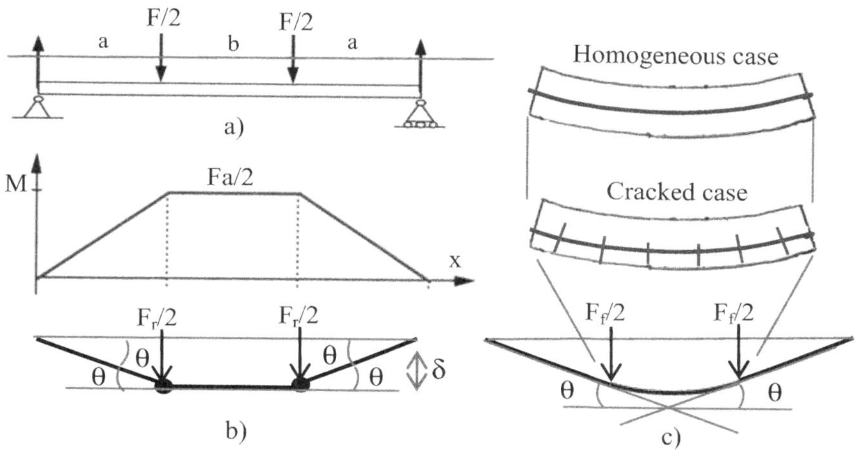

Figure 8.3. *Limit analysis applied to a four-point bending beam: a) characteristics of the beam and evolution of the bending moment; b) failure pattern with two plastic hinges; c) "plastic segment" concept*

On this same test, let us resume the reasoning but with the cracking moment M_f. As we have seen, the moment–curvature curve in Figure 8.1(a) shows a quasi-plateau at this point, which suggests a flow form during this period when cracks are formed and the transfer of forces from the cracked concrete to the rebars occurs.

Here, we can keep the idea, not of a localized hinge, but of a "plastic segment" over the entire area where the moment is constant. Thus, the pattern to be retained is that of Figure 8.3(c). The static approach gives $F_f = 2M_f/a$, and with the kinematic approach applied to this pattern, we obtain:

$$2F_f\delta/2 = M_f 2\theta, \text{ but as } \delta = a\theta, \text{ we find } F_f = 2M_f/a \qquad [8.5]$$

This result completely justifies the idea of the "plastic segment" pattern.

This plastic segment can be translated in two different ways:

– by a homogeneous description throughout the segment, which amounts to assuming that the concrete is damaged in a diffuse way while the resumption of forces by the rebar is also homogeneous over the entire length of the segment;

– by a description close to reality which is a progressive cracking, regularly spaced and a transfer of forces from concrete to rebar, which this time is non-homogeneous.

In both cases, the overall behavior satisfies the idea of a "flow" on a quasi-plateau corresponding to the cracking moment M_f. Both of these patterns are a representation of the same reality, and in section 8.2 we will see how to use them to make progress on the characterization of cracking of this plastic segment.

8.2. Tie rod cracking

Several authors have worked on the progressive cracking of reinforced concrete tie rods, mainly to study the main indicators of cracking (spacing, opening, etc., see CEOS.fr (2017)) and analyze crack formation details, such as strain evolution and progressive degradation of the reinforcement-concrete bond (Mayer and Eligehausen 1998). This type of study was mainly used to develop crack spacing and opening formulas presented in the model code (MC2010 2013). At the same time, finite element calculation, using concrete behavior models incorporating a damage diffusion procedure, was used to simulate the progressive cracking of reinforced concrete tie rods. However, these calculations are not trivial. Reproducing the distribution of cracks involves introducing, in addition to refined models, certain stochastic approaches (Mang 2015), which makes the analysis long and complex.

8.2.1. *Localized cracking and diffuse damage*

In the previous paragraph, we have shown the interest of the "plastic segment" concept in which, when the CLS is reached, a flow form appears, corresponding to the progressive cracking of concrete and, at the same time, to the transfer of forces from concrete to steel. We apply these limit analysis concepts to the case of tie rods.

A reinforced concrete tie rod can be considered as the superposition of two elements, one of concrete of section A_c and the other of steel of section A_s, in a totally symmetrical distribution within the section of the tie rod. However, these are not two independent solid elements; the contact interfaces between solid elements generate restrictions related to stresses and strains, which must be taken into account.

The first descriptive scheme of reaching the cracking limit, called the *localized scheme,* is presented in Figure 8.4. Cracks are evenly spaced throughout the tie rod (spacing s_c), and the length segment l considered for analysis will include at least two cracks. In this scheme, the static approach gives the critical cracking force F_c at the level of a section:

$$F_c = \int_{A_c} \sigma_c ds + \int_{A_s} \sigma_s ds, \text{ where ds = dy dz} \quad [8.6]$$

The kinematic approach leads to:

$$F_c\delta = A_c \int_0^l \sigma_c \varepsilon dx + A_s \int_0^l \sigma_s \varepsilon dx \rightarrow F_c\delta = \int_0^l (A_c\sigma_c + A_s\sigma_s)\varepsilon dx \quad [8.7]$$

As $\int_0^l \varepsilon dx = \delta$, the kinematic approach leads to the solution $A_c\sigma_c + A_s\sigma_s = F_c$, regardless of x. Specifically, this means that in each section of the tie rod:

– the stress is uniform in each of the two constituent elements of the tie rod;

– the transfer between concrete and steel must be done according to the rule:

$$\sigma_s = \frac{F_c - A_c\sigma_c}{A_s} \quad [8.8]$$

The interface laws between the two solid steel and concrete elements remain to be defined. For this, hypotheses are needed. Let us consider a tie rod segment $[0, s_c/2]$, $x = 0$ being located to the right of a crack:

– cracks are assumed to pass through, so at the right of a crack the stress $\sigma_c(0) = 0$ and by virtue of equation [8.8], $\sigma_s(0) = F_c/A_s$;

– the stress transfer between steel and concrete takes place over a length x_r, on either side of each crack, and the transfer law is assumed to be linear along this length;

– beyond x_r, stress is assumed to be constant in both materials and at the cracking limit of concrete:

$$\forall\, x \in [x_r, \tfrac{s_c}{2}],\ \sigma_c(x) = f_t \text{ and } \sigma_s(x) = \frac{F_c - A_c f_t}{A_s} \quad [8.9]$$

which reflects the fact that stress is constant in steel throughout the area.

To make progress on tie rod behavior, it is necessary to formulate other hypotheses on the behavior of the two materials and their interface:

– considering a percentage of steel ($\rho = A_s/A_c$) higher than the percentage of fragility (see (Eurocode 2 2004; MC2010 2013)), steel is assumed to remain linear elastic at the CLS level;

– concrete is considered elastic-damageable; in addition, the degradation of the interface, along the regularization length x_r, is assumed to occur such that strain remains locally the same on concrete and steel, $\varepsilon_s(x) = \varepsilon_c(x)$.

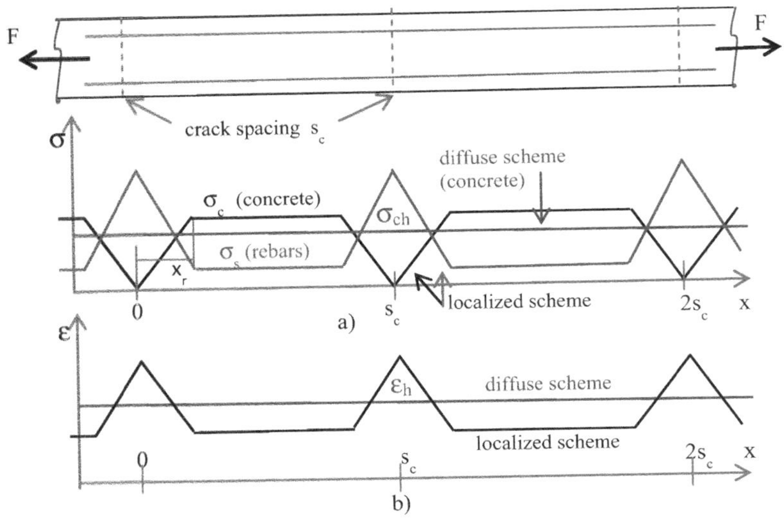

Figure 8.4. *Diffuse (red) and localized (black and blue) cracking patterns in a tie rod: a) stress distribution; b) strain distribution. For a color version of this figure, see www.iste.co.uk/mazars/damage.zip*

A second descriptive scheme of the tie rods at CLS is that of the *diffuse scheme*. As seen for the localized scheme, the tie rod is always assumed to consist of two elements, concrete and steel, but the whole is totally homogenized. Thus, in the diffuse scheme, we can take the following outcomes forward:

1) The static approach gives:

$$F_c = A_c\sigma_c + A_s\sigma_s \qquad [8.10]$$

The difference from the localized scheme is that σ_s and σ_c are constant, not only within each section, but also over the length of the beam ($\forall\, x$, see red curves of the diffuse damage diagram in Figure 8.4).

2) As in the localized scheme:

- steel is considered to remain elastic linear at CLS;

- concrete is elastic-damageable, but its behavioral law is to be considered "homogenized", because it includes the degradation of the entire section, as well as that of the interface along the length of regularization x_r; it is therefore not the usual concrete damage law that must be used; to distinguish it, we will name the damage d_h;

- interface degradation is assumed to occur such that strain is locally the same on concrete and steel, $\varepsilon_s(x) = \varepsilon_c(x)$.

From all of these considerations, at a given moment t of the "plastic" flow of cracking, we arrive at the following results.

1) *For the localized scheme*:

$$F(t) = \sigma_c(x)A_c + \sigma_s(x)A_s, \text{ or } \sigma_c(x) = (F(t) - \sigma_s(x)A_s)/A_c, \forall x \quad [8.11]$$

- at the level of a crack ($x = 0$):

$$F(t) = \varepsilon(0).E_sA_s \quad [8.12]$$

- in the regularization zone ($x = s_c/2$):

$$F(t) = \varepsilon\left(\frac{s_c}{2}\right).(E_cA_c + E_sA_s) \quad [8.13]$$

2) *For the diffuse or homogenized pattern*: in this pattern, indices relating to strains and stresses will be indicated as h. Thus, the elastic-damageable behavior of the homogenized part of concrete is written as:

$$\sigma_{ch}(t) = \varepsilon_h(t).E_c(1 - d_h)$$

which leads to:

$$F(t) = \varepsilon_h(t).(E_c(1 - d_h)A_c + E_sA_s) \quad [8.14]$$

8.2.1.1. *Mapping diffuse and localized schemes*

The two patterns are a representation of the same phenomenon: the CLS; they must be such that static equilibrium and kinematic coherence are ensured for the chosen limit state (given in Figure 8.4).

1) *Static equilibrium*: by integrating the value of $F(t)$, obtained from equations [8.12] and [8.13], in equation [8.14], we obtain:

- for $x = s_c/2$:

$$\sigma_c\left(\frac{s_c}{2}\right) = \varepsilon_h(t).(E_c(1 - d_h) + \rho E_s(\varepsilon_h(t) - \varepsilon\left(\frac{s_c}{2}\right)) \quad [8.15]$$

- for $x = 0$:

$$0 = \varepsilon_h(t).(E_c(1 - d_h) + \rho E_s(\varepsilon_h(t) - \varepsilon(0)) \quad [8.16]$$

2) *Kinematic coherence*: at a given time t of the cracking process, the overall displacement must be the same for the two schemes. For the interval [0, $s_c/2$], this is written as:

$$\left(\varepsilon(0) + \varepsilon(x_r)\right)\frac{x_r}{2} + \varepsilon\left(\frac{s_c}{2}\right).\left(\frac{s_c}{2} - x_r\right) = \varepsilon_h(t).\frac{s_c}{2} \quad [8.17]$$

- between x_r and $s_c/2$, $\varepsilon(x)$ is constant, so equation [8.17] leads to:

$$\varepsilon_h(t) = \varepsilon\left(\frac{s_c}{2}\right)\left(1 - \frac{x_r}{s_c}\right) + \varepsilon(0).\frac{x_r}{s_c} \quad [8.18]$$

- from equations [8.14], [8.16], and [8.18], we deduce:

$$\sigma_{ch}(t)/_{\rho E_s} = (1 - \frac{x_r}{s_c})(\varepsilon(0) - \varepsilon\left(\frac{s_c}{2}\right)) \quad [8.19]$$

- but equations [8.12] and [8.13] give:

$$\varepsilon(0) = \varepsilon\left(\frac{s_c}{2}\right).(\frac{E_c}{\rho E_s} + 1) \quad [8.20]$$

- finally, from equations [8.18] and [8.20], we obtain:

$$\varepsilon(0) = \varepsilon_h(t).\frac{1+E_c/\rho E_s}{1+x_r E_c/s_c \rho E_s} \quad [8.21]$$

- thus, equations [8.14] and [8.16] give:

$$\sigma_{ch}(t) = \varepsilon_h(t).E_c(1 - d_h) = \varepsilon_h(t).E_c\frac{1-x_r/s_c}{1+x_r E_c/s_c \rho E_s} \quad [8.22]$$

- from this, we obtain:

$$(1 - d_h) = \frac{1-x_r/s_c}{1+x_r E_c/s_c \rho E_s} \quad [8.23]$$

Equation [8.23] gives the evolution of the homogenized damage of the diffuse scheme, d_h. The latter is conditioned by the characteristics of the rebar-concrete composite by the parameters ρ, Ec and Es and is determined by the ratio $\frac{x_r}{s_c}$. E_c, E_s and ρ (rebar/concrete ratio) are known, and x_r is the stress regularization length, a characteristic of the rebar-concrete system, which must be determined from specific experiments, as shown in Figure 8.5 (Mayer and Eligehausen 1998).

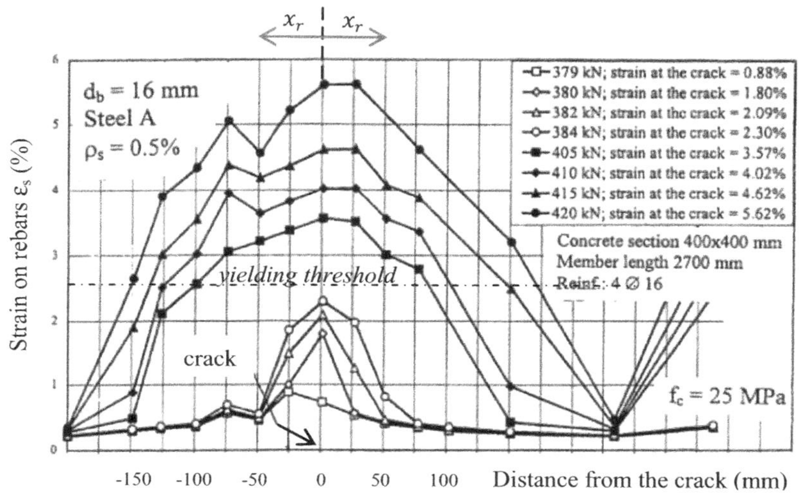

Figure 8.5. *Test on tie rod (Mayer and Eligehausen 1998): effects of decohesion between rebar and concrete around a crack*

This experiment, carried out on the basis of tests on a certain number of tie rods with a section of 0.4 × 0.4 m and comprising four HA16 reinforcements, leads to the determination of the strain evolution of the steel in the vicinity of a crack for different loading stages (see Figure 8.5). Two stages of rebar decohesion are shown:

– one, up to 384 kN, is related to crack formation;

– the other, above 405 kN, is linked to the crack opening and at the same time to the plasticization of rebars ($\varepsilon_s > 0.25\%$).

The CLS stages correspond to the first phase. In this phase, we find an almost fixed value for x_r during the cracking process. This value, which is reported in equation [8.23], leads to the following major result.

IMPORTANT.– *For a specific tie rod and at a given crack flow stage characterized by the strain $\varepsilon_h(t)$, a homogenized damage value d_h corresponds to a single mean crack spacing:*

$$s_c = x_r\left[\frac{1}{d_h}\left(\frac{E_c}{\rho E_s} - 1\right) - \frac{E_c}{\rho E_s}\right] \qquad [8.24]$$

There are two limitations here:

– when there is no damage ($d_h = 0$), there is therefore no crack and s_c does not exist;

– conversely, when the damage is total ($d_h = 1$), a minimum value appears $s_c = x_r$. We shall see later that this remains a limit that is never reached due in particular to the tension stiffening phenomenon, which we shall discuss again in section 8.2.2.

8.2.2. *Behavioral law for concrete in the diffuse scheme*

Just before the start of the cracking process, the concrete part is still in the elastic phase:

$$\sigma_{ch}(t) = E_c \varepsilon_h(t) \quad [8.25]$$

At this stage, the two stress curves (the diffuse curve $\sigma_{ch} - \varepsilon_h$ and the localized curve $\sigma_c - \varepsilon_c$) of the patterns shown in Figure 8.4 are merged, and the threshold value of this stress is the tensile strength f_t. We have seen in section 8.2.1 that the behavioral law to be adopted for the diffuse scheme was not the usual behavioral law of concrete because the homogenization carried out here leads to including the degradation of the entire section in this law, as well as the degradation of the interface along the regularization length x_r. Below, we will show that this law is implicitly included in the relationships established in section 8.2.1 and that it is totally determined by the formulated hypotheses and, in particular, the equivalence between the two schemes, diffuse and localized. In the localized scheme, during the cracking process and outside the stress regularization zones, the stress remains constant between two cracks and reaches the threshold ($f_t = \varepsilon_{t0} E_c$); the corresponding strain is ε_{t0}.

As a result, equation [8.15] becomes:

$$f_t = \varepsilon_{t0} E_c = \varepsilon_h(t).(E_c(1-d_h) + \rho E_s(\varepsilon_h(t) - \varepsilon_{t0}) \quad [8.26]$$

This gives us:

$$(1-d_h) = \frac{\varepsilon_{t0}}{\varepsilon_h(t)}(1+\rho\frac{E_s}{E_c}) - \rho\frac{E_s}{E_c} \quad [8.27]$$

Equation [8.27], introduced in equation [8.22], gives the following expression for the post-peak part of the $\sigma_{ch}(t) - \varepsilon_h(t)$ curve:

$$\sigma_{ch}(t) = E_c \varepsilon_{t0} - \rho E_s(\varepsilon_h(t) - \varepsilon_{t0}) \quad [8.28]$$

Thus, due to the chosen crack scheme (see Figure 8.4), the tensile behavior of the concrete part relative to the diffuse scheme is composed of two straight lines (see Figure 8.7):

– that before the peak corresponding to the elastic phase given by equation [8.25];

– that after the peak corresponding to the damage phase given by equation [8.28].

In this last phase, we find that $\sigma_{ch} = f_t$ when $\varepsilon_h = \varepsilon_{t0}$ and $\sigma_{ch} = 0$; when $\varepsilon_h = \varepsilon_r$, it is a strain limit value (of failure) for which equation [8.28] gives:

$$\varepsilon_r = \varepsilon_{t0}(1 + {}^{E_c}/_{\rho E_s}) \qquad [8.29]$$

It can be seen that this value, and therefore the post-peak phase of the behavior, is totally conditioned by the steel-concrete ratio ρ, the lower limit of which is the minimum percentage of brittleness, below which the failure becomes fragile and does not enter the progressive cracking process established here, and the upper limit of which is $\rho = 1$, which leads to a concrete-free element that is not the subject of the study. Typical ratios range from a few tenths of a percent to a few percent.

NOTE 8.1.– The approach proposed for tie rods will be the same for a cylindrical reinforced concrete reservoir under pressure, which can be considered, before cracking, to have a homogeneous stress distribution in the wall (thin tube hypothesis). This is the case in a large radius reservoir, for which a section can be schematically represented by a tie rod. Suppose this reservoir is subjected to the pressure p of a fluid. It is established that the tangential stress in the wall of radius R and thickness e is $\sigma = pR/e$. The idea of the two schemes (localized and diffuse) is to be taken up again beyond reaching the crack threshold (when $\sigma = f_t$), and the relationships established for tie rods can be used directly.

NOTE 8.2.– The tie rod approach can also be applied to the case of reinforced concrete beams subjected to circular bending (case of Figure 8.3 for the part under constant moment). The idea of the diffuse scheme and the localized scheme with regularly spaced cracks on the lower side are to be taken up again. However, since the cracks only open over part of the height of the beam, additional hypotheses must be made to describe the local equilibrium states and the kinematic compatibility between the two patterns. Some regulations propose the concept of an equivalent tie rod within the lower area of the beam. We leave it to the interested reader to work on this issue, which should be resolved without great complexity.

8.2.3. *Application to an experiment on tie rods carried out at EPFL*

An experimental campaign on reinforced concrete tie rods was carried out at the EPF Lausanne by Mivelaz (1996), the test principles of which are given in Figure 8.6. Here, we use some of these results to validate the proposed approach.

Within the framework of the CLSs that take place in a "flow" phase leading to a force–displacement evolution in a quasi-plateau (see Figure 8.8), we seek to determine the characteristics of this cracking, in particular the average spacings between cracks.

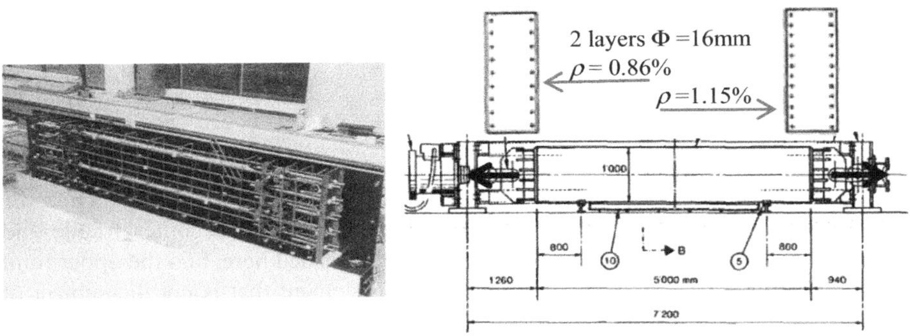

Figure 8.6. *Experimentation on tie rods (Mivelaz 1996): view of the test set-up and details of rebars for the two types of tie rods considered*

From the relationships established in section 8.2.1, three principal ones will be needed to solve the problem: equations [8.22], [8.23] and [8.24]:

$$\left.\begin{aligned} &\sigma_{ch}(t) = E_c\varepsilon_{t0} - \rho E_s(\varepsilon_h(t) - \varepsilon_{t0}) \\ &\sigma_{ch}(t) = \varepsilon_h(t).E_c(1 - d_h) \\ &s_c = x_r\left[\frac{1}{d_h}\left(\frac{E_c}{\rho E_s} - 1\right) - \frac{E_c}{\rho E_s}\right] \end{aligned}\right\} \quad [8.30]$$

The first makes it possible to define the post-peak of the bilinear behavior of the concrete part of the diffuse scheme. The second associates a damage value with any point of this homogenized curve. The third gives the mean spacing of cracks (s_c) in the localized scheme corresponding to a level of damage of the diffuse scheme (d_h).

Element	Rebar/concrete ratio (ρ)	Elastic modulus (GPa)	Tensile strength (MPa)
R4	0.0086	30	2.5
E4	0.0086	45	3
E5	0.0115	45	3

Table 8.1. *EPFL tie rods: characteristics of materials (for steel E_S = 200 GPa)*

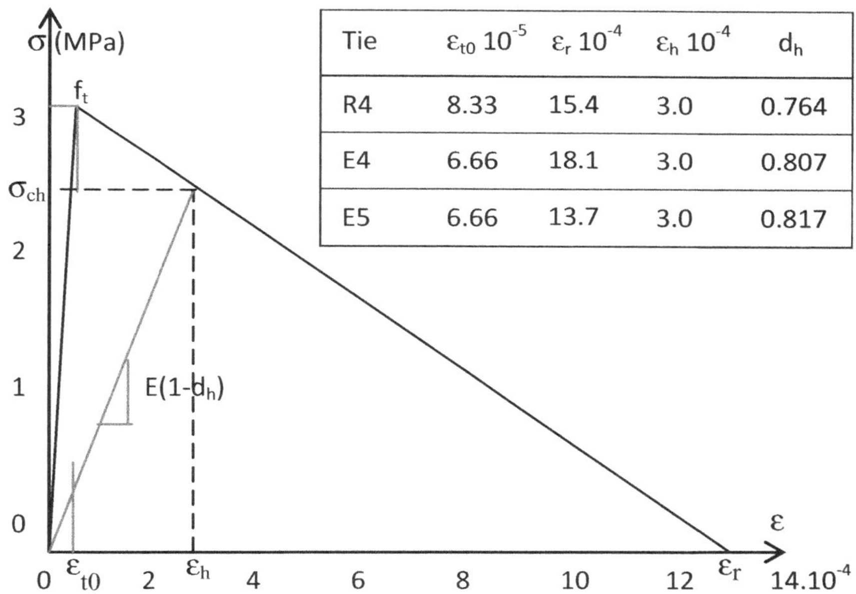

Tie	$\varepsilon_{t0}\ 10^{-5}$	$\varepsilon_r\ 10^{-4}$	$\varepsilon_h\ 10^{-4}$	d_h
R4	8.33	15.4	3.0	0.764
E4	6.66	18.1	3.0	0.807
E5	6.66	13.7	3.0	0.817

Figure 8.7. *Concrete behavior for tie rod E5 and characteristics table for the three tie rods, as well as damage for a fixed CLS ($\varepsilon_h = 3 \times 10^{-4}$)*

From equation [8.30] and from the characteristics given in Table 8.1, the post-peak for the concrete behavior of the different tie rods can be determined:

$$\left.\begin{array}{l} -\text{ R4: } \sigma_{ch}(t) = -1{,}720\varepsilon_h(t) + 2.64 \text{ MPa} \\ -\text{ E4: } \sigma_{ch}(t) = -1{,}720\varepsilon_h(t) + 3.12 \text{ MPa} \\ -\text{ E5: } \sigma_{ch}(t) = -2{,}300\varepsilon_h(t) + 3.155 \text{ MPa} \end{array}\right] \qquad [8.31]$$

The example of one of the three behavior curves (E5) that result from there is given in Figure 8.7. This curve shows the main characteristics of the behavior (ε_{t0}, ε_h, ε_r), the values of which are given in the associated table. During the tests, crack measurements were carried out at a given value of the mean strain of the tie rod, $\varepsilon_h = 3 \times 10^{-4}$, a situation for which cracking is considered to be stabilized. At this stage, and from previous equations, we can calculate σ_{ch}, and from the second equation [8.30], we can deduce d_h. The table in Figure 8.7 provides this value for each of the three tie rods.

Finally, the third equation [8.30] makes it possible to give the corresponding value of the mean spacing cracks (s_c) as a function of x_r, the regularization length of stress from each crack. Assuming that x_r is within the same range as the value given by the tests presented in Figure 8.5 (x_r # 0.05 m), we can infer the results given in Table 8.2, which are in good agreement with the experimental results.

Tie rod	Experiment (m)	Calculation (m)
R4	0.27	0.33
E4	0.25	0.27
E5	0.24	0.21

Table 8.2. *Crack spacings measured and calculated for the three tie rods*

Figure 8.8(b) shows (dotted lines) the experimental force–displacement curve of tie rod E5 and the force–displacement straight line relating to the single rebars (dash-dotted lines). This leads to two observations:

– progressive cracking leads to a very discontinuous experimental curve, each jump corresponding to the formation of at least one crack;

– the cracking being stabilized, a residual difference is observed between the curve measured and that corresponding to the effect of the rebars alone: it is the tension stiffening which reflects the effect of concrete which continues to act between the cracks.

The concept used for the homogenized tie rod gives an “average” account of the progressive formation of cracks (flow on a quasi-plateau); however, the bilinear behavior curve leads to a progressive exhaustion of the strength of the concrete and cannot therefore account for the tension stiffening. To this end, in Figure 8.8(a) we propose a σ–ε curve, parabolic in nature, respecting in the first phase the linear progression of damage and leading beyond a value ε_m (peak of the parabola) to a residual stress capable of describing the phenomenon of tension stiffening.

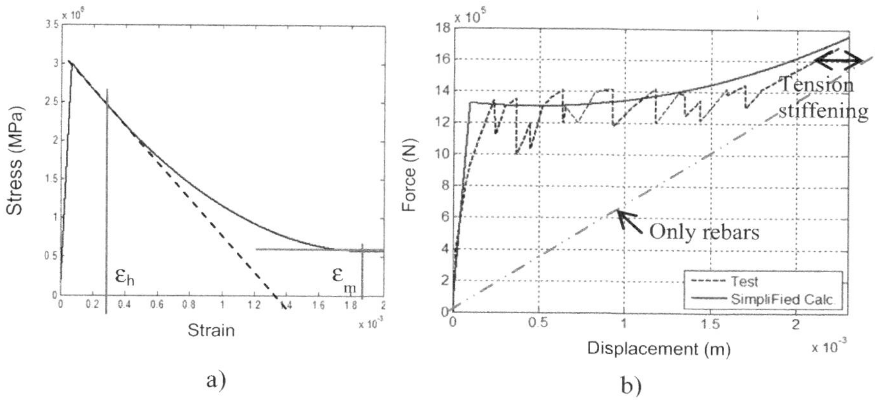

Figure 8.8. *Simulation of the force–displacement behavior of the tie rod: a) σ–ε curve for the diffuse scheme relating to tie rod E5; solid line, curve modified to take into account the tension stiffening in the established cracking phase; b) force–displacement for tie rod E5, simplified calculation-experiment comparison*

From this curve, the overall force-displacement behavior of the tie rod (concrete force + rebar force) is compared (see Figure 8.8(b)) with the experimental curve. This figure shows the consistency of the proposed modeling, which, as seen, is extremely simple to implement.

As we are confident in this approach, we propose to use it in section 8.3 to anticipate cracking at an early age of a circular gusset of a confinement structure.

8.3. Homogeneous field created by concrete maturation within a cylindrical wall

The application proposed here has the particularity of being a real case in relation to the construction and study of the behavior over time of a one-third scale model, a faithful reproduction of a plant containment enclosure of the French nuclear fleet for double-wall technologies (see VeRCoRs (2018)).

8.3.1. *VeRCoRs program and model*

The principal objectives of the program were to:

– study the structural behavior of the internal enclosure at an early age and during aging;

– study the behavior during pressure tests and under conditions where the thermomechanical load (vapor pressure) is maintained for several days;

– study the effects of aging on the sealing of the structure.

The internal enclosure is a cylindrical object of prestressed concrete constructed in successive layers. The base slab is cast in a single phase on its pedestal, and the gusset located at the interface between the base slab and the wall of the enclosure is made of a single layer placed on the already hardened concrete base slab. Above the gusset, the second layer of the cylindrical wall is poured several days later (see Figure 8.9(b)).

The purpose of this study is to address the prediction of gusset cracking at an early age, primarily before the second layer is applied. Cracking is due to both the massiveness of the object and to the restrained shrinkage due to the presence of the base slab, which is a solid and very massive reinforced concrete cylinder.

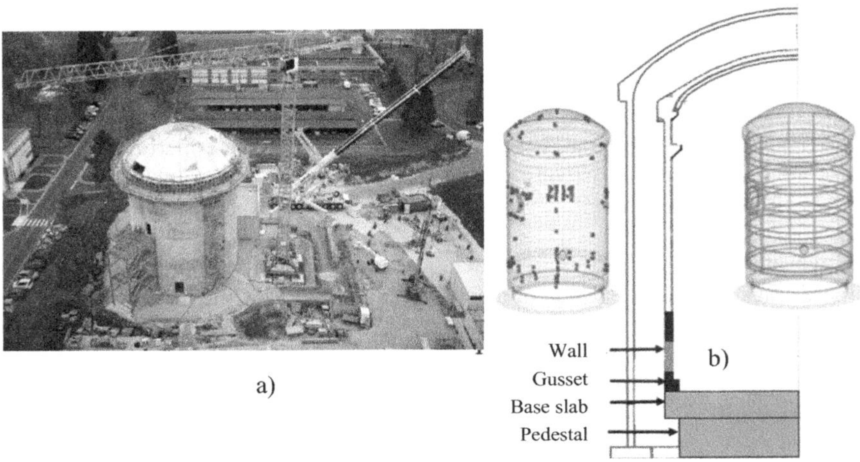

Figure 8.9. *a) VeRCoRs model under construction; b) cross-section of the double wall to locate the gusset and the location of the various measurement sensors (temperature, strains, optical fibers, etc.). For a color version of this figure, see www.iste.co.uk/mazars/damage.zip*

Figure 8.9 gives an overview of all VeRCoRs facilities and a cross-section of the model. During construction, measurements (temperatures, strains, etc.) were carried out just after concrete pouring and at intervals of 1 h thereafter. Subsequently, during aging and pressure tests, many measurements are collected every day on countless sensors (temperature, strain, geometric variations, humidity, etc.). During construction, hundreds of concrete samples were prepared and tested to determine material properties and behavioral parameters, including moduli, strengths, shrinkage, creep, permeability, etc. All of these measurements make it possible to follow different evolutions and, consequently, to improve models that relate to each of the phenomena analyzed.

With regard to the gusset behavior at an early age, many models have been tested, with success in some cases, for predicting changes in temperature and strain, but the prediction of cracking characteristics is still an open problem (Corbin and Garcia 2016). Indeed, finite element calculations of gussets mainly stumble on two points:

– the appearance of cracking destroys the axisymmetry of the problem and requires a 3D calculation;

– the calculations carried out, most often with a damage model for concrete, lead to the damage threshold to almost homogeneous stress and strain fields throughout the gusset, and the generation of the localization necessary to represent cracking zones goes through probabilistic approaches that complicate the procedure and do not always lead to realistic schemes (Bouhjiti 2018).

Here, the proposal is to consider the problem differently and, in the manner of the approach used for tie rods, extract the characteristics of this cracking from the diffuse scheme-localized scheme dialog.

However, in comparison with the case dealt with in section 8.2 for tie rods, the problem differs for two main reasons:

1) the gusset is a circular element, which can be considered axisymmetric;

2) loading is exclusively due to the combined effects of young age and boundary conditions, particularly hindering the establishment of strains due to the presence of the base slab, which is a particularly rigid body.

In accordance with point 1, the choice was made to treat the problem axisymmetrically (2D): discretization is performed in a section of the gusset, and the results are considered to be identical around the entire perimeter of the cylinder.

Point 2 requires treating thermo-hydro-mechanical problems from the date of casting the gusset to the formation of all the cracks (experiments have shown that they are almost stabilized after 12 days). To do this, we have resumed the work already done to simulate the behavior of the RG8 experiment (see section 6.2.4), which is a large beam whose shrinkage was restrained by metal struts (see Mazars et al. 2018). As it was a beam, the problem was treated using a multifiber Timoshenko beam discretization (1D modeling). In this case, the modeling is in 2D in the section of the gusset and axisymmetric on the perimeter.

Axisymmetric calculation leads to a homogeneous result and is therefore the "diffuse scheme" of the approach presented in section 8.2. Note that the geometry of the gusset respects the assumptions of thin tubes:

– the hypothesis is made that the homogeneity can be extended to the section, in particular for damage (d_h assumed to be almost constant in the section of the gusset);

– under these conditions, the use of equations [8.30] established in the diffuse scheme-localized scheme dialog for tie rods can be extended to the gusset case.

8.3.2. *Mesh of the gusset and temperature conditions*

In an axisymmetric problem, the mesh is fully described by a vertical section of the structure and assumptions of axial symmetry (see section A.1.3).

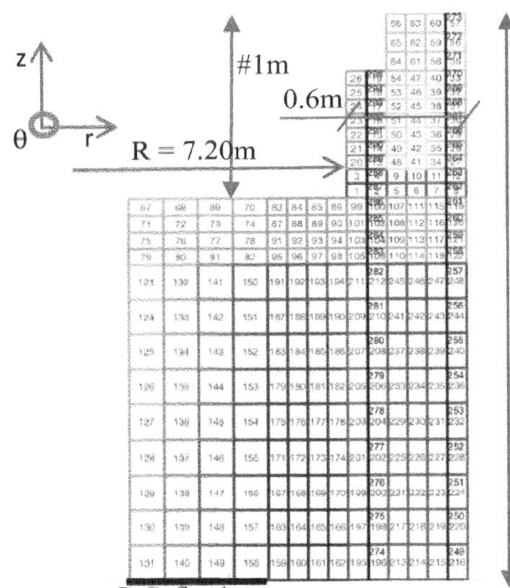

a)

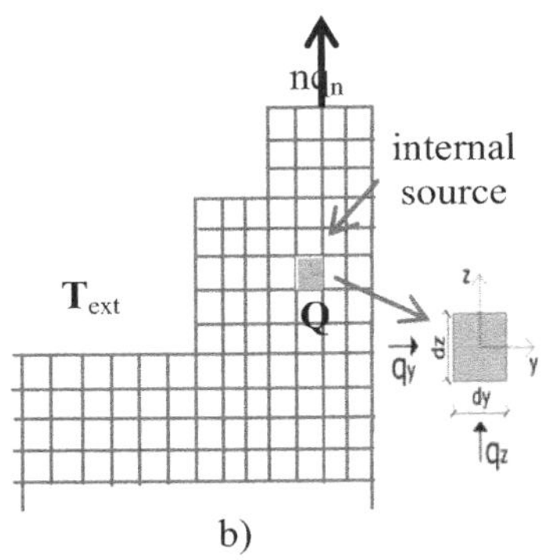

b)

Time	0-3h	3h-7h	17h-35h	>35h
T_{ext}°C	20	38	48	15

c)

Figure 8.10. *a) Mesh size of gusset and base slab; b) thermal equilibria: internal source of heat production, internal and external flows, and external conditions (T_{ext}); c) table of values for outdoor temperatures during the 12 days of analysis. For a color version of this figure, see www.iste.co.uk/mazars/damage.zip*

The concrete of the base slab is hardened when the gusset is poured. It is a very massive object (full cylinder with an average diameter of 15 m and a thickness of 2 m). To simplify the problem, the central part of the base slab is not meshed, and the mesh is blocked at the connection with the foundation (see Figure 8.10(a)). In total, 250 Quad4 are used for the gusset. They are 10 × 10 cm, and their dimensions are modulated in the base slab to manage the gusset-base slab bond and the distance to the gusset. Rebars parallel to the cross-sections are bar elements, their presence on the perimeter is homogenized (each bar represents the generator of a cylinder), and the stiffness of reinforcements perpendicular to the cross-sections (longitudinal rebars for the gusset) is integrated into the axisymmetric analysis.

To recreate the thermal conditions observed during the casting of a gusset in an actual confinement structure, a specific external temperature (T_{ext}) was applied by a heating process acting under tarpaulin during the first two days. The table in Figure 8.10(c) shows the evolution of temperatures considered for calculations during the 12 days of the analysis.

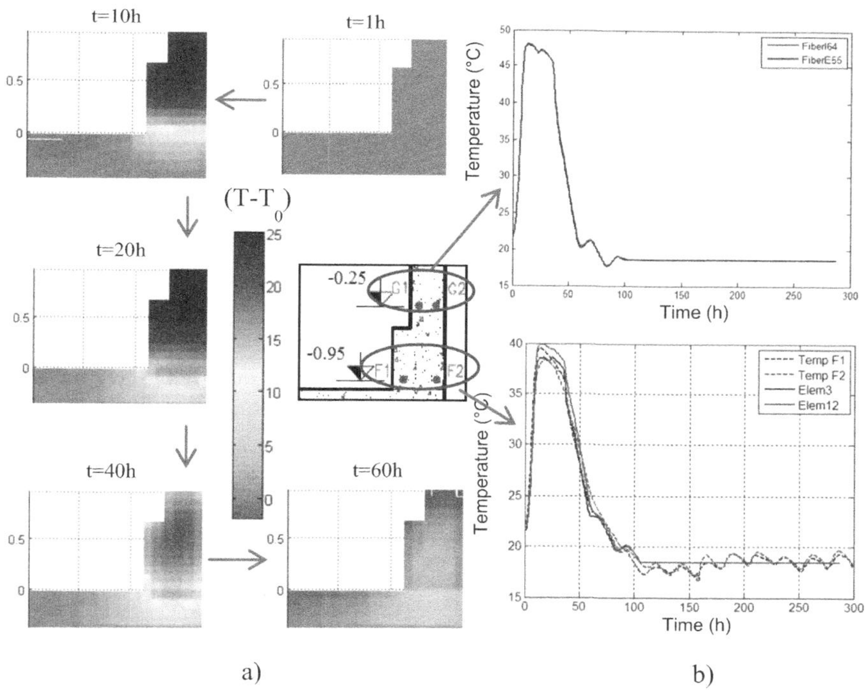

Figure 8.11. *Temperature in the gusset: a) evolution of the field in the current section of the gusset from 1 to 60 h; b) evolution as a function of time in four measuring points, at the top and bottom of the gusset (solid lines: calculation; dotted lines: measurements). For a color version of this figure, see www.iste.co.uk/mazars/damage.zip*

As shown in the THM calculations conducted for the RG8 beam (see section 6.2), determining temperatures requires resolving a group of three equations: the heat equation [6.6], the equation giving the degree of hydration as a function of time [6.7] and that relating to heat flows in exchange with the outside [6.11]. The calculation was carried out with Cast3M on a mesh identical to that given for the mechanical part in Figure 8.10(a).

Figure 8.11 shows the results obtained, after the identification of parameters, from laboratory tests on samples. It includes:

– the evolution of the temperature field between 1 h and 60 h (see Figure 8.11(a)); it is noted that it is the end of the gusset that reaches the highest temperatures 20 h after casting;

– temperature changes as a function of time, at the top of the gusset (narrowest part, points G1-G2) where the calculation curves are totally superimposed and near the connection with the base slab, points F1-F2, for which temperatures measured in situ (dotted curves) are given (see Figure 8.11(b)).

It can be seen how significant the temperature differences between the gusset and base slab are ($\Delta T > 20°C$), with temperature peaks at the top of the gusset reaching 48°C. Moreover, the calculation–measurement comparison appears to be excellent (points F1-F2).

8.3.3. *Creep, shrinkage and mechanical properties*

Due to the massiveness of the structure and the relative brevity of the experiment (12 days), concrete drying is not taken into account in the analysis.

Steel/concrete ratio (ρ)	Steel elastic modulus (GPa)	Concrete elastic modulus (GPa)	f_t at 28 days (MPa)	f_c at 28 days (MPa)
0.007	210	30	2.8*	44

*Estimated from cylinder splitting tests (Φ = 11cm) (see proposed principles in section 4.3).

Table 8.3. *Material characteristics of gusset*

In this context and using the principle of the calculation carried out for the RG8 beam (see section 6.2):

– the autogenous shrinkage was calibrated from tests previously carried out on samples (κ_∞ = 150 μm/m);

– the basic creep model was also calibrated from previous laboratory tests and, to cover the entire duration of the test in an optimized way, the characteristic times of the three Kelvin–Voigt models taken into account are: $\tau_1 = 0.1$ day, $\tau_2 = 1$ day and $\tau_3 = 10$ days;

– knowing the evolution of the degree of hydration (function of temperature and time: see equations [6.7] to [6.9]), the evolution of the maturity of concrete $M(t)$ is known at all points of the gusset, as well as the mechanical characteristics:

$$E\,(M) = E.M,\ f_t(M) = E(M).\,\varepsilon_{t0},\ \sigma_{c0}(M) = E(M).\,\varepsilon_{c0} \tag{8.32}$$

ε_{t0} and ε_{c0} are concrete damage thresholds under tension and compression, respectively, which are considered constant regardless of the maturity value, and are therefore determined from the 28 days characteristics measured on laboratory samples (see Table 8.3).

8.3.4. *Mechanical calculation*

The gusset is an axisymmetric object. As such, the stress, strain and damage fields will be homogenized around its perimeter. The analogy with tie rods of section 8.2 leads us to be able to use equations [8.30] and the characteristics of Table 8.3 to obtain, in particular, the behavior curve of concrete to be used at 28 days for the diffuse scheme. This behavior is bilinear for the pre- and post-peak phases, respectively:

$$\sigma_{ch} = E\varepsilon_{t0} \text{ and } \sigma_{ch} = -1470\varepsilon_{ch} + 2.925 \text{ MPa} \tag{8.33}$$

This curve is given in Figure 8.12(a) (dashed lines for post-peak), but as discussed about tie rods, taking into account the tension stiffening for the significant strains led to using, for calculations, the curve presented in Figure 8.12(a) in solid line.

More generally, the complete tensile-compression curve is given in Figure 8.12(b). This is used in the μ model at the base of the gusset calculation, which respects a damage evolution law that is different from that in the usual exponential because, as explained for tie rods (see Figures 8.7 and 8.8(a)), it accounts for the damage of the concrete section, as well as of the concrete–rebar connection. Its expression is given as:

$$\boldsymbol{\sigma}_{ch} = \Lambda(1 - d_h) : \boldsymbol{\varepsilon}_h, \text{ where } (1 - d_h) = \frac{Y_0}{Y}\left\{1 + \frac{Y^2 - 2A(Y - Y_0) - Y_0^2}{2(Y_0 - B)(Y_0 - A)}\right\} \tag{8.34}$$

The two equations [8.34] reflect a 3D behavior in the spirit of the μ model. Y is the pilot damage variable ($Y = rY_{th} + (1 - r)Y_{ch}$), Y_0 is the value of this variable at the damage threshold ($Y_0 = r\varepsilon_{t0h} + (1 - r)\varepsilon_{c0h}$); A and B are the shape variables of the behavior curves, as defined in the model presentation in section 3.2, varying in the intervals $[A_t, A_c]$ and $[B_t, B_c]$.

Since the curve of the diffuse scheme is not the usual exponential law, the values of the parameters are very different:

– $A_t = \varepsilon_m$, ε_m ($= 2.6 \times 10^{-3}$) vertex of the parabola under tension; the stress is constant beyond this point;

– $B_t = \varepsilon_r$, ε_r ($= 2 \times 10^{-3}$), point of convergence with the axis $\sigma = 0$, of the line of the post-peak behavior under tension given by equation [8.33];

– $A_c = |\varepsilon_M|$, ε_M ($= -2.6 \times 10^{-3}$), vertex of the parabola under compression. Finally, $B_c = 0$. Furthermore, $\varepsilon_{t0h} = 9.3 \times 10^{-5}$ and $\varepsilon_{c0h} = 3.33 \times 10^{-4}$ are derived from sample data (see Table 8.3).

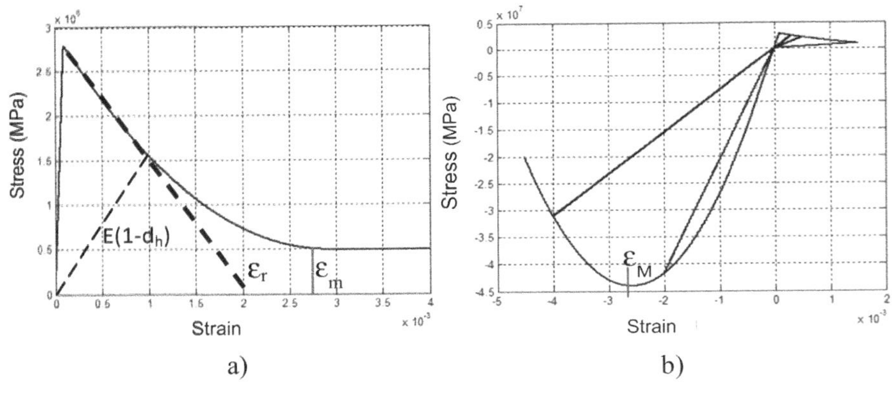

Figure 8.12. *Response of the μ model identified for mature concrete of the gusset (28 days). a) Response under tension; b) uniaxial cyclic response*

8.3.5. *Principal results and comparisons with in situ measurements*

8.3.5.1. *Strains*

During the first hours, the strain combines thermal and shrinkage effects. Measurements were made at points of interest in three directions: (tangential (θθ), vertical (zz) and radial (rr)). Figure 8.13(a) shows the calculation–measurement comparisons for tangential strain, which acts on the vertical cracking of the gusset.

Comparisons are made on four specific measurement points: G1 and F1 in intrados of the gusset and G2 and F2 in extrados. Overall, these results are of high quality.

Figure 8.13(b) shows the displacement fields (amplified) at different times after gusset casting, which greatly instructs on the combined effects of temperature and shrinkage. During the exothermic phase of maturation, the gusset, hindered by the base slab that does not heat up, strains outwards into a corolla (t = 25 h). Conversely, when the cooling is complete and autogenous shrinkage is already progressing well (t = 105 h), the gusset strains inwards. These movements will have consequences on damage at an early age, as we will see below.

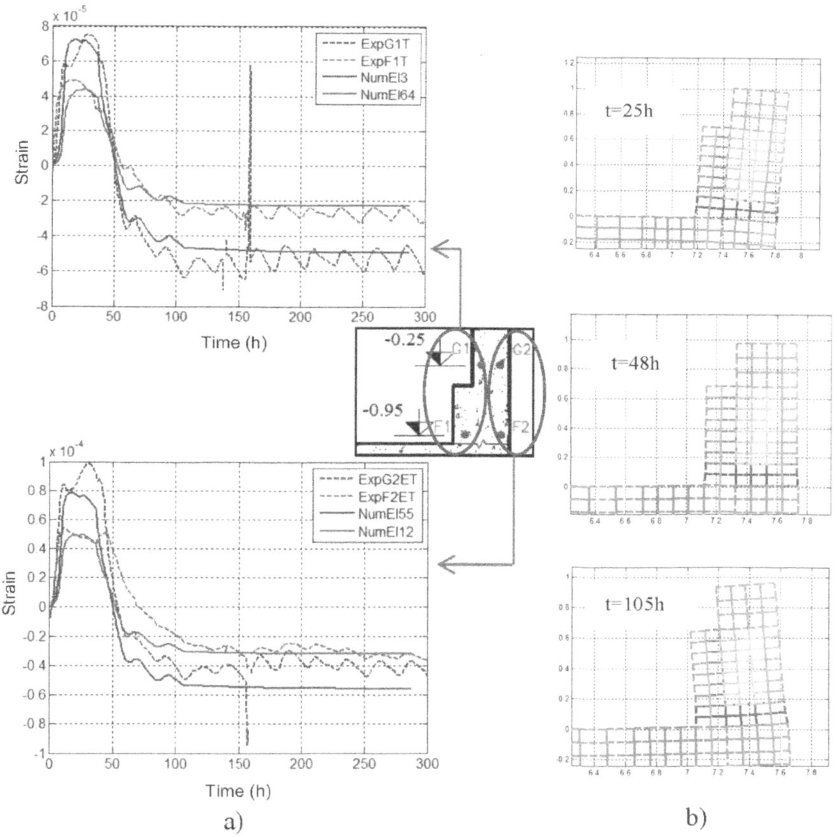

Figure 8.13. *Evolution of strain in the gusset during the first 12 days: comparison measurements (dotted lines), calculations (solid lines). a) Tangential strain at four specific measuring points; b) displacement field at different times after casting at t = 0. For a color version of this figure, see www.iste.co.uk/mazars/damage.zip*

8.3.5.2. *Damage*

These results are the culmination of the procedure set up in this chapter. As presented for tie rods in section 8.2, the calculation gives a diffuse damage d_h over the entire perimeter of the gusset. This is the case, with the calculation carried out in axisymmetric 2D, and because of the observed movements of the gusset in Figure 8.13(b), localized damage appears in the plane of the section at the base of the gusset (see Figure 8.14(a)): first inwards when the overall strain is outwards and then outwards when the overall strain is inwards.

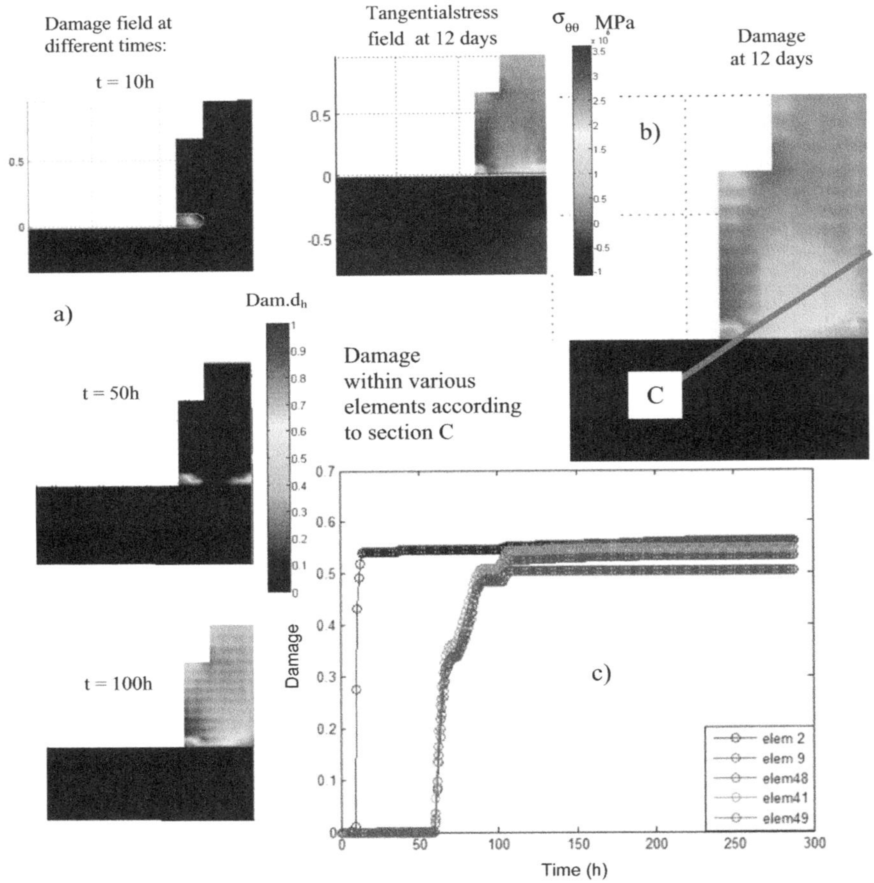

Figure 8.14. *Damage in the gusset due to concrete maturation: a) damage field within the gusset at different times since pouring; b) stress state (tangential in MPa) and damage state at 12 days; c) evolution of damage as a function of time along the C cut within the gusset. For a color version of this figure, see www.iste.co.uk/mazars/damage.zip*

These localizations are the beginnings of horizontal cracking at the base of the gusset, but the intensity of this damage has not been until the opening of a crack, which has not been observed experimentally. Beyond the first hours, during which this localization takes place, damage gradually spreads throughout the gusset (see Figure 8.14(a) at t = 100 h). The calculation continued up to 12 days (288 h); at this stage, it can be seen that the stress and damage fields are almost uniform within the gusset.

Curves (see Figure 8.14(c)) show the evolution of damage on a section of the gusset, and it is noted that it peaks at a value d_h # 0.55 from approximately 120 h (5 days). For the points of this cut at the left base of the gusset, the rise of damage is faster (# 10 h), a phenomenon that is related to the early localization observed in this area.

8.3.5.3. *Cracking indicators*

The last equation of group [8.30] gives the link between d_h, x_r, and s_c. Considering that the conditions of the test presented in Figure 8.5 are not very different from those relating to the gusset (principal rebars of the same diameter) and without specific results for x_r, we have taken into account the same value as that used for the tie rods of section 8.2: x_r # 0.05 m. Thus, with $d_h = 0.55$, the mean crack spacing at 12 days is estimated to be $s_c = 1.06$ m.

Figure 8.15(b) reproduces, in a developed form on an extrados half-gusset (between two ribs, used to position prestressing anchors), cracks found to be present on three dates after casting: 5 days, 12 days and 8 months. We note that:

– cracks are almost all formed at 5 days, a finding which is consistent with that made in the calculation results for damage that stabilizes within the gusset at about 120 h (5 days (see Figure 8.14(c)));

– these increase little in number after 5 days, but rather in length.

Table 8.4 provides a comparison of the mean measured and calculated spacings. Calculated at 12 days, the mean value of the spacing between cracks estimated by the procedure presented here is quite relevant.

Table 8.4 also gives estimates of the opening of mean cracks w_c that have been determined using the localized scheme Figure 8.4.

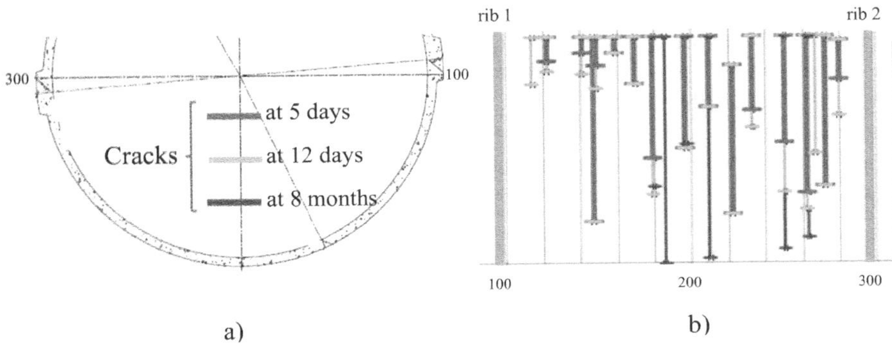

Figure 8.15. *Cracks observed on the outer cylindrical wall of the gusset: a) observation zone between two ribs; b) state of the three-period crack, presented in a developed form from 100 to 300 grades. For a color version of this figure, see www.iste.co.uk/mazars/damage.zip*

Time	Measured spacing	Calculated spacing	Measured opening	Calculated opening
t_0 + 5 days	# 1.2 m	–	<0.1 mm	–
t_0 + 12 days	# 1.2 m	1.06 m	0.1 mm	0.095 mm
t_0 + 8 months	# 1.2 m	–	0.2 mm	–

Table 8.4. *Measurement-calculation comparisons on mean spacing and mean crack opening within the gusset*

The reasoning to do this is as follows:

– if we assume that any strain beyond the peak of the behavior curve σ–ε will contribute to cracks opening, then this strain that is established in the stress regularization zone ($x \in [-x_r, x_r]$) will lead to an opening:

$$w_c = 2\int_0^{x_r}(\varepsilon(x) - \varepsilon_{0t})dx \quad [8.35]$$

– from the use of the localized scheme and in relation to the transition zone of the stresses of a given crack, we have $x = 0$ ($\sigma_c = 0$ and $\varepsilon_c = \varepsilon_r$) and $x = x_r$ ($\sigma_c = f_t$ and $\varepsilon_c = \varepsilon_{t0}$). Thus, the bilinear behavior curve in Figure 8.12(a) gives:

$$w_c = \frac{1}{2}2(\varepsilon_r - \varepsilon_{t0})x_r = (2 10^{-3} - 9.3 \times 10^{-5})0.05 = 0.095\ 10^{-3}\ m \quad [8.36]$$

This result leads to almost the same value as that measured at 12 days (0.1 mm, see Table 8.4).

8.4. Conclusion

The proposed simplified method is based on an approach where the calculation operates in a scheme generating diffuse damage. A second localized cracking scheme derived from limit analysis concepts is also considered. The dialog between the two schemes, diffuse and localized, leads to mean cracking indicators (spacing and opening).

It is simple to apply to treat tie rods because while it is essentially analytical (three relationships to use), this method is also effective for more physically complex cases (effects of concrete maturation at an early age) and mechanically simple cases (axisymmetric problem). Applied to the gusset of the VeRCoRs model, it is found that the different fields (strain, damage, etc.) are almost homogeneous, and the dialog between the diffuse scheme and the localized scheme leads to very good values indicating cracking.

Controlling cracking in concrete structures remains a very topical issue because it is in tune with the needs of safety and durability. This is a subject on which the sophisticated modeling tools, developed in recent decades, are still struggling because generating so-called "complex" resolution systems involves numerical difficulties and a multitude of parameters. The use of limit analysis, developed by our "great elders" when the means of numerical investigations were reduced, shows that a deep reflection established with pragmatism before any calculation still retains its supremacy on subjects as complex as cracking.

9

Exercises and Supplements

This chapter proposes both exercises and supplements related to certain themes covered in the different chapters. The aim is to become familiar with some particular aspects of modeling.

9.1. Determining mechanical characteristics from experimental curves

In Chapter 1, we analyzed the experimental behavior of concrete and gave the curves in Figure 1.9(b) for uniaxial behavior under compression. These are the $|\sigma_1|$–$|\varepsilon_1|$ and $|\sigma_1|$–ε_2 curves (reproduced in Figure 9.1), obtained by repeated loadings (with unloadings), thus showing the changes in rigidity.

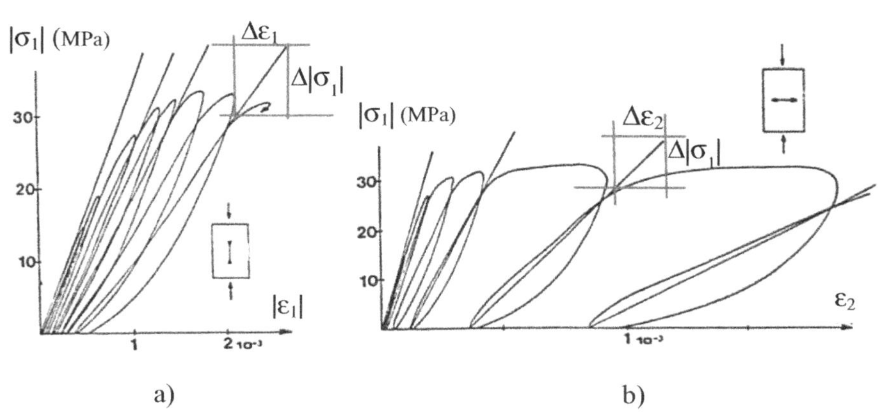

Figure 9.1. *Behavior of concrete under simple compression: a) longitudinal behavior $|\sigma_1|$–$|\varepsilon_1|$ and b) transversal behavior $|\sigma_1|$–ε_2*

The question raised here is how to extract the elastic modulus E from these curves and its evolution, and the Poisson coefficient ν and its evolution.

– From equation [A.1], for an initially isotropic elastic material in uniaxial compression, we get:

$$\varepsilon_1 = \frac{\sigma_1}{E} \text{ and } \varepsilon_2 = -\frac{\vartheta}{E}\sigma_1 = \varepsilon_3 \qquad [9.1]$$

– These equations lead to:

$$E = \frac{\sigma_1}{\varepsilon_1} \text{ and } \vartheta = -\frac{E\varepsilon_2}{\sigma_1} = -\frac{\varepsilon_2}{\varepsilon_1} \qquad [9.2]$$

With the curve shown in Figure 9.1(a), for a given loading state, the measured modulus E_m is directly derived from the ratio $|\Delta\sigma_1|/|\Delta\varepsilon_1|$. With the curve shown in Figure 9.1(b), the slopes of the unloading lines give, according to equation [9.1], $\Delta\sigma_1/\Delta\varepsilon_2 = -E_m/\nu_m$. Thus, at a given loading level, we obtain the Poisson coefficient: $\Delta\sigma_1/\Delta\varepsilon_2 = -\Delta\sigma_1/\Delta\varepsilon_1\,\nu_m$, that is, $\nu_m = -\Delta\varepsilon_2/\Delta\varepsilon_1 = \Delta\varepsilon_2/|\Delta\varepsilon_1|$, as given in equation [9.2].

Thus, considering that the damage is related to the ratio between the original modulus E and the measured modulus E_m ($E_m/E = 1 - D_m$), we can deduce from the two curves the evolution of E_m and ν_m as a function of the measured damage D_m and find that:

– D_m changes from 0 to 0.55 at the end of the curve; however, by forming the hypothesis, formulated in the models presented in Chapter 3, of the absence of residual strain at the end of the unloading (unloading lines are all supposed to pass through the origin), the calculation damage, D, then changes from 0 to 0.7 at the end of the curve.

– Damage is not isotropic: if it were, ν_m would be constant and therefore so would $\Delta\varepsilon_2/|\Delta\varepsilon_1|$, which is not the case in Figure 9.1(b). Indeed, we note that from one loop to another, especially during the post-peak, at fixed $\Delta\sigma_1$, $\Delta\varepsilon_2$ grows much faster than $|\Delta\varepsilon_1|$, which means that ν_m increases from 0.2 initially to values that exceed 0.5 at the end of the process.

In a material, when $\nu = 0.5$, the strain occurs at a constant volume. Indeed, the volume variation here is:

$$\frac{\Delta V}{V} = \varepsilon_1 + \varepsilon_2 + \varepsilon_3 = \varepsilon_1 + 2\varepsilon_2 = \frac{\sigma_1}{E}(1 - 2\vartheta) \qquad [9.3]$$

Thus, $\nu = 0.5$ leads to $\Delta V/V = 0$, and $\nu > 0.5$ would lead to $\Delta V/V > 0$ (σ_1 here is < 0). This would mean that during a compression strain, there would be production of material, which obviously does not make sense. In fact, it is an "illusion" related to the fact that when vertical cracks are formed, they open at the end of the process, and this opening generates an apparent and sharp increase in transverse strains ε_2. It is therefore necessary to be cautious in modeling these phenomena. The choice made in the models presented in Chapter 3 is that of isotropic damage. This leads to considering that ν is constant and, therefore, leads to underestimating the prediction of transverse strains in a compression test. This is the price to pay for using this simplifying hypothesis.

9.2. Mazars model: axisymmetric triaxial loading under compression

Let us consider a cylindrical specimen of the X_1 axis. It receives a vertical compression load, $\sigma_1 < 0$, and a lateral stress such that $\sigma_2 = \sigma_3 = \beta\sigma_1$ (radial loading) where $0 < \beta < 1$.

From equation [A.3] written in the principal coordinate system (X_1, X_2, X_3), for $i = 1, 2, 3$ we get:

$$\varepsilon_i = \frac{1+\vartheta}{E}\sigma_i - \frac{\vartheta}{E}\sigma_1(1+2\beta) \qquad [9.4]$$

that is:

$$\varepsilon_1 = \frac{\sigma_1}{E}(1-2\beta\vartheta),\ \varepsilon_2 = \frac{\sigma_1}{E}\big(\beta - \vartheta(1+\beta)\big) = \varepsilon_3 \qquad [9.5]$$

It can be seen that with the hypotheses made ($\beta < 1$), the axial stress is greater than the lateral one, then $\varepsilon_1 < 0$ and, if moreover $\beta > \frac{\vartheta}{(1-\vartheta)}$, then $\varepsilon_2 = \varepsilon_3 < 0$. Under these conditions, according to equation [3.2], the equivalent strain $\varepsilon_{eq} = 0$ and therefore D_t, D_c and, consequently, D are all zero, regardless of the loading level. This condition prevents any transverse extension, and as a result, the material does not become damaged. We come across one of the limitations of the model that favors the presence of extension. In reality, the absence of extension leads, for a high level of stress, to a consolidation if β is close to 1 and a consolidation-mode II combination if β is different from 1. These situations are characterized as loading types referenced as C and B in Table 1.2.

Here, let us consider the case where extensions are possible (type of loading referenced as A in Table 1.2), then: $\beta < \frac{\vartheta}{(1-\vartheta)}$, assuming that $\nu = 0.2$, current value for concrete, $\beta < 0.25$.

Thus, ε_1 is always negative and $\varepsilon_2 = \varepsilon_3$ are positive. We then obtain the equivalent strain:

$$\varepsilon_{eq} = \sqrt{\varepsilon_2^2 + \varepsilon_3^2} = \varepsilon_2\sqrt{2} \quad [9.6]$$

From equations [3.1] and [9.5], for the damaged material, we obtain:

$$\varepsilon_1 = \frac{\sigma_1}{E(1-D)}(1-2\beta\vartheta),\ \varepsilon_2 = \frac{\sigma_1}{E(1-D)}\big(\beta - \vartheta(1+\beta)\big) = \varepsilon_3 \quad [9.7]$$

that is:

$$\varepsilon_{eq} = \frac{\sigma_1}{E(1-D)}\big(\beta - \vartheta(1+\beta)\big)\sqrt{2} = \frac{\varepsilon_1}{(1-2\beta\vartheta)}\big(\beta - \vartheta(1+\beta)\big)\sqrt{2} \quad [9.8]$$

This is an expression from which we can determine, for a given β, $\varepsilon_{eq} = f(\varepsilon_1)$.

The damage is given by equation [3.5]: $D = \alpha_t D_t + \alpha_c D_c$, and, in the particular case where all the principal stresses are negative, then equations [3.6] become: $\boldsymbol{\sigma} = \langle\boldsymbol{\sigma}\rangle_-$ and $\boldsymbol{\varepsilon} = \boldsymbol{\varepsilon}_c$. From equation [3.7], we get $\alpha_t = 0$ and $\alpha_c = 1$ and therefore, according to equations [3.5] and [3.10]:

$$D = D_c = 1 - \frac{(1-A_c)\varepsilon_0}{\varepsilon_{eq}} - A_c \exp\{-B_c(\varepsilon_{eq} - \varepsilon_0)\} \quad [9.9]$$

D being known, the behavior is thus completely defined from equation [9.7]:

$$\sigma_1 = \frac{\varepsilon_1 E(1-D)}{(1-2\beta\vartheta)} \text{ and } \varepsilon_2 = \frac{\varepsilon_1}{(1-2\beta\vartheta)}\big(\beta - \vartheta(1+\beta)\big) = \varepsilon_3 \quad [9.10]$$

Considering the numerical values E = 30,000 MPa, ν = 0.2, A_t = 0.8, $B_t = 2.10^{+4}$, A_c = 1.4, and B_c = 1,850, Figure 9.2 shows, for several values of β, the axial behavior $\sigma_1 = f(\varepsilon_1)$ of the sample (Mazars 1986).

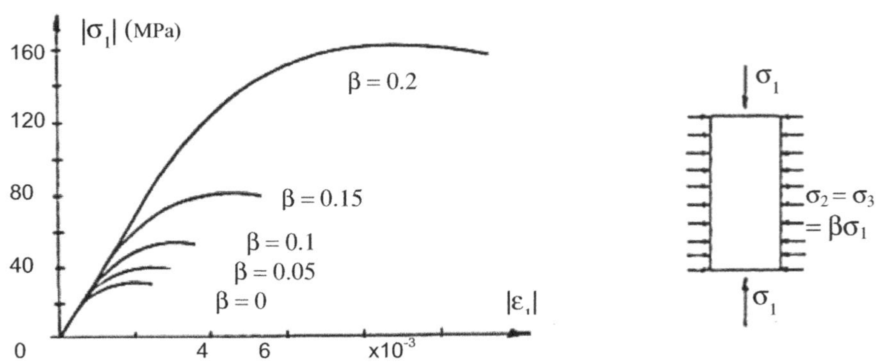

Figure 9.2. *Axisymmetric triaxial behavior given by the Mazars model for different $\beta = \sigma_2/\sigma_1$ ratios*

It is found that the intensity of the lateral stress has an extremely favorable effect on the material's behavior to cope with a compression action. In practice, this lateral stress can be achieved in columns, by a metal tube in which concrete is poured (this is a widespread practice in offshore engineering), by performing binding via transverse reinforcements whose decrease in spacing intensifies the lateral stress (Cusson and Paultre 1994), or by externally sheathing the column with a carbon fiber fabric (Desprez et al. 2013).

9.3. Local and non-local damage

Here, the idea is to present the peculiarities of calculations that generate localization or non-localization damage through the example of a concrete bar loaded under tension. We will then see how to approach finite element calculations to respect their objectivity (mesh independence) by associating in particular local calculation and the Hillerborg method.

9.3.1. *Example of a concrete bar under direct tension*

Geometric characteristics of the bar: diameter $\Phi = 10$ cm, length L = 60 cm; elastic-damageable behavioral law respecting the constitutive laws of the Mazars model (or the μ model, which under monotonous tension gives the same results). The characteristics of the material are given in the table in Figure 9.3.

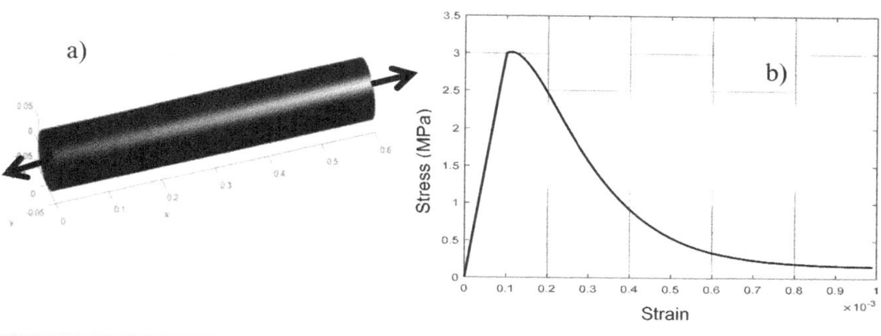

Elastic modulus E	Tensile strength f_t	Damage threshold ε_0	A_t	B_t
30 GPa	3 MPa	1×10^{-4}	0.95	9,000

Figure 9.3. *a) Cylindrical bar under tension; b) behavioral law and table of characteristics for modeling (Mazars model). For a color version of this figure, see www.iste.co.uk/mazars/damage.zip*

9.3.2. *Local model response: impact of the number of elements*

To carry out calculations, we operate on the same bar cut into several 1D bar elements whose number is variable (from 1 to 10); to locate the failure, we weakened one of the elements close to the middle of the beam, reducing its cross-section by 0.1%. Each of the bars is then loaded under tension, with imposed displacement until it reaches local strains close to 4×10^{-4}.

Figure 9.4 shows the results obtained. It can be seen that the localization of damage is carried out on a single element: from the entire bar for N = 1 to 1/10 of its length for N = 10 (see Figure 9.4(a)). Moreover, the overall behavior appears very different depending on the number of elements (Figure 9.4(b)). For N = 10, it leads to a brutal failure beyond the peak, which is characteristic of a very fragile behavior. It is shown that if N is large, the post-peak results in a snap-back, which means that both the effort and the displacement return, and, if the number of elements tends toward infinity, the dissipated energy (area under the curve) then tends toward 0.

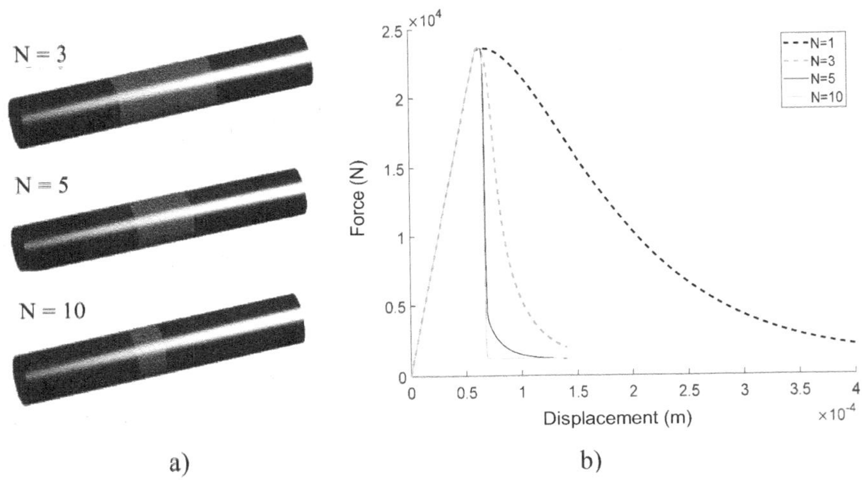

Figure 9.4. *Bar under tension: local calculation: a) localized damage on a single element (total number of elements, N); b) overall behavior force–displacement as a function of N. For a color version of this figure, see www.iste.co.uk/mazars/damage.zip*

It is clear that under these conditions, precautions must be taken before approaching a "local" calculation if the calculation results are not dependent on the size of the elements used.

In the third part of this exercise we will see that the method advocated by Hillerborg (see section 4.4.2), consisting of adapting the behavioral law to the size of the elements, is a good answer to this important problem.

9.3.3. *Non-local damage problem*

For details on the concept of non-local damage (Pijaudier-Cabot and Bazant 1987), see section 4.5. In this exercise, we choose l_c = 0.06 m as the characteristic length (l_c is related to the size of the largest aggregates, assumed here to be d_a = 20 mm and $l_c = 3d_a$ is recommended).

As in section 9.3.2, we will conduct several calculations here by changing the number of elements. For a fixed number N, it is first necessary to determine the vicinity of the element where damage will appear. For the local calculation, we will weaken the section of a central element of the x-axis (no. 11 for a mesh with 20 elements, see Figure 9.5) and the calculation of the vicinity is as follows:

– before any mechanical calculation, we calculate the weight function (g) for all the elements neighboring the weakened element by storing this result:

$$g(x-s) = \exp\left[-\left(\frac{2(x-s)}{l_c}\right)^2\right], \text{ if } |x-s| < l_c \text{ and } g(x-s) = 0 \text{ if } |x-s| > l_c \quad [9.11]$$

– during loading, the distribution of non-local equivalent strains ($\bar{\varepsilon}_{eq}$) shall be determined at every step from local equivalent strains (ε_{eq}):

$$\bar{\varepsilon}_{eq} = \frac{\int_{V_r} g(x-s)\varepsilon_{eq} dV}{\int_{V_r} g(x-s) dV} \quad [9.12]$$

– damage is then calculated, not with ε_{eq}, but with $\bar{\varepsilon}_{eq}$, which leads to the determination of non-local damage D.

After all the calculations are performed, we obtain the results presented in Figure 9.6. The observation is that, beyond 15 elements, the damaged areas and global force–displacement curves are identical, making the process independent of the mesh. Fifteen elements on a bar of 60 cm in length are elements of 4 cm; it is therefore deduced from these results that it is necessary to have three minimum elements in the representative volume V_r (such that $V_r = 2l_c$ = 12 cm here) to ensure objectivity with respect to the mesh. For other applications, reference may be made to Saouridis and Mazars (1989, 1992).

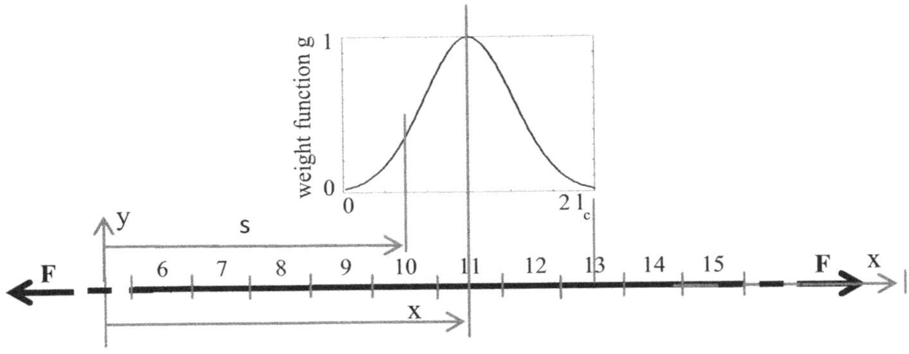

Figure 9.5. *Determination of the vicinity of the element that is damaged, here no. 11 of the x-axis is damaged (case where N = 20). For a color version of this figure, see www.iste.co.uk/mazars/damage.zip*

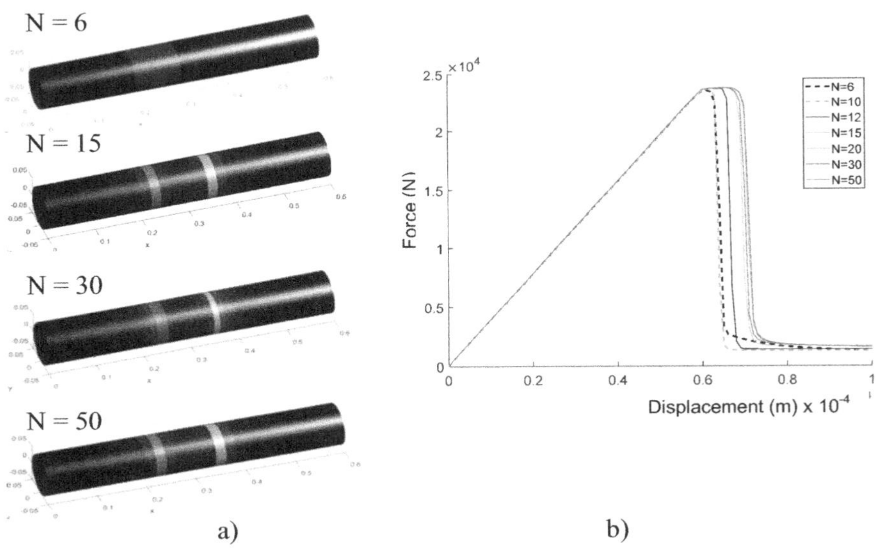

Figure 9.6. *Bar under tension: non-local calculation: a) independence of the extent of damage with the number of elements; b) force–displacement behavior as a function of N, showing near perfect identity beyond 15 elements. For a color version of this figure, see www.iste.co.uk/mazars/damage.zip*

9.3.4. *Objective calculation with a local model: Hillerborg method*

As we have seen, local calculation leads to a dependence on the mesh and therefore to a non-objective calculation. However, it can be imagined that an objective calculation can be obtained by choosing the correct size of elements (in the zone where there is damage)

From the calculations made in sections 9.3.2 and 9.3.3, it can be seen that, all other things being equal (same behavioral law and same geometrical characteristics), by choosing an element of size $h = 2l_c = 12$ cm, that is, here N = 5, the compared local–non-local results are close (see Figure 9.7).

Since controlling the size of elements is not always easy, Hillerborg et al. (1976) proposed a method that allows the alternative:

– either we can control the size of elements and adapt it to the chosen behavioral law;

– or the size cannot be chosen and it is the behavioral law that is adapted.

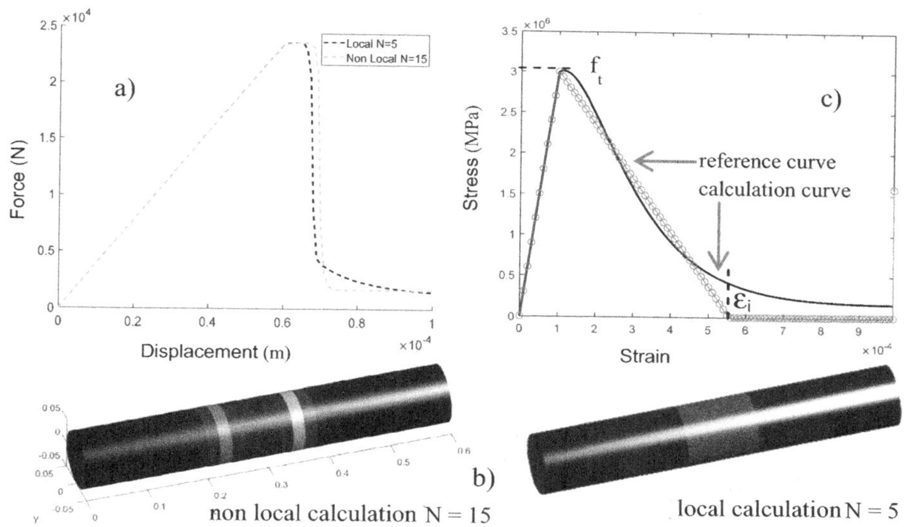

Figure 9.7. *Bar under tension: (a and b) concordance between a non-local calculation and a local calculation performed with an element size $h = 2l_c$; c) setting of the behavior curve in order to respect a fixed value of G_f/h (area under the triangle $= \varepsilon_i f_t/2$). For a color version of this figure, see www.iste.co.uk/mazars/damage.zip*

However, at the basis of the method, the fracture energy during the damage process must be respected in all cases. We commented on this in section 4.4 and proposed the use of the intrinsic part of this energy: G_f ($G_f = G_F$/k, G_F is the total fracturing energy and k # 2). The procedure is shown in Figure 4.5 to set the tension curve σ–ε to meet this requirement (see equations [4.11] and [4.12]).

This point has been anticipated here, and the initial behavior curve is based on the following data: G_f = 100 N/m, h = 0.12 m (= $2l_c$). Compliance with these data is ensured by proceeding, as shown in Figure 9.7(c), as follows:

– the triangle is drawn, which shows a bilinear curve respecting G_f (vertex f_t, base $0 - \varepsilon_i$), where:

$$\varepsilon_i = \frac{2G_f}{hf_t} \qquad [9.13]$$

– the σ–ε curve of the calculation model is adjusted so that:

$$\int_0^{\varepsilon_i} \sigma \varepsilon d\varepsilon = G_f/h \qquad [9.14]$$

In the Mazars model (or μ model), A_t and B_t should be fixed to comply with equation [9.14]. Section 4.4.1 states that it is prudent to take an A_t value of less than 1, both for numerical facilitation and for extrinsic energy ($G_F - G_f$). To do this, we have chosen A_t= 0.95, only B_t remains to be adjusted, for which we have chosen B_t = 9,000 (see Figure 9.8(a), where Gf/h # 833).

More generally, the problem is dealt with as follows:

– the material data are the problem data: E, f_t (or ε_0), G_f; moreover, A_t is fixed (# 0.9);

– on the basis of an a priori determined mesh, knowing h, it is easy to deduce G_f/h and consequently ε_i from equation [9.13];

– the numerical processing of equation [9.14] leads, if A_t is fixed, to the plotting of the curve $B_t = f(G_f/h)$, as shown in Figure 9.8(a). As G_f/h is known, B_t is also known, and the behavioral law is also known.

Figure 9.8(b) gives, starting from the case studied with h = 12 cm, some examples for other h values (15 and 20 cm), and Figure 9.8(c) shows the proximity of the responses obtained for a local calculation carried out on a bar under tension with this procedure.

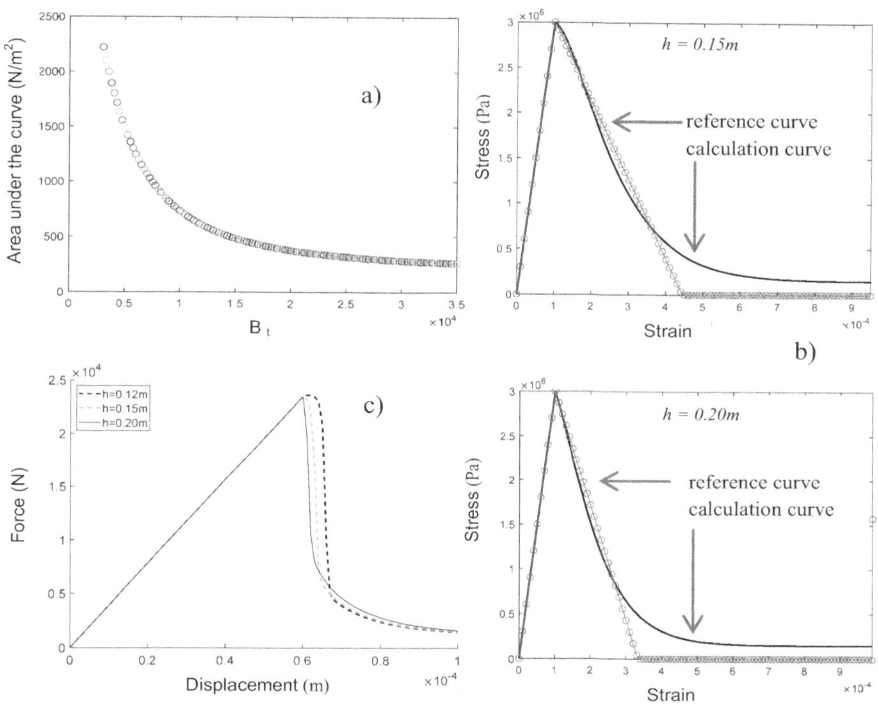

Figure 9.8. *The Hillerborg method was applied to local calculation. a) Evolution of the parameter B_t as a function of G_f/h (area under the σ–ε curve limited to ε_l); b) adjustment of the behavioral law for two values of h; c) concordance between the different calculations (h = 0.12 m, 0.15 m and 0.20 m). For a color version of this figure, see www.iste.co.uk/mazars/damage.zip*

9.3.5. *Conclusion*

The objectivity of finite element calculations is a major point for the simulation of the behavior of structures. We have just shown that, in the case of non-reinforced concrete, without any particular procedure, the localization is systematic and leads to a strong dependence on the results at the fineness of the mesh. As we have shown, the concept of non-local damage is one way of dealing with this type of problem. This concept is more generally part of so-called regularization methods (regularization by kinematic enrichment, gradient regularization, or non-local regularization) and leads to the use of a characteristic length (l_c) to calibrate the procedure, which, despite everything, gives the calculation a complementary heaviness (the vicinity of each Gauss point needs to be determined beforehand by assigning it a weight function, depending on the distance to this point).

Hillerborg regularization is simpler to apply from this point of view. Most likely, less efficient for concrete, it is a good compromise for reinforced concrete because the localization is much less severe because of the presence of rebars and their regulatory role. Furthermore, we have shown, starting from Chapter 5, that the discretization of reinforced concrete elements by beam elements automatically introduces diffuse damage for bending-type loadings. Section 5.2.3 provides details on the management of this particularity.

9.4. On the μ model

9.4.1. *Reaching the damage threshold, load-unload criterion*

With the μ model (Mazars et al. 2015), on this subject, the problem differs from what was seen for the Mazars model (see section 3.1.1) by the fact that two threshold surfaces are to be considered here because two equivalent strains are involved, one for cracking (ε_t) and the other for crushing (ε_c).

In this case, the load–unload criterion depends on the triaxial factor value r (see section 9.4.2 for its calculation), and it is expressed mathematically as follows:

– If r = 1, only the threshold surface $f(\varepsilon_t, Y_t) = 0$ is likely to be reached, then only Y_t (= $\max(\varepsilon_{t0}, \varepsilon_t)$) must be updated and three situations arise:

- if $f(\varepsilon_t, Y_t) < 0$, the point is inside the surface, then $\dot{d} = \frac{\mathrm{d}d}{\mathrm{d}t} = 0$ (d does not progress);

- if $f(\varepsilon_t, Y_t) = 0$ and $df = 0$, the point is on the surface and drags it with it, then $\dot{d} > 0$ (d increases);

- if $f(\varepsilon_t, Y_t) = 0$ and $df < 0$, the point is on the surface and returns inwards (unloading), then $\dot{d} = 0$.

– If r = 0, only the threshold surface $f(\varepsilon_c, Y_c) = 0$ is likely to be reached, then only Y_c (= $\max(\varepsilon_{c0}, \varepsilon_c)$) must be updated and three situations arise:

- if ($f(\varepsilon_c, Y_c) < 0$, the point is inside the surface, then $\dot{d} = 0$ (d does not progress);

- if $f(\varepsilon_c, Y_c) = 0$ and $df = 0$, the point is on the surface and drags it with it, then $\dot{d} > 0$ (d increases);

- if $f(\varepsilon_c, Y_c) = 0$ and $df < 0$, the point is on the surface and returns inwards (unloading), then $\dot{d} = 0$.

– If $0 < r < 1$, both threshold surfaces are likely to be reached, then Y_t and Y_c are updated as presented above, and the damage-driven variable $Y = rY_t + (1-r)Y_c$ is calculated:

- if Y increases, then d increases ($\dot{d} > 0$);
- if Y remains constant, then $\dot{d} = 0$.

These particularities must be integrated into the flowchart presented in section 4.2.2:

– if $r = 1$, only Y_t is updated and $Y = Y_t$;

– if $r = 0$, only Y_c is updated and $Y = Y_c$;

– if $0 < r < 1$, both Y_t and Y_c must be updated and $Y = rY_t + (1-r)Y_c$.

Thus, all cases lead to a progression of Y, pilot variable of damage, which once determined, allows access to the material behavior given by equations [3.20]:

$$\boldsymbol{\sigma} = (1-d)\boldsymbol{\Lambda}: \boldsymbol{\varepsilon} \tag{9.15}$$

It is also written as:

$$\boldsymbol{\sigma} = (1-d)\lambda Tr(\boldsymbol{\varepsilon})\mathbf{1} + 2(1-d)\mu$$

or from equation [A.3] as:

$$\boldsymbol{\varepsilon} = \frac{1+\vartheta}{E(1-d)}\boldsymbol{\sigma} - \frac{\vartheta}{E(1-d)}\mathrm{Tr}(\boldsymbol{\sigma})\mathbf{1} \tag{9.16}$$

Furthermore, we have seen in Chapter 3, in equation [3.27], that the laws of evolution are of the form:

$$d = f(Y, A, B) \tag{9.17}$$

As described earlier, this model (like the Mazars model) is said to be explicit because there is a one-to-one relationship between the value of the Y variable and the damage value. Knowing the value of the pilot variable, it is sufficient to use the expressions given for d in equation [3.27] to completely solve the problem.

9.4.2. *On the stress triaxiality factor*

We have seen, in section 3.2.1, the definition of the pilot variable of the effective damage d:

$$Y = r\,Y_t + (1 - r)Y_c, \qquad [9.18]$$

Y_t and Y_c are the maximum values of the equivalent cracking and crushing strains, respectively.

r, the triaxiality factor, is defined as:

$$r = \frac{\sum_i \langle \tilde{\sigma}_i \rangle_+}{\sum_i |\tilde{\sigma}_i|} \qquad [9.19]$$

$\langle \tilde{\sigma}_i \rangle_+$ is the positive value of the effective stress, which according to equation [2.9] in Chapter 2 is defined by:

$$\tilde{\sigma}_i = \frac{\sigma_i}{(1-d)} \qquad [9.20]$$

$|\tilde{\sigma}_i|$ is the absolute value of this effective stress.

Some observations relating to r:

– as $\boldsymbol{\sigma} = \boldsymbol{\Lambda}(1 - d) : \boldsymbol{\varepsilon}$ then, $\tilde{\boldsymbol{\sigma}} = \boldsymbol{\Lambda} : \boldsymbol{\varepsilon}$ is independent of d; r is also independent, and its calculation therefore does not depend on the damage state;

– if the loading is only a combination of tensile stresses or zero stresses, then all $\tilde{\sigma}_i$ are ≥ 0, which implies that $\sum_i \langle \tilde{\sigma}_i \rangle_+ = \sum_i |\tilde{\sigma}_i|$ and leads to $r = 1$;

– if the loading is only a combination of compressive stress or zero stress, then all $\tilde{\sigma}_i$ are ≤ 0, which implies that $\sum_i \langle \tilde{\sigma}_i \rangle_+ = 0$ and leads to $r = 0$;

– if the loading is a combination of tensile and compressive stresses, then $\sum_i \langle \tilde{\sigma}_i \rangle_+ \neq \sum_i |\tilde{\sigma}_i|$ and $0 < r < 1$.

For example, in the case of a biaxial shear stress, $\tilde{\sigma}_3 = 0$ and $\tilde{\sigma}_1 = -\tilde{\sigma}_2 > 0$, then:

$$r = \frac{\tilde{\sigma}_1}{\tilde{\sigma}_1 - \tilde{\sigma}_2} = 0.5 \qquad [9.21]$$

Thus, more generally, in the case of a biaxial stress ($\tilde{\sigma}_3 = 0$), if the sector concerned is included inside the positive part of the axes ($\tilde{\sigma}_1, \tilde{\sigma}_2$), then $r = 1$; if it is inside the negative part of the same axes, then $r = 0$; and for the other two sectors that mix tension and compression, $0 < r < 1$. It is within these sectors that the points in simple shear are found on the bisector $\tilde{\sigma}_1 = -\tilde{\sigma}_2$.

9.4.3. *Response to triaxial axisymmetric compression loading*

The triaxial situation pushes the modeling to the limits of their capacity. In this context, we propose this type of challenge for the μ model in this section. With respect to this model, a triaxial compression load covers, according to the relative intensities of axial stresses, all the types of stresses listed in Table 1.2:

– type A, allowing extensions and favoring crushing damage in mode I;

– type B without extension, but with damage in mode II, for which a correction of the pilot variable of crushing damage makes it possible to respond;

– types C and C′, where the strong confinements generate a so-called "consolidation" damage, corresponding to the collapse of porous medium and which is to be coupled with that of crushing. All of these aspects are discussed here, and we will conclude that, in most cases (low and medium confinements), the use of crushing-type modeling is sufficient to obtain good results.

As in section 9.2, the specimen is a cylinder with axis X_1 (see Figure 9.2). It receives a vertical compression load $\sigma_1 < 0$ and a lateral stress such that:

– in the first phase, $\sigma_2 = \sigma_3 = \beta\sigma_1$ (radial loading) with $0 < \beta < 1$;

– in a second phase, $\sigma_2 = \sigma_3 = s$, s remaining constant while the intensity of σ_1 increases until failure.

The loading parameters are $\beta = 0.8$, and s takes the following values: 0, 20, 35, 50, and 100 MPa. Table 9.1 gives the material characteristics used.

E (MPa)	ν	Y_{c0}	A_c	B_c	$I_{\varepsilon 0}$	$I_{\varepsilon 1}$	*a*	*b*
30,000	0.2	$3.5\ 10^{-4}$	1.2	490	–0.006	–0.06	–0.78	0.00132

Table 9.1. *Modeling parameters*

For the first phase, equations [9.4] and [9.5] are also valid here, as well as the observations that have been made on the value of β: if $\beta < \frac{\vartheta}{(1-\vartheta)}$, then the extensions are possible ($I_{\varepsilon+} > 0$), and as we are in the case where the triaxiality factor r is equal to 0, as seen in section 3.3:

$$C = \frac{\sum_i \langle \varepsilon_i \rangle_+}{\sum_i \langle \varepsilon_{io} \rangle_+} = \frac{I_{\varepsilon+}}{I_{\varepsilon 0+}} > 0 \qquad [9.22]$$

Thus, the pilot variable to be used is $Y_c^* = CY_c$, and the effective crushing damage d is activated when $Y_c^* = CY_{c0}$. Otherwise, if $\beta > \frac{\vartheta}{(1-\vartheta)}$, (i.e. with a current value for the concrete of ν = 0.2, if $\beta > 0.25$), then $I_{\varepsilon+} = 0$, $C = 0$ and d is not activated. In the case considered here, $\beta = 0.8$, the first phase starts in a linear elastic manner, but we will see in section 9.4.3.2 that under certain conditions, consolidation damage can be activated during this phase.

On the other hand, the second phase can be complex with activation of both damage patterns, as we will see in sections 9.4.3.1 and 9.4.3.2.

9.4.3.1. *Activation of crushing damage*

Except in the case where $s = 0$, which is the particular situation of uniaxial compression for which the damage will be exclusively of the crushing type, in all other cases, beyond the threshold $CY_{c0} = C\varepsilon_{c0}$ (see Table 9.2), the crushing damage mode cannot be activated in the first phase because $\beta > 0.25$.

On the other hand, in the second phase where $\sigma_2 = \sigma_3 = s$ (constant value) and where the intensity of σ_1 increases, the value of $\beta = s/\sigma_1$ decreases progressively. When it reaches 0.25, then $I_{\varepsilon+}$ becomes positive and d is activated. Table 9.2 shows the values corresponding to $\sigma_1 = s/0.25$, for the various confinements considered, but this assumes that crushing appears first (otherwise, we will see that the combination of modes complicates the case).

The problem can also be addressed by strains. In this case:

$$\varepsilon_1 = \frac{1}{E_d}(\sigma_1 - 2s\vartheta) \text{ and } \varepsilon_2 = \varepsilon_3 = \frac{1}{E_d}(s - \vartheta(\sigma_1 + s)) \qquad [9.23]$$

Crushing is activated if $I_{\varepsilon+} > 0$ or, because ε_1 is always negative, if $\varepsilon_2 > 0$.

Thus, if:

$$\varepsilon_2 = \frac{1}{E_d}(s - \vartheta(s + \varepsilon_1 E_d + 2\vartheta s) > 0$$

or if:

$$-\varepsilon_1 > \frac{s}{-\vartheta E_d}(1 - \vartheta - 2\vartheta^2), \text{ then d is activated} \qquad [9.24]$$

Assuming that this threshold appears first ($E_d = E$ at the time of activation), the various values of this crushing threshold are given in Table 9.2. If consolidation appears first, then $E_d = E\,(1 - d_{hc})$ and the value of $-\varepsilon_1$ increases, but moderately as d_{hc} remains low.

9.4.3.2. *Activation of "consolidation" damage*

Consolidation damage d_{hc} is activated if I_ε exceeds the $I_{\varepsilon 0}$ threshold.

If activation takes place in the first phase, starting from equations [9.7], we get:

$$I_\varepsilon = \varepsilon_1 + \varepsilon_2 + \varepsilon_3 = \frac{\sigma_1}{E_d}(1 - 2\vartheta + 2\beta - 4\beta\vartheta)$$
$$= \frac{\sigma_1}{E_d}(1 - 2\vartheta)(1 + 2\beta) \qquad [9.25]$$

With $\beta = 0.8$, at the threshold we obtain $\frac{2.6\sigma_1}{E_d}(1 - 2\vartheta) = I_{\varepsilon 0}$, or $\sigma_1 = \frac{E_d I_{\varepsilon 0}}{2.6(1-2\vartheta)}$. Taking into account the numerical values in Table 9.1 and, as in section 9.4.3.1, considering that $E_d = E$ (valid if consolidation appears first), then, $\sigma_l = -115$ MPa when it is triggered.

If activation takes place in the second phase:

$$\varepsilon_1 = \frac{1}{E_d}(\sigma_1 - 2s\vartheta) \text{ and } \varepsilon_2 = \varepsilon_3 = \frac{1}{E_d}(s - \vartheta(\sigma_1 + s)) \qquad [9.26]$$

Or, $I_\varepsilon = \frac{(1-2\vartheta)}{E_d}(\sigma_1 + 2s)$; we then obtain the following at the threshold:

$$\sigma_1 = \frac{E_d I_{\varepsilon 0}}{(1-2\vartheta)} - 2s \qquad [9.27]$$

Taking the numerical values given in Table 9.1, it can be seen (see Table 9.2) that consolidation never appears first in the second phase; either it appears in the first phase ($s = -100$ MPa), or failure is achieved by crushing before it is possible. In the case where E is previously affected by crushing, $E_d = E(1 - d)$ and the intensity of σ_l decreases. This is the case when $s = -20$ MPa or -35 MPa, for which failure appears at -106 MPa and -160 MPa, respectively, which does not allow consolidation to be activated.

The $s = -50$ MPa case is particular. The consolidation threshold is -115 MPa if it appears in the first phase, but it has not been activated because the second phase started earlier (-62.5 MPa); in this second phase, since β decreases, the consolidation threshold is postponed and it appears only for -200 MPa, at the same time as the crushing threshold.

Table 9.2 summarizes the different possible situations in the combination of the two modes of damage and also gives the results obtained for failure simulations of the various confinement solutions proposed.

<table>
<tr><th>-s (MPa)</th><th>0</th><th>20</th><th>35</th><th>50</th><th>100</th></tr>
<tr><td>-σ_1 at the beginning of the second phase</td><td rowspan="5">Special case of simple compression, crushing begins at $-\sigma_1 = 10.5$ MPa and $-\varepsilon_1 = 3.5 \times 10^{-4}$</td><td>25</td><td>43.75</td><td>62.5</td><td>125</td></tr>
<tr><td>-σ_1 at the threshold if consolidation appears in the first phase</td><td>115</td><td>115</td><td>115</td><td>115</td></tr>
<tr><td>-ε_1 at the threshold if crushing appears in the second phase</td><td>0.0024</td><td>0.0042</td><td>0.006</td><td>>0.012</td></tr>
<tr><td>-σ_1 at the threshold if crushing appears in the second phase</td><td>80</td><td>140</td><td>200</td><td><400</td></tr>
<tr><td>-σ_1 at the threshold if consolidation appears in the second phase</td><td><260</td><td><230</td><td>200</td><td>–</td></tr>
<tr><td colspan="6">Maximum stress achieved (failure)</td></tr>
<tr><td>Experimental results</td><td>–32</td><td>–107</td><td>–152</td><td>–214</td><td>–368</td></tr>
<tr><td>Num. results: coupled μ model</td><td>–32</td><td>–106</td><td>–160</td><td>–211</td><td>–376</td></tr>
<tr><td>Num. results: crushing μ model</td><td>–32</td><td>–106</td><td>–163</td><td>–215</td><td>–428</td></tr>
</table>

Table 9.2. *Achieving different thresholds (damage and failure) depending on the intensity of the confinement and experimental – numerical results comparison. The thresholds reached first are indicated in italics. For a color version of this table, see www.iste.co.uk/mazars/damage.zip*

This table shows that, depending on the intensity of the second phase of confinement, damage modes are triggered differently:

– for s = –20 MPa, only crushing appears from –80 MPa (second phase) and causes failure at –106 MPa;

– for s = –35 MPa, we enter the second phase at –43.75 MPa, crushing begins at approximately –140 MPa and causes failure at –160 MPa;

– for s = –50 MPa, consolidation and crushing begin at the same time during the second phase at approximately –200 MPa, which causes failure at approximately –211 MPa;

– for $s = -100$ MPa, consolidation begins during the first phase at –115 MPa, and crushing actually begins at 375 MPa, which causes a very rapid failure at approximately 376 MPa (see Table 9.2 for the results given by the coupled μ model).

In the latter case, there is a high incidence of consolidation damage, which has significantly reduced the failure. Calculations have shown that in the absence of consolidation, crushing would then have started for $\sigma_1 = -400$ MPa and failure would have occurred at approximately –428 MPa (see Table 9.2 and Figure 9.9(c) for the results given by the crushing μ model).

The impact of consolidation is much lower for low or medium confinement and can be overlooked for confinements below 50 MPa.

Figure 9.9(b) gives the results obtained experimentally by two researchers on a similar concrete and on similar loading trajectories (Ramtani 1990; Gabet 2008). A comparison is then made with the results of calculations obtained with the μ model in two situations:

– Figure 9.9(c) gives the results obtained by only considering the effects of crushing damage ($E_d = E(1 - d)$);

– Figure 9.9(d) gives the results obtained by taking into account the coupling effects of crushing and consolidation damage ($E_d = E\,(1 - d)\,(1 - d_{hc})$).

The observation is that, overall, the coupled μ model, with d and d_{hc}, gives good results for both low and high confinement values. However, Figure 9.9(c) shows that, except for confinement at 100 MPa, the model with crushing damage alone (crushing μ model) works very well.

However, in both cases, we find the same weakness as that observed in uni- and biaxial loadings, the underestimation of strains in transverse directions (ε_2, ε_3), because the Poisson coefficient remains invariable throughout the phase of damage; it is the price to pay with the use of an isotropic model.

In conclusion, except for high confinement (>50 MPa), we can ignore consolidation damage. The interest is a simplification because let us remember that the basic model requires only uniaxial tests to identify parameters, while the laws of evolution of d_{hc} require four complementary parameters whose identification is based on triaxial tests of a high technicality and which few laboratories are able to carry out.

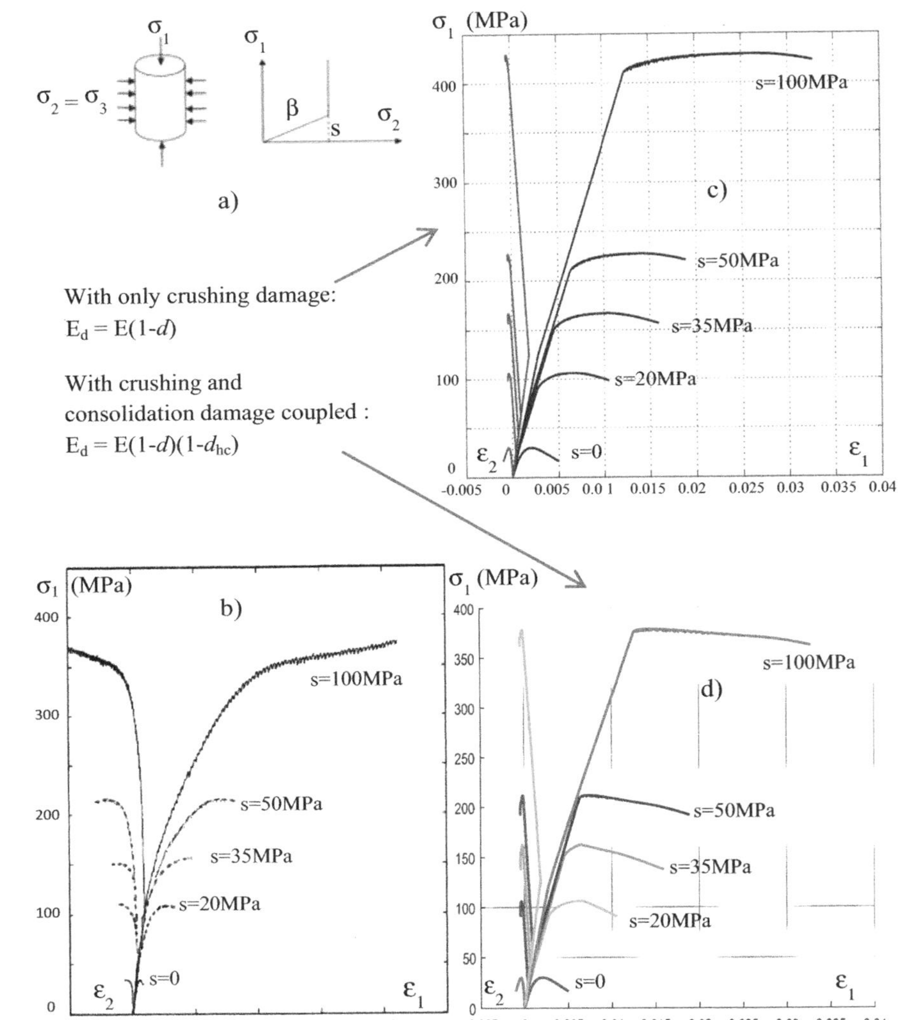

Figure 9.9. *Triaxial test. For a color version of this figure, see www.iste.co.uk/mazars/damage.zip*

COMMENT ON FIGURE 9.9.– *(a) Sample and loading trajectory: after a radial path ($\sigma_2 = \sigma_3 = \beta\sigma_1$), the lateral stress is fixed at s and σ_1 increases until failure; (b) experimental results (Ramtani 1990; Gabet 2006); (c) responses to the μ model taking into account only crushing damage d; (d) responses of the model by coupling crushing damage d and consolidation damage d_{hc}.*

NOTE 9.1. There are a variety of behaviors during different phases.

– *Crushing*: calculations are carried out in the same way as in exercise 9.2, adapting the problem to the use of the pilot damage variable $Y_c^* = CY_c$ instead of ε_{eq} of the Mazars model. The procedure to be carried out is then given.

– Calculation of $C = \frac{\sum_i \langle \varepsilon_i \rangle_+}{\sum_i \langle \varepsilon_{io} \rangle_+} = \frac{I_{\varepsilon+}}{I_{\varepsilon 0+}}$, $C \neq 0$ if $I_{\varepsilon+} = 2\langle \varepsilon_2 \rangle_+ > 0$. We have seen that this can only be verified in the second phase, for which we get $I_{\varepsilon+} = 2\frac{1}{E_d}(s - \vartheta(\sigma_1 + s))$. Moreover, it was seen in Chapter 3 that $I_{\varepsilon 0+}$ is the projection of $I_{\varepsilon+}$ on the closest $\sigma_i = 0$ plane, calculated by setting the compression of lower intensity to 0. Here, $|\sigma_1| > |\sigma_2| = |\sigma_3|$, we choose $|\sigma_3| = 0$ as an example, and we get $I_{\varepsilon+0} = \frac{1}{E_d}(s - \vartheta(2\sigma_1 + s))$, that is:

$$C = 2\frac{s-\vartheta(\sigma_1-\vartheta(\sigma_1+s)}{s-\vartheta(\sigma_1-\vartheta(\sigma_1+s)}, \text{ or } C = 2\frac{s(1-\vartheta-2\vartheta^2)-\vartheta E_d\varepsilon_1}{s(1-\vartheta-4\vartheta^2)-2\vartheta E_d\varepsilon_1} \qquad [9.28]$$

from which we deduce $Y_c^* = CY_c = C\varepsilon_c$ max, with $\varepsilon_c = \frac{I_\varepsilon}{5(1-2\vartheta)} + \frac{6\sqrt{J_\varepsilon}}{5(1+\vartheta)}$, which is the pilot damage variable calculated as given by equations [3.25] and [3.26].

– *Consolidation*: this calculation is independent of that of crushing. There is consolidation if $I_\varepsilon < 0$, and equations [9.25] and [9.27] give the expressions for I_ε as a function of the phase considered. Knowing the high and low thresholds, $I_{\varepsilon 0}$ and $I_{\varepsilon l}$, respectively, we get:

$$\text{if } I_\varepsilon < I_{\varepsilon 0},\ d_{hc} = 0;\ \text{if } I_{\varepsilon 0} < I_\varepsilon < I_{\varepsilon 1},\ d_{hc} = 1 + a + b/I_\varepsilon \text{ and if } I_\varepsilon > I_{\varepsilon 1},\ d_{hc} = d_{hc1} \qquad [9.29]$$

– *Determination of behavior*: this is done conventionally with the law:

$$\boldsymbol{\sigma} = \Lambda(1-d)(1-d_{hc}) : \boldsymbol{\varepsilon}$$

using, depending on the phase, equations [9.7] and [9.23] and $E_d = E(1-d)(1-d_{hc})$.

9.5. On the restraint degree R in situations of restrained shrinkage

This coefficient R, known as the restraint degree, reflects the fact that a structure is prevented from straining with respect to its supports. $0 \le R \le 1$, when $R = 0$, there is no restraint, and when $R = 1$, the restraint is total.

Two types of restraint can be considered:

– external restraint caused by a connection with the supports of the concerned element; coefficient R depends on the stiffness of the element itself and its support, and the effects depend on the phenomenon that generates the restraint to the strain (expansion, shrinkage, effects related to early age, etc.);

– internal restraint, connected in particular to the presence of rebars that restrain the establishment of strain to which the concrete is subjected, during drying, for example.

Here, we are interested in external restraint, which may, for example, correspond to the case of a wall with concrete poured vertically on a very stiff foundation block. At an early age, this wall will be subjected to maturation effects, which, as we have seen in Chapters 6 and 8, will cause "large" volume variations, and the restraint caused by the stiff substratum will lead to the development of cracks. This cracking will depend on the geometrical and material characteristics of the structure (quality of concrete, geometry, presence of rebars, casting conditions, etc.).

Here, we propose determining R in the case of the application presented in section 6.2 concerning the RG8 beam subject to restrained shrinkage by interposing metal struts between beam ends (Mazars et al. 2018).

In this case, the definition of the restraint degree is:

$$R = \frac{\text{Stiffness of struts}}{\text{Total stiffness}} = \frac{\text{Ke}}{\text{Ke+Kc}} = \frac{1}{1+{}^{\text{Kc}}/_{\text{Ke}}} \qquad [9.30]$$

K_e and K_c are the stiffnesses of the struts and the concrete element, respectively. The following expressions are used to calculate them:

$$\text{K}_\text{e} = \frac{F_e}{u} = \frac{A_S\sigma_S}{\varepsilon_S L} = \frac{A_S\sigma_S E_S}{\sigma_S L} = \frac{A_S E_S}{L}$$

Similarly:

$$\text{K}_\text{c} = \frac{A_C E_C}{L} \text{ and } \frac{\text{K}_\text{c}}{\text{K}_\text{e}} = \frac{A_C E_C}{A_S E_S} \qquad [9.31]$$

In the example of the RG8 beam under restrained shrinkage (see section 6.2), struts are tubes 32.4 cm in the outside diameter and 0.55 cm in thickness; the cross-section of the beam in its running part is 0.8 × 0.5 m. With E_s = 210 GPa and E_c = 30 GPa, we get:

$$R = \frac{1}{1+\frac{A_C E_C}{A_S E_S}} = \frac{1}{1+\frac{0.8\ 0.5\ 30}{2\ 0.0464\ 210}} = 0.62 \qquad [9.32]$$

The restraint degree is therefore 62%.

If we take the case of a beam embedded at these two ends in an infinitely stiff rocky massif, then the stiffness of struts and the total stiffness merge and $R = 1$. Conversely, if there is no strut, $R = 0$.

In the same vein, Engström (2011) studied the case of vertical walls cast on a supposedly infinitely stiff support. The restraint degree then varies within the structure with the distance from the embedded base in the vicinity of which $R = 1$. He showed that the change in height depended on L/H and established a system of charts giving the value of R as a function of y/L (y is the position considered at the center of the wall with respect to the base; see Figure 9.10). R represents a risk and not a cracked state and serves in particular to provide a minimum rebar percentage adapted to each zone of the wall.

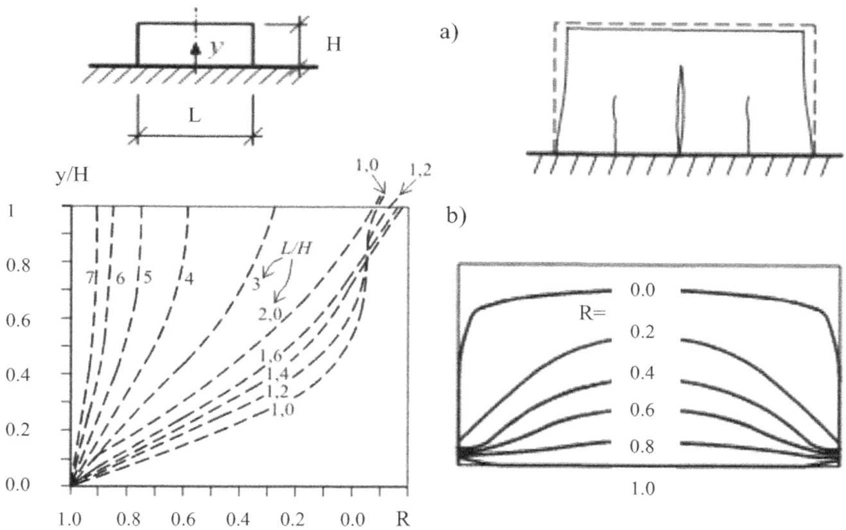

Figure 9.10. *a) Vertical wall cast on a stiff support and cracking created at an early age; b) value of the restraint degree R = f (y/H) parameterized by L/H, reflecting the hinderance in establishing strains and the risk of cracking (according to Engström (2011))*

Codes and regulations echoed this idea. For example, EC2 (2004) and ACI (1995) recommend the following:

$$R = \left(\frac{L/_{H}-2}{L/_{H}+1}\right)^{y/H} \text{ if } L/_{H} \geq 2.5 \text{ and } R = \left(\frac{L/_{H}-1}{L/_{H}+10}\right)^{y/H} \text{ if } L/_{H} < 2.5 \qquad [9.33]$$

9.6. Solving a simple structure using the PVP*

This simple example shows how the principle of virtual powers (PVP*) is used to then formulate the finite element method.

9.6.1. *Problem position*

1) Force applied in A_2, such that $\boldsymbol{F} = -F\boldsymbol{y}$.

2) Boundary conditions (BC):

$$\begin{cases} u_{1_x} = 0 \\ u_{1_y} = 0 \\ u_{3_x} = 0 \\ u_{3_y} = 0 \end{cases} \qquad [9.34]$$

3) Unknowns to calculate: $\boldsymbol{U_2} = u_{2_x}\boldsymbol{x} + u_{2_y}\boldsymbol{y}$.

4) Question: calculate displacements at node A_2.

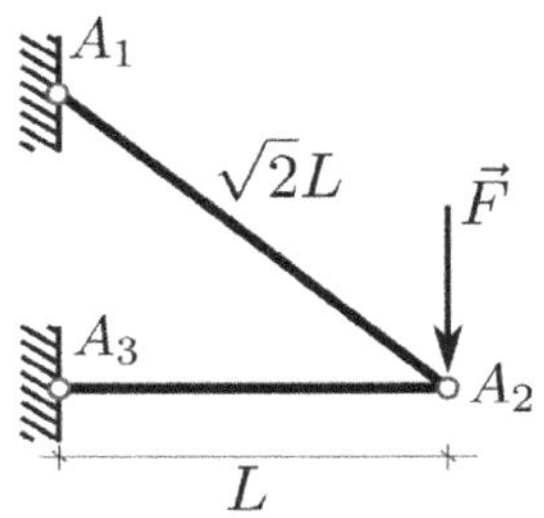

Figure 9.11. *Example of two bar truss structure. For a color version of this figure, see www.iste.co.uk/mazars/damage.zip*

9.6.2. *Using PVP* (assembling the contributions of the elements): preliminary comments for solving*

Truss elements orientation is noted $\boldsymbol{n}_{ij}$.

Internal forces in bars:

– are axial only;

– are noted q_i and q_j for each bar.

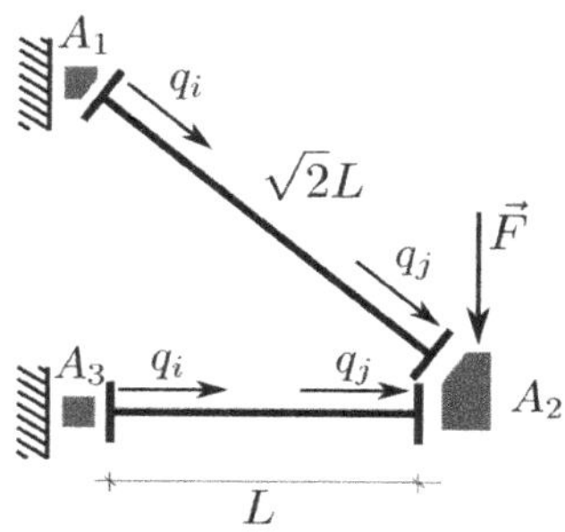

Figure 9.12. *Minimal decomposition of structure into elements and into nodes. For a color version of this figure, see www.iste.co.uk/mazars/damage.zip*

9.6.2.1. *Writing the terms of the PVP**

For each bar, we can develop the PVP* expression [A.18] as follows:

$$\begin{aligned}\mathcal{P}^{\star^{el}}_{\text{int}} &= \int_0^{l_{el}} \frac{ES}{l_{el}^2}\left[{}^t\boldsymbol{n}_{ij}\left(\boldsymbol{u}^\star{}_j - \boldsymbol{u}^\star{}_i\right)\right]\left[{}^t\boldsymbol{n}_{ij}\left(\boldsymbol{u}_j - \boldsymbol{u}_i\right)\right]dx \\ \Leftrightarrow \mathcal{P}^{\star^{el}}_{\text{int}} &= \int_0^{l_{el}} \frac{ES}{l_{el}^2}\,{}^t\left(\boldsymbol{u}^\star{}_j - \boldsymbol{u}^\star{}_i\right)\underbrace{\boldsymbol{n}_{ij}\,{}^t\boldsymbol{n}_{ij}}_{\boldsymbol{n}_{ij}\otimes\boldsymbol{n}_{ij}}\left(\boldsymbol{u}_j - \boldsymbol{u}_i\right)dx \\ \Leftrightarrow \mathcal{P}^{\star^{el}}_{\text{int}} &= \frac{ES}{l_{el}}\,{}^t\left(\boldsymbol{u}^\star{}_j - \boldsymbol{u}^\star{}_i\right)\boldsymbol{n}_{ij}\otimes\boldsymbol{n}_{ij}\left(\boldsymbol{u}_j - \boldsymbol{u}_i\right)\end{aligned}$$

Separating the degrees of freedom gives:

$$\mathcal{P}^{\star^{el}}_{\text{int}} = \left[\boldsymbol{u}^\star{}_i, \boldsymbol{u}^\star{}_j\right]\underbrace{\frac{ES}{l_{el}}\begin{bmatrix}\boldsymbol{n}_{ij}\otimes\boldsymbol{n}_{ij} & -\boldsymbol{n}_{ij}\otimes\boldsymbol{n}_{ij} \\ -\boldsymbol{n}_{ij}\otimes\boldsymbol{n}_{ij} & \boldsymbol{n}_{ij}\otimes\boldsymbol{n}_{ij}\end{bmatrix}}_{\boldsymbol{k}_{el}}\begin{bmatrix}\boldsymbol{u}_i \\ \boldsymbol{u}_j\end{bmatrix}$$

9.6.2.2. *Application of the PVP* to the structure in a virtual velocity field*

$$\forall \boldsymbol{u}^\star \text{ KA } 0, \mathcal{P}^\star_{\text{int}} = \mathcal{P}^\star_{\text{ext}}$$

$$\mathcal{P}^\star_{\text{ext}} = {}^t\boldsymbol{F}\boldsymbol{u}^\star_2 = -Fu^\star_{2_y}$$

$$\mathcal{P}^\star_{\text{int}} = \left[\boldsymbol{u}^\star{}_1, \boldsymbol{u}^\star{}_2\right]\boldsymbol{k}_{①}\begin{bmatrix}\boldsymbol{u}_1 \\ \boldsymbol{u}_2\end{bmatrix} + \left[\boldsymbol{u}^\star{}_2, \boldsymbol{u}^\star{}_3\right]\boldsymbol{k}_{②}\begin{bmatrix}\boldsymbol{u}_2 \\ \boldsymbol{u}_3\end{bmatrix}$$

or:

$$\boldsymbol{k}_{①} = \frac{E_1 S_1}{2l_1}\begin{bmatrix} 1 & -1 & -1 & 1 \\ -1 & 1 & 1 & -1 \\ -1 & 1 & 1 & -1 \\ 1 & -1 & -1 & 1 \end{bmatrix} \text{ and } \boldsymbol{k}_{②} = \frac{E_2 S_2}{l_2}\begin{bmatrix} 1 & 0 & -1 & 0 \\ 0 & 0 & 0 & 0 \\ -1 & 0 & 1 & 0 \\ 0 & 0 & 0 & 0 \end{bmatrix}$$

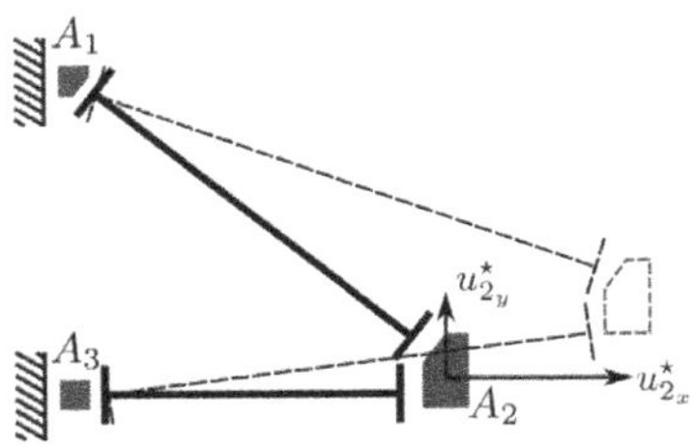

Figure 9.13. *Virtual velocity field chosen to solve the problem. For a color version of this figure, see www.iste.co.uk/mazars/damage.zip*

By gathering the degrees of freedom, we can finally write as follows:

$\forall \boldsymbol{u}^\star$ KA at 0 (means, virtual field = 0 at the boundary conditions),

$$^t\begin{bmatrix} \boldsymbol{u}_1^\star \\ \boldsymbol{u}_2^\star \\ \boldsymbol{u}_3^\star \end{bmatrix} \begin{bmatrix} \frac{E_1S_1}{2l_1} & -\frac{E_1S_1}{2l_1} & -\frac{E_1S_1}{2l_1} & \frac{E_1S_1}{2l_1} & 0 & 0 \\ -\frac{E_1S_1}{2l_1} & \frac{E_1S_1}{2l_1} & \frac{E_1S_1}{2l_1} & -\frac{E_1S_1}{2l_1} & 0 & 0 \\ -\frac{E_1S_1}{2l_1} & \frac{E_1S_1}{2l_1} & \frac{E_1S_1}{2l_1} + \frac{E_2S_2}{l_2} & -\frac{E_1S_1}{2l_1} & -\frac{E_2S_2}{l_2} & 0 \\ \frac{E_1S_1}{2l_1} & -\frac{E_1S_1}{2l_1} & -\frac{E_1S_1}{2l_1} & \frac{E_1S_1}{2l_1} & 0 & 0 \\ 0 & 0 & -\frac{E_2S_2}{l_2} & 0 & \frac{E_2S_2}{l_2} & 0 \\ 0 & 0 & 0 & 0 & 0 & 0 \end{bmatrix}$$

$$\times \begin{bmatrix} \boldsymbol{u}_1 \\ \boldsymbol{u}_2 \\ \boldsymbol{u}_3 \end{bmatrix} = -F u^\star_{2_y}$$

KA: kinematically admissible.

KA field at 0, $\Rightarrow \boldsymbol{u}^\star{}_1 = \boldsymbol{u}^\star{}_3 = \boldsymbol{0} \Rightarrow$.

$$\forall u^\star_{2_x}, u^\star_{2_y},$$

$$^t\begin{bmatrix} u^\star_{2_x} \\ u^\star_{2_y} \end{bmatrix} \begin{bmatrix} -\frac{E_1S_1}{2l_1} & \frac{E_1S_1}{2l_1} & \frac{E_1S_1}{2l_1} + \frac{E_2S_2}{l_2} & -\frac{E_1S_1}{2l_1} & -\frac{E_2S_2}{l_2} & 0 \\ \frac{E_1S_1}{2l_1} & -\frac{E_1S_1}{2l_1} & -\frac{E_1S_1}{2l_1} & \frac{E_1S_1}{2l_1} & 0 & 0 \end{bmatrix}$$

$$\times \begin{bmatrix} \boldsymbol{u}_1 \\ \boldsymbol{u}_2 \\ \boldsymbol{u}_3 \end{bmatrix} = {}^t\begin{bmatrix} u^\star_{2_x} \\ u^\star_{2_y} \end{bmatrix} \begin{bmatrix} 0 \\ -F \end{bmatrix}$$

Which gives us:

$$\begin{bmatrix} -\frac{E_1S_1}{2l_1} & \frac{E_1S_1}{2l_1} & \frac{E_1S_1}{2l_1}+\frac{E_2S_2}{l_2} & -\frac{E_1S_1}{2l_1} & -\frac{E_2S_2}{l_2} & 0 \\ \frac{E_1S_1}{2l_1} & -\frac{E_1S_1}{2l_1} & -\frac{E_1S_1}{2l_1} & \frac{E_1S_1}{2l_1} & 0 & 0 \end{bmatrix} \begin{bmatrix} \boldsymbol{u}_1 \\ \boldsymbol{u}_2 \\ \boldsymbol{u}_3 \end{bmatrix} = \begin{bmatrix} 0 \\ -F \end{bmatrix}$$

And finally, by considering the BC (boundary conditions) of the problem $\boldsymbol{u}_1 = \boldsymbol{u}_3 = \boldsymbol{0}$, we then get the system to solve:

$$\begin{bmatrix} \frac{E_1S_1}{2l_1}+\frac{E_2S_2}{l_2} & -\frac{E_1S_1}{2l_1} \\ -\frac{E_1S_1}{2l_1} & \frac{E_1S_1}{2l_1} \end{bmatrix} \begin{bmatrix} u_{2_x} \\ u_{2_y} \end{bmatrix} = \begin{bmatrix} 0 \\ -F \end{bmatrix}$$

This simple example also shows two things in the case of this structure intrinsically discretized into elements and nodes:

– the equilibrium of the structure can be expressed as the sum of contribution of elementary parts (here, truss bar elements);

– there is a stiffness matrix (and possibly an internal force vector) for each of these elements that can be added through an assembly technique on the degrees of freedom of the problem.

Appendix

Prerequisites in Solid Mechanics and Finite Element Methods

This appendix has been divided into three parts:

– a part relating to the mechanics of deformable solids and the basics of linear elasticity;

– a part relating to mechanical equilibrium in the weak form;

– a part relating to the finite element method (FEM) practice.

A.1. Prerequisites in mechanics of deformable solids

This is not a solid mechanics course. Only the principles and main relationships that flow from it and may be useful throughout this book are referenced. Readers who would like clarification can refer to the books by Batoz and Dhatt (1990a, 1990b, 1990c); Zienkiewicz et al. (2013) or Simo and Hughes (1998).

A.1.1. *Bases in linear elasticity*

It is customary to present the "intrinsic" equations of a material (those that characterize the mechanical properties) by using a priori a complete set of coefficients linking each stress component to each strain component. This results in a tensor relationship connecting the stress tensor to that of strains $\boldsymbol{\sigma} = \boldsymbol{\Lambda} : \boldsymbol{\varepsilon}$:

$$\begin{bmatrix} \sigma_{11} & \sigma_{12} & \sigma_{13} \\ \sigma_{21} & \sigma_{22} & \sigma_{23} \\ \sigma_{31} & \sigma_{32} & \sigma_{33} \end{bmatrix} = [\Lambda_{ijkl}] \begin{bmatrix} \varepsilon_{11} & \varepsilon_{12} & \varepsilon_{13} \\ \varepsilon_{21} & \varepsilon_{22} & \varepsilon_{23} \\ \varepsilon_{31} & \varepsilon_{32} & \varepsilon_{33} \end{bmatrix} \quad [A.1]$$

The various $\mathbf{\Lambda}$ coefficients are obtained by permutation of the indices $ijkl$. From considerations of symmetry and orientation, the number of these coefficients is reduced. For example, in linear isotropic elasticity, E and ν are, respectively, the elastic modulus (called Young's modulus) and the coefficient of contraction (called Poisson's coefficient), the relationship $\boldsymbol{\sigma} = \mathbf{\Lambda}\varepsilon$ is reduced to:

$$\boldsymbol{\sigma} = \lambda Tr(\boldsymbol{\varepsilon})\mathbf{1_d} + 2\mu\boldsymbol{\varepsilon} \qquad \text{[A.2]}$$

or for the component ij; $\sigma_{ij} = \lambda\varepsilon_{kk}\delta_{ij} + 2\mu\varepsilon_{ij}$.

λ and μ are Lamé coefficients: $\lambda = \frac{\nu E}{(1+\nu)(1-2\nu)}$ and $\mu = \frac{E}{2(1+\nu)}$ and $\mathbf{1_d}$ is the identity tensor. $Tr(\varepsilon)$ is the trace of the strain tensor. δ_{ij} equals 1 if $i = j$ and equals 0 if not.

The dual form is:

$$\boldsymbol{\varepsilon} = \frac{1+\nu}{E}\boldsymbol{\sigma} - \frac{\nu}{E}Tr(\boldsymbol{\sigma})\mathbf{1_d} \qquad \text{[A.3]}$$

For the component ij, let: $\varepsilon_{ij} = \frac{1+\nu}{E}\sigma_{ij} - \frac{\nu}{E}\sigma_{kk}\delta_{ij}$.

The symmetry conditions cited above ($\varepsilon_{ij} = \varepsilon_{ji}$ and $\sigma_{ij} = \sigma_{ji}$) lead to only six components out of nine of the stress and strain tensors being considered. In this case, the following equations completely define the isotropic elastic linear behavior of a solid body. Thus, we are talking about Voigt notation. The presence of the terms in $\sqrt{2}$ makes it possible to obtain the same result by the dot product between the stresses and strains in the form of a vector or by the doubly contracted product that the matrix writes. This will be particularly useful for writing the energy balances needed to balance the system, as we will see in section A.2. In the local coordinate system, this leads to:

$$\begin{bmatrix} \varepsilon_{11} \\ \varepsilon_{22} \\ \varepsilon_{33} \\ \sqrt{2}\varepsilon_{23} \\ \sqrt{2}\varepsilon_{31} \\ \sqrt{2}\varepsilon_{12} \end{bmatrix} = \begin{bmatrix} \frac{1}{E} & -\frac{\nu}{E} & -\frac{\nu}{E} & 0 & 0 & 0 \\ -\frac{\nu}{E} & \frac{1}{E} & -\frac{\nu}{E} & 0 & 0 & 0 \\ -\frac{\nu}{E} & -\frac{\nu}{E} & \frac{1}{E} & 0 & 0 & 0 \\ 0 & 0 & 0 & \frac{1}{2G} & 0 & 0 \\ 0 & 0 & 0 & 0 & \frac{1}{2G} & 0 \\ 0 & 0 & 0 & 0 & 0 & \frac{1}{2G} \end{bmatrix} \begin{bmatrix} \sigma_{11} \\ \sigma_{22} \\ \sigma_{33} \\ \sqrt{2}\sigma_{23} \\ \sqrt{2}\sigma_{31} \\ \sqrt{2}\sigma_{12} \end{bmatrix} \qquad \text{[A.4]}$$

or in the principal coordinate system:

$$\begin{bmatrix} \varepsilon_1 \\ \varepsilon_2 \\ \varepsilon_3 \end{bmatrix} = \begin{bmatrix} \frac{1}{E} & -\frac{\nu}{E} & -\frac{\nu}{E} \\ -\frac{\nu}{E} & \frac{1}{E} & -\frac{\nu}{E} \\ -\frac{\nu}{E} & -\frac{\nu}{E} & \frac{1}{E} \end{bmatrix} \begin{bmatrix} \sigma_1 \\ \sigma_2 \\ \sigma_3 \end{bmatrix} \qquad \text{[A.5]}$$

The shear modulus is given by:

$$G = \mu = \frac{E}{2(1+\nu)} \qquad \text{[A.6]}$$

ε_{ii} and σ_{ii} are normal strains and normal stresses, respectively, and ε_{ij} and σ_{ij} are shear strains and stresses, respectively. In the prinicipal coordinate system (X,Y, Z), the $\boldsymbol{\sigma}$ and ε matrices diagonalize and lead to defining the three principal strains and stresses $\varepsilon_1, \varepsilon_2, \varepsilon_3$ and $\sigma_1, \sigma_2, \sigma_3$, respectively. This situation simplifies the matrix equation as seen above.

NOTE A.1.– The elastic coefficient matrix is complicated if the material does not have the same properties in all directions. In the particular case of a transverse isotropy (identical characteristics for any pair of directions symmetrical with respect to an axis, which is the case for wood or a unidirectional composite), then five independent coefficients appear: two moduli (E_1 and E_3), two contraction coefficients (ν_{12} and ν_{13}) and a shear modulus (G_{13}). This number rises to nine independent coefficients in orthotropic elasticity (three orthogonal axes of particular mechanical properties) and to 21 in the most general case. We will limit ourselves here to the case of complete isotropy, which therefore leads to the use of only two independent coefficients in linear elasticity, E and ν.

A.1.2. *Invariants and special relationships*

A.1.2.1. *Stress and strain tensor invariants*

Due to the isotropy hypothesis, invariant magnitudes appear for stress and strain tensors whatever the axis system considered. There are three invariants:

$$\begin{cases} \sigma_I = Tr(\sigma) = \sigma_{11} + \sigma_{22} + \sigma_{33} \\ \sigma_{II} = \frac{1}{2}Tr(\sigma^2) = \frac{1}{2}\sigma_{ij}\sigma_{ij} \\ \sigma_{III} = \frac{1}{3}Tr(\sigma^3) = \frac{1}{3}\sigma_{ij}\sigma_{jk}\sigma_{ki} \end{cases} \qquad \text{[A.7]}$$

and:

$$\begin{cases} \varepsilon_I = Tr(\varepsilon) = \varepsilon_{11} + \varepsilon_{22} + \varepsilon_{33} \\ \varepsilon_{II} = \frac{1}{2}Tr(\varepsilon^2) = \frac{1}{2}\varepsilon_{ij}\varepsilon_{ij} \\ \varepsilon_{III} = \frac{1}{3}Tr(\varepsilon^3) = \frac{1}{3}\varepsilon_{ij}\varepsilon_{jk}\varepsilon_{ki} \end{cases} \qquad \text{[A.8]}$$

i and j can take all the values from 1 to 3 and Tr means Trace, which is the arithmetic sum of the components of the diagonal of the tensor considered.

These expressions are simplified in the principal coordinate system:

$$\begin{cases} \sigma_I = \sigma_1 + \sigma_2 + \sigma_3 \quad \sigma_{II} = \frac{1}{2}(\sigma_1\sigma_2 + \sigma_2\sigma_3 + \sigma_3\sigma_1) \quad \sigma_{III} = \frac{1}{3}(\sigma_1\sigma_2\sigma_3) \\ \varepsilon_I = \varepsilon_1 + \varepsilon_2 + \varepsilon_3 \quad \varepsilon_{II} = \frac{1}{2}(\varepsilon_1\varepsilon_2 + \varepsilon_2\varepsilon_3 + \varepsilon_3\varepsilon_1) \quad \varepsilon_{III} = \frac{1}{3}(\varepsilon_1\varepsilon_2\varepsilon_3) \end{cases} \qquad [A.9]$$

A.1.2.2. *Hydrostatic pressure, spherical tensor and deviatoric tensor*

The mean stress or hydrostatic pressure p is the arithmetic mean of normal stresses:

$$p = -\frac{Tr(\sigma)}{3} = -\frac{\sigma_{11} + \sigma_{22} + \sigma_{33}}{3} \qquad [A.10]$$

It is independent of the chosen coordinate system and allows us to define a stress tensor partition:

$$\boldsymbol{\sigma} = -\begin{bmatrix} p & 0 & 0 \\ 0 & p & 0 \\ 0 & 0 & p \end{bmatrix} + \begin{bmatrix} s_{11} & s_{12} & s_{13} \\ s_{21} & s_{22} & s_{23} \\ s_{31} & s_{32} & s_{33} \end{bmatrix} = -p\mathbf{1_d} + \boldsymbol{s} \qquad [A.11]$$

where $s_{ij} = \sigma_{ij}$ if $i \neq j$ and $s_{ii} = \sigma_{ii} - \frac{Tr(\sigma)}{3}$; $-p\mathbf{1_d}$ is the spherical part and $\boldsymbol{s}$ the deviatoric part.

The same applies to the strain tensor for which we find:

$$\boldsymbol{\varepsilon} = -e_h\mathbf{1_d} + \boldsymbol{e}$$

where for the spherical strain: $e_h = -\frac{Tr(\varepsilon)}{3} = -\frac{\varepsilon_{11}+\varepsilon_{22}+\varepsilon_{33}}{3}$ and for the deviatoric strain $e_{ij} = \varepsilon_{ij}$ if $i \neq j$ and $e_{ii} = \varepsilon_{ii} - \frac{Tr(\varepsilon)}{3}$.

A.1.2.3. *Deviator invariants*

As before, it may be necessary to consider the invariants of the deviators s and e, which in the principal coordinate system gives:

$$\begin{cases} s_I = s_1 + s_2 + s_3 \;\; s_{II} = \frac{1}{2}(s_1s_2 + s_2s_3 + s_3s_1) \;\; s_{III} = \frac{1}{3}(s_1s_2s_3) \\ e_I = e_1 + e_2 + e_3 \;\; e_{II} = \frac{1}{2}(e_1e_2 + e_2e_3 + e_3e_1) \;\; e_{III} = \frac{1}{3}(e_1e_2e_3) \end{cases} \qquad [A.12]$$

NOTE A.2.– It is easily shown that the second invariants of the deviator can be written according to the components of the tensor. For example, for strains, we find:

$$e_{II} = -\frac{1}{12}\left[(\varepsilon_1 - \varepsilon_2)^2 + (\varepsilon_1 - \varepsilon_3)^2 + (\varepsilon_2 - \varepsilon_3)^2\right] \qquad [A.13]$$

A.1.3. *Case of axisymmetric problems*

The transition from a three-dimensional frame to an axisymmetric formulation of the problems passes into the calculation of the terms of the strain tensor, which can be reduced to a number of 4. The classic strains of the plane $(r - z)$ are accompanied by the orthoradial term:

$$\varepsilon_{\theta\theta} = \frac{1}{r}\boldsymbol{U}_r \qquad \text{[A.14]}$$

The strain tensor is then written as:

$$\begin{bmatrix} \varepsilon_{rr} \\ \varepsilon_{zz} \\ \varepsilon_{rz} \\ \varepsilon_{\theta\theta} \end{bmatrix} \qquad \text{[A.15]}$$

In linear elasticity, the constitutive law in axisymmetry can be deduced simply from the three-dimensional expression, since all the terms of normal strains exist and it is therefore sufficient to scratch the rows and columns of the shear strains out of the plane (and which are decoupled from the others). No projection of the Hooke tensor is therefore required as is the case for plane stresses:

$$\begin{bmatrix} \sigma_{rr} \\ \sigma_{zz} \\ \sigma_{rz} \\ \sigma_{\theta\theta} \end{bmatrix} = \begin{bmatrix} \lambda + 2\mu & \lambda & 0 & \lambda \\ \lambda & \lambda + 2\mu & 0 & \lambda \\ 0 & 0 & 2\mu & 0 \\ \lambda & \lambda & 0 & \lambda + 2\mu \end{bmatrix} \begin{bmatrix} \varepsilon_{rr} \\ \varepsilon_{zz} \\ \varepsilon_{rz} \\ \varepsilon_{\theta\theta} \end{bmatrix} \qquad \text{[A.16]}$$

In the case of the FEM, the discretization can therefore be done in a plane $(r - z)$ with the aid of planar elements (triangles with three nodes, quadrangles with four nodes, etc.) adapted to the axisymmetric case in order to calculate orthoradial strains.

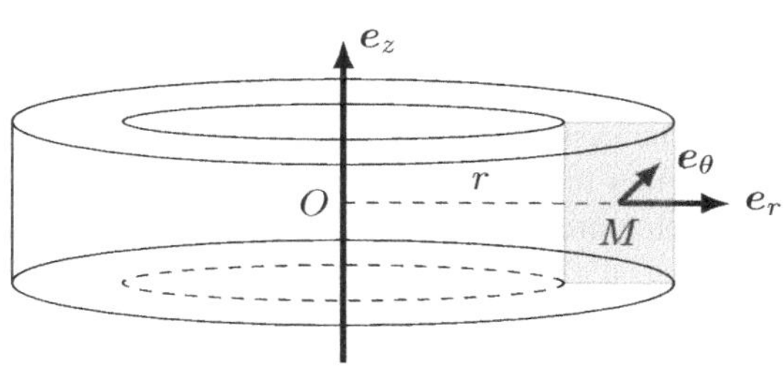

Figure A.1. *Representation of axes used for axisymmetric problems and the equilibrium of a material element around the point M*

A.2. Basics of finite element calculations

A.2.1. *Framework of solid mechanics and variable formulation*

A.2.1.1. *Equilibrium equation in solid mechanics*

The mechanics of continuous media makes it possible to write the equilibrium of a solid by means of an equation with partial derivatives (see equation [A.17]) relating to stresses and with boundary conditions (BC) on these stresses, as well as on the kinematics. In dynamics, initial conditions (IC) on the kinematics are also required.

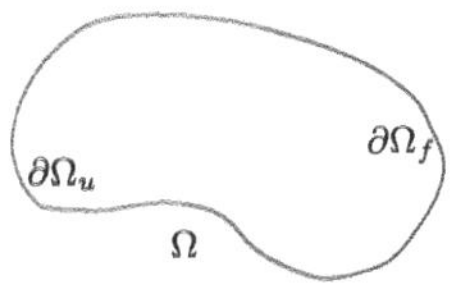

$$div(\boldsymbol{\sigma}) + \boldsymbol{f} = \rho \ddot{\boldsymbol{u}}$$
$$+\text{CL}: \begin{cases} \boldsymbol{u} = \boldsymbol{u_b} \text{ sur } \partial\Omega_u \\ \boldsymbol{\sigma}.\boldsymbol{n} = \boldsymbol{t} \text{ on } \partial\Omega_f \\ +\text{IC, on: } \boldsymbol{u}(0), \dot{\boldsymbol{u}}(0) \text{ and } \ddot{\boldsymbol{u}}(0) \end{cases} \quad \text{[A.17]}$$

For a color version of this equation, see www.iste.co.uk/mazars/damage.zip.

In this equation, $\boldsymbol{f}$ represents a force density, $\boldsymbol{u}$ are the displacements, $\dot{\boldsymbol{u}}$ are the velocities, $\ddot{\boldsymbol{u}}$ are the accelerations, and IC are initial conditions.

A.2.1.2. *Variational formulation (or "weak form")*

The system formulation [A.17] is valid at any point in a system. It is called strong formulation. However, its use may prove to be complex and it is preferred to write it in the form of a projected equation, also called a variational formulation (or weak form) given by equation [A.18] (see Salençon 1996):

$$\forall \boldsymbol{u}^\star, \underbrace{\int_\Omega {}^t\boldsymbol{u}^\star.\rho\ddot{\boldsymbol{u}} d\Omega}_{\mathcal{P}^\star_{\text{acc}}} + \underbrace{\int_\Omega \boldsymbol{\varepsilon}^\star : \boldsymbol{\sigma} d\Omega}_{\mathcal{P}^\star_{\text{int}}} = \underbrace{\int_\Omega {}^t\boldsymbol{u}.^\star \boldsymbol{f} d\Omega + \int_{\partial\Omega_f} {}^t\boldsymbol{u}^\star.\boldsymbol{t} dS}_{\mathcal{P}^\star_{\text{ext}}} \quad \text{[A.18]}$$

This variational form is completely equivalent to the mechanical problem of the system [A.17]. On the other hand, it makes it possible to write the equilibrium of each part and subpart of the system (according to the way in which the virtual field is chosen, which is none other than a field of projection of the equilibrium equation). This form is therefore much more interesting and efficient than the writing of the local equilibrium.

It is also called the virtual work principle or the principle of virtual powers (PVP*) since the $\boldsymbol{u}^{\star}$ has the mathematical structure of a velocity field (which can also be seen as a variation in displacement in some approaches).

A.2.2. *General framework of the FEM*

The FEM consists of spatially discretizing a domain by decomposing it minimally in the form of nodes and elements connecting these nodes. Over the entire area of these elements, the displacement field is interpolated by means of functions of interpolations on the displacements of nodes (also called degrees of freedom (dof)) because of equation [A.19]). The number of nodes being finite, it is then possible to interpolate the entire displacement field by means of a finite number of dof.

$$\boldsymbol{u}(M) = \sum_{i=1}^{N_{\text{nodes}}} u_i N_i(M) = \boldsymbol{N}_e \boldsymbol{u}_e \qquad \text{[A.19]}$$

The set of dof of the system is then combined in the form of a vector. The vector $\boldsymbol{U}$ then contains all of the unknowns of the system. The dof attached to the element Ω_e are then attached to a position in these tables whose numbering is specified in a connectivity table. The latter can also be noted in expanded form ($\hat{\boldsymbol{u}}_e$). This expanded vector is written in a dof base that corresponds to the total number of dof of the system, which can prove to be very practical for the construction of the overall system. However, these displacements can also be put in compact form ($\boldsymbol{u}_e$) expressed in the dof base of the element considered. One or the other of these entries is correct and makes it possible to handle the assembly of the contributions of each of the elements differently.

Numbering of tables:

$$\boldsymbol{U} = \begin{bmatrix} u_1 \\ \vdots \\ \vdots \\ u_i \\ \vdots \\ u_n \end{bmatrix} = \begin{bmatrix} \vdots \\ u_{e_i} \\ u_{e_j} \\ \vdots \\ u_{e_k} \\ \vdots \end{bmatrix}, \hat{\boldsymbol{u}}_e = \begin{bmatrix} 0 \\ u_{e_i} \\ u_{e_j} \\ \vdots \\ 0 \\ u_{e_k} \\ 0 \end{bmatrix}, \text{ or } \quad \boldsymbol{u}_e = \begin{bmatrix} u_{e_i} \\ u_{e_j} \\ u_{e_k} \end{bmatrix}$$

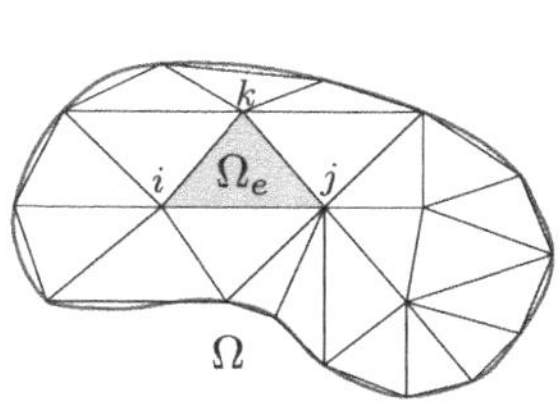

For a color version of this equation, see www.iste.co.uk/mazars/damage.zip.

Strains can be calculated from these displacements. In the case of a small strain hypothesis, we can write:

$$\varepsilon = \frac{1}{2}\left(\boldsymbol{Grad}(\boldsymbol{u}) + {}^t\boldsymbol{Grad}(\boldsymbol{u})\right) \text{ et } \varepsilon^{\star} = \frac{1}{2}\left(\boldsymbol{Grad}(\boldsymbol{u}^{\star}) + {}^t\boldsymbol{Grad}(\boldsymbol{u}^{\star})\right) \quad \text{[A.20]}$$

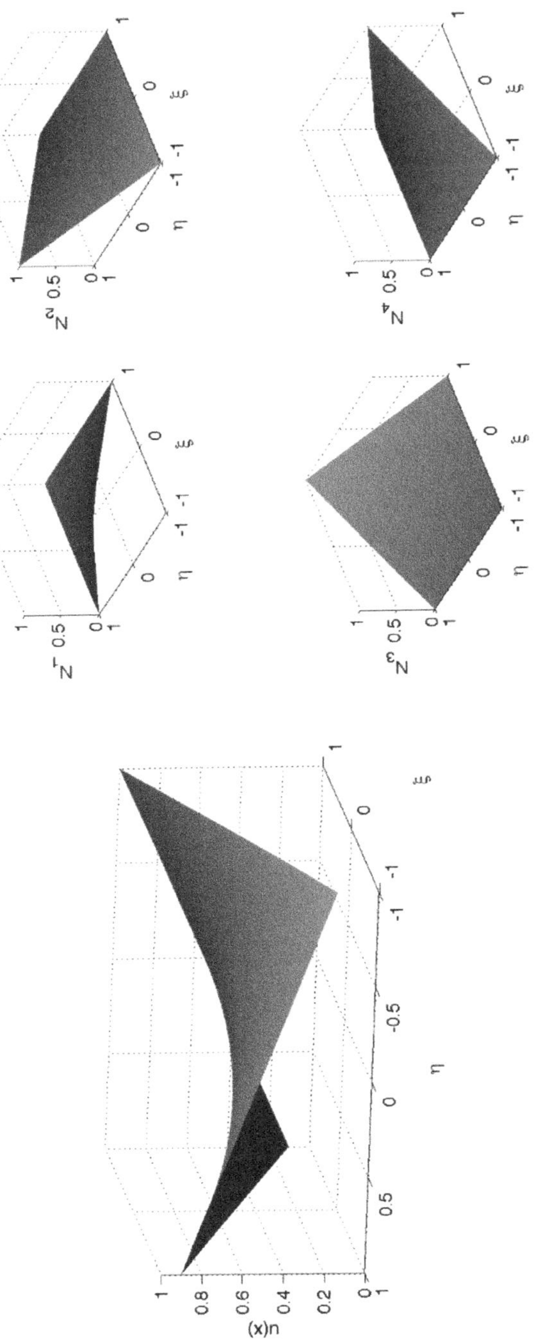

Figure A.2. *Interpolation of a displacement field using four interpolation functions* N_1, N_2, N_3 *and* N_4

The terms of this strain tensor can be recorded in vector form (for easier numerical manipulation).

A Voigt notation introduced by equation [A.4] for writing stresses and strains in the form of a vector can then be used so as to keep the dot product of these two vectors in the PVP* expression in the power term of internal forces (so that it is equal to the same double dot product contracted between the matrices). Thus, by calculating each of the terms, it appears that the matrix $\boldsymbol{B}$ contains the derivatives of the form functions:

$$\boldsymbol{\varepsilon} = \boldsymbol{B}_e \boldsymbol{u}_e = \hat{\boldsymbol{B}}_e \hat{\boldsymbol{u}}_e \text{ and } \boldsymbol{\varepsilon}^\star = \boldsymbol{B}_e \boldsymbol{u}_e^\star = \hat{\boldsymbol{B}}_e \hat{\boldsymbol{u}}_e^\star \qquad [A.21]$$

With the linear elasticity hypothesis (Hooke's law), we can write:

$$\boldsymbol{\sigma} = \boldsymbol{\Lambda}\boldsymbol{\varepsilon} = \boldsymbol{\Lambda}\boldsymbol{B}_e \boldsymbol{u}_e = \boldsymbol{\Lambda}\hat{\boldsymbol{B}}_e \hat{\boldsymbol{u}}_e \qquad [A.22]$$

A.2.2.1. *Discretized equilibrium*

All of the kinematic and stress variables calculated previously are then reinjected into the equilibrium equation [A.18]. We then get (for a color version of this equation, see www.iste.co.uk/mazars/damage.zip):

$$\forall \boldsymbol{u}^\star \text{KA } 0 \qquad [A.23]$$

$$\sum_{e=1}^{n} {}^t\hat{\boldsymbol{u}}_e^\star \underbrace{\int_{\Omega_e} {}^t\hat{\boldsymbol{N}}_e \rho \hat{\boldsymbol{N}}_e d\Omega}_{\hat{\boldsymbol{m}}_e} \ddot{\hat{\boldsymbol{u}}}_e + \sum_{e=1}^{n} {}^t\hat{\boldsymbol{u}}_e^\star \underbrace{\int_{\Omega_e} {}^t\hat{\boldsymbol{B}}_e \boldsymbol{\Lambda} \hat{\boldsymbol{B}}_e d\Omega}_{\hat{\boldsymbol{k}}_e} \hat{\boldsymbol{u}}_e$$

$$= \sum_{e=1}^{n} {}^t\hat{\boldsymbol{u}}_e^\star \underbrace{\left[\int_{\Omega_e} {}^t\hat{\boldsymbol{N}}_e \boldsymbol{f} d\Omega + \int_{\partial\Omega_e} {}^t\hat{\boldsymbol{N}}_e \boldsymbol{t} dS\right]}_{\hat{\boldsymbol{f}}_e}$$

This leads to an expression of the assembly of contributions elements between nodes in the form of a sum:

$$\forall \boldsymbol{u}^\star \text{KA } 0, \sum_{e=1}^{n} {}^t\hat{\boldsymbol{u}}_e^\star \hat{\boldsymbol{m}}_e \ddot{\hat{\boldsymbol{u}}}_e + \sum_{e=1}^{n} {}^t\hat{\boldsymbol{u}}_e^\star \hat{\boldsymbol{k}}_e \hat{\boldsymbol{u}}_e = \sum_{e=1}^{n} {}^t\hat{\boldsymbol{u}}_e^\star \hat{\boldsymbol{f}}_e \qquad [A.24]$$

$$\Leftrightarrow \boldsymbol{M}\ddot{\boldsymbol{U}} + \boldsymbol{K}\boldsymbol{U} = \boldsymbol{F}$$

The dynamic equation thus obtained appears to be a matrix of mass $\boldsymbol{M}$, a matrix of stiffness $\boldsymbol{K}$ and an external force vector $\boldsymbol{F}$ of the system.

$\boldsymbol{M}$ and $\boldsymbol{K}$ are symmetrical matrices and $\boldsymbol{M}^{-1}\boldsymbol{K}$ has N positive eigenvalues ω_i^2.

NOTE A.3.– Matrices and vectors written in their expanded form in the global degrees of freedom base are represented by:

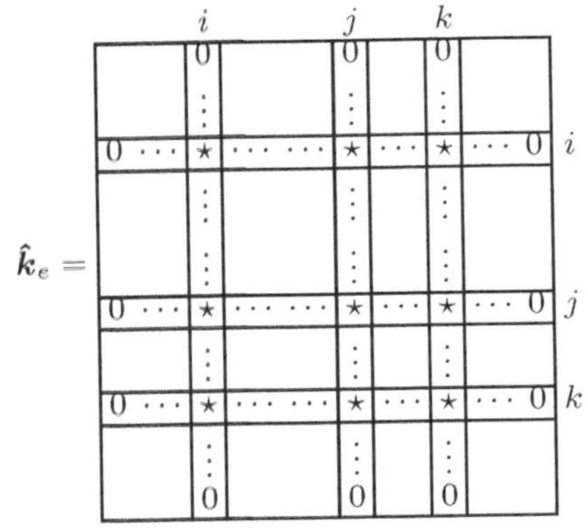

$$\Rightarrow \boldsymbol{k}_e = \begin{bmatrix} \star & \star & \star \\ \star & \star & \star \\ \star & \star & \star \end{bmatrix}$$

Same for the mass matrix $\hat{\boldsymbol{m}}_e$ the force vector $\hat{\boldsymbol{f}}_e$

The mass $\boldsymbol{M}$, stiffness $\boldsymbol{K}$ and the vector force matrices $\boldsymbol{F}$ are exactly the sum of contributions of each element, respectively, $\hat{\boldsymbol{m}}_e$, $\hat{\boldsymbol{k}}_e$ and $\hat{\boldsymbol{f}}_e$.

NOTE A.4.– Physical interpretation of PVP* (or weak form) application.

The application of the PVP* in the virtual degrees of freedom actually allows for the equilibrium of each node of the problem to be written. Each equation in the assembled system is the static equilibrium of a node.

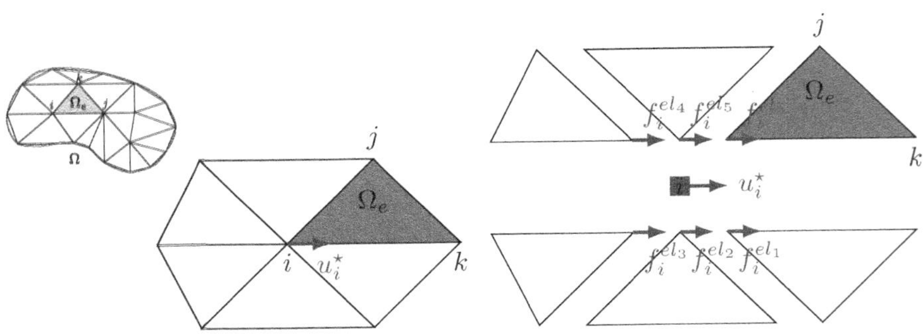

Figure A.3. *Representation of the equilibrium of a node in the direction of $u_i^\star$ using discretization from the finite element method. For a color version of this figure, see www.iste.co.uk/mazars/damage.zip*

The PVP* term in factor of $u_i^\star$, allows for the equilibrium of the node i according to $\vec{X}$ to be written:

$$... + u_i^\star \left(f_i^{el_1} + f_i^{el_2} + f_i^{el_3} + f_i^{el_4} + f_i^{el_5} + f_i^{el} \right) + ... = 0 \qquad \text{[A.25]}$$

NOTE A.5.– A simple PVP* application exercise is proposed in section 9.5.

A.2.3. *Solving the discretized equation*

A.2.3.1. *Linear elasticity static problem*

In the case of statics, the problem to be solved is:

$$\boldsymbol{K}\boldsymbol{U} = \boldsymbol{F}_{ext} \tag{A.26}$$

The resolution is done simply by inverting the matrix $\boldsymbol{K}$ by $\boldsymbol{U} = \boldsymbol{K}^{-1}\boldsymbol{F}_{ext}$.

A.2.3.2. *Nonlinear static problem*

A.2.3.2.1. Writing the equilibrium equation via PVP*

If the behavior of the material is nonlinear, then Hooke's law no longer applies. A nonlinear constitutive law of the type:

$$\boldsymbol{\sigma} = \boldsymbol{f}\left(\boldsymbol{\varepsilon}\right) = \boldsymbol{f}\left(\boldsymbol{u}_e\right) = \boldsymbol{f}\left(\hat{\boldsymbol{u}}_e\right) \tag{A.27}$$

The PVP* is written as:

$$\sum_{e=1}^{n} {}^t\hat{\boldsymbol{u}}_e^\star \underbrace{\int_{\Omega_e} {}^t\hat{\boldsymbol{B}}_e\boldsymbol{\sigma}d\Omega}_{\hat{\boldsymbol{p}}_e} = \sum_{e=1}^{n} {}^t\hat{\boldsymbol{u}}_e^\star \underbrace{\left[\int_{\Omega_e} {}^t\hat{\boldsymbol{N}}_e\boldsymbol{f}d\Omega + \int_{\partial\Omega_e} {}^t\hat{\boldsymbol{N}}_e\boldsymbol{t}dS\right]}_{\hat{\boldsymbol{f}}_e} \tag{A.28}$$

Then, finally:

$$\boldsymbol{P}\left(\boldsymbol{U}\right) = \boldsymbol{F} \tag{A.29}$$

This vector equation is nonlinear. A Newton–Raphson or loading factor increment (arc length) method is then used to solve it.

For this, the residue is then defined as:

$$\boldsymbol{R} = -\boldsymbol{P}\left(\boldsymbol{U}\right) + \boldsymbol{F} \tag{A.30}$$

The purpose of these resolution methods is to find $\boldsymbol{U}$ such that $\boldsymbol{R} = \boldsymbol{0}$.

A.2.3.2.2. Newton–Raphson method

To simply explain the approach, let us take a 1D problem.

Let the nonlinear equation $f(x) = 0$ be solved.

The principle is to linearize this equation to obtain the equation of the tangent T_f to the curve C_f. The following equation is then resolved: T_f^k: $y(x^k) = f'(x^k)\Delta x^k + f(x^k) = 0$.

In practice, from an initial condition x^0, the value of the function $f(x^0)$ is calculated and then the equation of the tangent. We deduce Δx^0 and therefore a new value $x^1 = x^0 + \Delta x^0$. This procedure is repeated successively at the point x^k (current), then by updating by $x^{k+1} = x^k + \Delta x^k$. We then gradually converge toward the solution (see Figure A.4) (when $|f(x^{k+1})| \leq \varepsilon^r$, where ε^r is the convergence criterion).

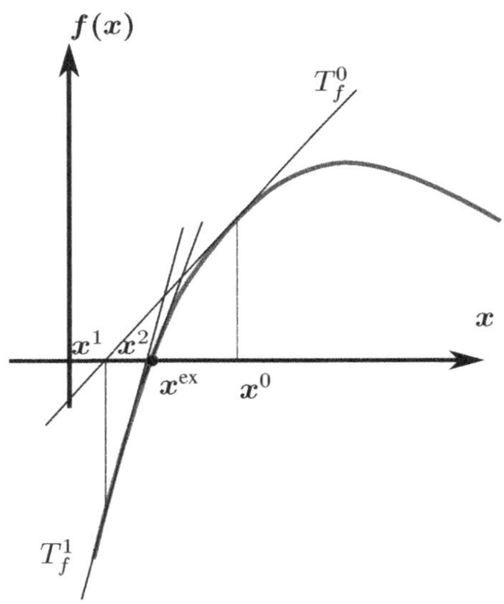

Figure A.4. *Principle of the Newton method; starting from the initial condition x^0, we converge toward the exact solution x^{ex}. For a color version of this figure, see www.iste.co.uk/mazars/damage.zip*

In mechanics, the function f is the residue of equation [A.30]:

$$\boldsymbol{R}(\boldsymbol{U}) = 0 \qquad \text{[A.31]}$$

The algorithm below generalizes the Newton–Raphson method to multidimensional cases.

1) Initialization on the loading step: $i = 0$, extraction of $\boldsymbol{F}_{i+1}$.

2) Initialization on iteration: $k = 1$.

3) Calculation of kinematics: $\boldsymbol{U}_{i+1}^{(k)} = \boldsymbol{U}_i$ and of the residual: $\boldsymbol{R}_{i+1}^{(k)}$.

4) Such that: $\|\boldsymbol{R}_{i+1}^{(k)}\| > \varepsilon^r$.

calculation of:

$$\left(\frac{d\boldsymbol{R}}{d\boldsymbol{U}}\right)_{i+1}^{(k)} = -\boldsymbol{K}^t \text{ (stiffness matrix)}$$

$$\delta \boldsymbol{U}_{i+1}^{(k)} = \frac{-\boldsymbol{R}_{i+1}^{(k)}}{\left(\frac{d\boldsymbol{R}}{d\boldsymbol{U}}\right)_{i+1}^{(k)}}$$

$$\boldsymbol{U}_{i+1}^{(k+1)} = \boldsymbol{U}_{i+1}^{(k)} + \delta \boldsymbol{U}_{i+1}^{(k)}$$

$k = k + 1$

calculation of: $\boldsymbol{R}_{i+1}^{(k)}$ and allocation $\boldsymbol{U}_{i+1} = \boldsymbol{U}_{i+1}^{(k)}$

end.

5) $i = i + 1$, back to (3).

6) End.

A.2.3.3. *Solving a dynamic problem using a Newmark time scheme*

In the case of a linear system, the dynamic equation can be written as follows:

$$\boldsymbol{M}\ddot{\boldsymbol{U}}(t) + \boldsymbol{C}\dot{\boldsymbol{U}}(t) + \boldsymbol{K}\boldsymbol{U}(t) = \boldsymbol{F}(\boldsymbol{t}) \qquad [A.32]$$

In this differential equation, the unknown is the displacement $\boldsymbol{U}$ and all its derivatives, velocities and accelerations $\dot{\boldsymbol{U}}$ and $\ddot{\boldsymbol{U}}$.

The Newmark method consists of discretizing the calculation of the derivatives over time. Thus, the second and first derivatives of the displacement are related to the displacement calculated at the time steps i and $i + 1$.

To do this, the method uses a set of displacements, velocities and accelerations $(\boldsymbol{U}_i = \boldsymbol{U}(t_i), \dot{\boldsymbol{U}}_i = \dot{\boldsymbol{U}}(t_i), \ddot{\boldsymbol{U}}_i = \ddot{\boldsymbol{U}}(t_i))$ at different time steps.

In addition, we note that $\Delta t = t_{i+1} - t_i$.

The Newmark formulas that make it possible to approximate the calculation of derivatives are as follows:

$$\begin{cases} \boldsymbol{U}_{i+1} = \boldsymbol{U}_i + \Delta t \dot{\boldsymbol{U}}_i + \Delta t^2 \left[\left(\frac{1}{2} - \beta\right) \ddot{\boldsymbol{U}}_i + \beta \ddot{\boldsymbol{U}}_{i+1} \right] \\ \dot{\boldsymbol{U}}_{i+1} = \dot{\boldsymbol{U}}_i + \Delta t \left[(1 - \gamma) \ddot{\boldsymbol{U}}_i + \gamma \ddot{\boldsymbol{U}}_{i+1} \right] \end{cases} \quad \text{[A.33]}$$

Depending on the values of γ and β, the problem to be solved is either explicit or implicit.

The inversion of this system allows $\ddot{\boldsymbol{U}}_{i+1}$ and $\dot{\boldsymbol{U}}_{i+1}$ to be expressed according to the converged terms $\boldsymbol{U}_i$, $\dot{\boldsymbol{U}}_i$, $\ddot{\boldsymbol{U}}_i$ and the unknown $\boldsymbol{U}_{i+1}$.

Finally, the dynamic equation at step $i+1$ is written as:

$$\underbrace{\left(\frac{1}{\beta \Delta t^2} \boldsymbol{M} + \frac{\gamma}{\beta \Delta t} \boldsymbol{C} + \boldsymbol{K} \right)}_{\tilde{\boldsymbol{K}}} \boldsymbol{U}_{i+1} = \quad \text{[A.34]}$$

$$\underbrace{\boldsymbol{F}_{i+1} + \left[\frac{\boldsymbol{C}\gamma}{\beta \Delta t} + \frac{\boldsymbol{M}}{\beta \Delta t^2} \right] \boldsymbol{U}_i + \left[\boldsymbol{C} \left(\frac{\gamma}{\beta} - 1 \right) + \frac{\boldsymbol{M}}{\beta \Delta t} \right] \dot{\boldsymbol{U}}_i + \left[\boldsymbol{C} \Delta t \left(\frac{\gamma}{\beta} (\frac{1}{2} - \beta) - (1 - \gamma) \right) + \frac{\boldsymbol{M}(\frac{1}{2} - \beta)}{\beta} \right] \ddot{\boldsymbol{U}}_i}_{\boldsymbol{f}_{i+1}}$$

The problem is solved by inverting the system $\tilde{\boldsymbol{K}} \boldsymbol{U}_{i+1} = \boldsymbol{f}_{i+1}$, and in particular by calculating $\tilde{\boldsymbol{K}}^{-1}$.

If the above equation has nonlinear terms (due to material or geometric nonlinearity), then a Newton–Raphson method is applied to each time step $i+1$.

A.2.4. *Modal analysis of a structure*

The modal analysis of a structure can be interesting when determining its dynamic signature (the frequencies at which it naturally vibrates associated with their modal shapes). This information is of interest in the context of dimensioning under dynamic loading of the structure.

A.2.4.1. *Calculation of eigenmodes*

The vibration modes of a structure are standing waves. So, we can decouple time and space as follows:

$$\boldsymbol{U}(x,t) = \sum_i u_i(t) \boldsymbol{\Phi}_i(x) \quad \text{[A.35]}$$

The contribution of each mode can therefore be written as (for mode i):

$$\boldsymbol{U}_i(x,t) = u_i(t)\boldsymbol{\Phi}_i(x) \qquad \text{[A.36]}$$

We will first consider this calculation on a single mode.

We will solve the equation of the dynamics in free oscillations and without damping. The response in un-damped free oscillations oscillates harmonically over time. So, we can write that:

$$u_i(t) = A\cos\omega_i t$$

This form of solution is then replaced in the equation of the dynamics in free oscillations.

We get:

$$\left(-\omega_i^2\boldsymbol{M} + \boldsymbol{K}\right)\boldsymbol{\Phi_i}(x) = \boldsymbol{0} \qquad \text{[A.37]}$$

This equation accepts a non-zero solution $\Phi_i(x)$ only if the determinant of $\left(-\omega_i^2\boldsymbol{M} + \boldsymbol{K}\right)$ is equal to 0.

Calculating the determinant of this matrix is not practical.

Let us rewrite the previous system such that:

$$\left(-\omega_i^2\boldsymbol{Id} + \boldsymbol{M}^{-1}\boldsymbol{K}\right)\boldsymbol{\Phi_i}(x) = \boldsymbol{0} \qquad \text{[A.38]}$$

(where $\boldsymbol{Id}$ is the identity matrix $n_{dim} \times n_{dim}$).

There is a problem with eigenvalues. The eigenmode squares ω_i^2 are the eigenvalues of the matrix $\boldsymbol{M}^{-1}\boldsymbol{K}$.

The modal deformations $\boldsymbol{\Phi_i}(x)$ are the eigenvectors of $\boldsymbol{M}^{-1}\boldsymbol{K}$

NOTE A.6.– The pulsation of the mode i has the following property:

$$-\omega_i^2\boldsymbol{M}\boldsymbol{\Phi}_i + \boldsymbol{K}\boldsymbol{\Phi}_i = 0$$

as $\boldsymbol{K}\boldsymbol{\Phi}_i = \omega_i^2\boldsymbol{M}\boldsymbol{\Phi}_i$ and so:

$$\omega_i^2 = \frac{{}^t\boldsymbol{\Phi}_i\boldsymbol{K}\boldsymbol{\Phi}_i}{{}^t\boldsymbol{\Phi}_i\boldsymbol{M}\boldsymbol{\Phi}_i} \qquad \text{[A.39]}$$

NOTE A.7.– The number of modes N is in practice the degree of freedom with a mass. If $\boldsymbol{M}$ is not invertible (if there are no masses associated with some dof, the reverse problem is solved. We then seek the eigenvalues of $\boldsymbol{K}^{-1}\boldsymbol{M}$, which are $\lambda_i = \frac{1}{\omega_i^2}$. N is then the number of $\lambda_i \neq 0$.

A.2.4.2. *Projection of equations into the modal base in forced regime by seismic action*

For each mode, let us replace $\boldsymbol{U}_i\,(x,t) = u_i(t)\boldsymbol{\Phi}_i(x)$ in the dynamic equation [A.26], we get N decoupled equations:

$$\ddot{u}(t) + \omega_i^2 u(t) = \frac{{}^t\boldsymbol{\Phi}_i \boldsymbol{P}(t)}{{}^t\boldsymbol{\Phi}_i \boldsymbol{M}\boldsymbol{\Phi}_i} = -\frac{{}^t\boldsymbol{\Phi}_i \boldsymbol{M}\boldsymbol{\Delta}}{{}^t\boldsymbol{\Phi}_i \boldsymbol{M}\boldsymbol{\Phi}_i} a_{\text{earthquake}}(t) \qquad [A.40]$$

where $\boldsymbol{\Delta}$ is the projection vector of the seismic acceleration $a_e(t)$ in the direction imposed in the coordinate system of the structure and $\omega_i^2 = \frac{{}^t\boldsymbol{\Phi}_i \boldsymbol{K}\boldsymbol{\Phi}_i}{{}^t\boldsymbol{\Phi}_i \boldsymbol{M}\boldsymbol{\Phi}_i}$, the system's fundamental pulsation for the mode i.

Let us note $Y_i(t)$ the solution of equation [A.41]:

$$\ddot{u}(t) + \omega_i^2 u(t) = -a_e(t)\text{earthquake}(t) \qquad [A.41]$$

NOTE A.8.– Property of orthogonality of modes: if $i \neq j$ then the pulsations $\omega_i^2 \neq \omega_j^2$ and therefore necessarily: ${}^t\boldsymbol{\Phi}_i \boldsymbol{M}\boldsymbol{\Phi}_j = 0$. This also implies that ${}^t\boldsymbol{\Phi}_i \boldsymbol{K}\boldsymbol{\Phi}_j = 0$. It is then said that the modes are orthogonal.

So, we get:

$$\boldsymbol{U}_i\,(x,t) = Y_i(t)\frac{{}^t\boldsymbol{\Phi}_i \boldsymbol{M}\boldsymbol{\Delta}}{{}^t\boldsymbol{\Phi}_i \boldsymbol{M}\boldsymbol{\Phi}_i}\boldsymbol{\Phi}_i \qquad [A.42]$$

The designing maximum displacement (in elastic) is then:

$$\boldsymbol{U}_i^{max} = S_d\frac{{}^t\boldsymbol{\Phi}_i \boldsymbol{M}\boldsymbol{\Delta}}{{}^t\boldsymbol{\Phi}_i \boldsymbol{M}\boldsymbol{\Phi}_i}\boldsymbol{\Phi}_i \qquad [A.43]$$

In this last expression, S_d is the spectral displacement. It represents the maximum displacement of a structure at 1 dof for a given natural frequency ω_i of the structure. This magnitude is generally given by seismic regulation across the elastic response spectrum.

A.2.4.3. *The different modal parameters*

We can then set the following variables (in the mode i):

– the equivalent static force on the mode i: $\boldsymbol{F}_i(t) = \boldsymbol{K}\boldsymbol{U}_i = \omega_i^2 Y_i(t) \frac{{}^t\boldsymbol{\Phi}_i \boldsymbol{M}\boldsymbol{\Delta}}{{}^t\boldsymbol{\Phi}_i \boldsymbol{M}\boldsymbol{\Phi}_i} \boldsymbol{M}\boldsymbol{\Phi}_i$ (indeed $\boldsymbol{K}\boldsymbol{\Phi}_i = \omega_i^2 \boldsymbol{M}\boldsymbol{\Phi}_i$);

– total shear force T_{o_i} (basically): $T_{o_i}(t) = {}^t\boldsymbol{\Delta}\boldsymbol{F}_i = \omega_i^2 Y_i(t) \frac{\left({}^t\boldsymbol{\Phi}_i \boldsymbol{M}\boldsymbol{\Delta}\right)^2}{{}^t\boldsymbol{\Phi}_i \boldsymbol{M}\boldsymbol{\Phi}_i}$;

– the maximum designing shear force: $T_i^{max} = S_{pa} \frac{\left({}^t\boldsymbol{\Phi}_i \boldsymbol{M}\boldsymbol{\Delta}\right)^2}{{}^t\boldsymbol{\Phi}_i \boldsymbol{M}\boldsymbol{\Phi}_i}$.

In the expressions in the previous list, we can also define some remarkable terms:

– the participation factor of mode i: $\Gamma_i = \frac{{}^t\boldsymbol{\Phi}_i \boldsymbol{M}\boldsymbol{\Delta}}{{}^t\boldsymbol{\Phi}_i \boldsymbol{M}\boldsymbol{\Phi}_i}$;

– the effective modal mass $\tilde{M}_i$ of the mode i: $\tilde{M}_i = \frac{\left({}^t\boldsymbol{\Phi}_i \boldsymbol{M}\boldsymbol{\Delta}\right)^2}{{}^t\boldsymbol{\Phi}_i \boldsymbol{M}\boldsymbol{\Phi}_i}$ (the sum of all actual modal masses relative to direction $\boldsymbol{\Delta}$ is equal to the total mass in that direction).

The participation coefficient is calculated as follows:

– the actual modal mass $\tilde{M}_i$ of mode i^e is determined by the relationship:

$$\tilde{M}_i = \frac{\left({}^t\Phi_i \boldsymbol{M}\boldsymbol{\Delta}\right)^2}{{}^t\Phi_i \boldsymbol{M}\Phi_i} \qquad \text{[A.44]}$$

– this value is then divided by the total mass to obtain the participation coefficient ν_i, i.e.:

$$\nu_i = \frac{\tilde{M}_i}{{}^t\boldsymbol{\Delta}\boldsymbol{M}\boldsymbol{\Delta}} \qquad \text{[A.45]}$$

A.3. Elaboration of the FEM for a continuous structure

A.3.1. *Case of 1D problems*

In order to illustrate the finite element calculation, it is proposed to solve a simple continuous structure: a continuous bar under tension.

Indeed, this example is unidirectional and therefore since its kinematics is simple, it makes it possible to manipulate the variational equations, as well as the discretization in an easy manner.

The bar is subjected to a concentrated force. The purpose of this problem is to calculate the displacements at any point on the bar.

Figure A.5. *Example of a unidirectional bar under tension*

A.3.1.1. *Problem discretization*

The FEM consists first of all of making an approximation of the actual geometry of the continuous bar by subdividing it into a finite number of nodes (which will carry the unknowns of the problem) and of elements (which connect these nodes).

In Figure A.6, three additional intermediate nodes are therefore considered in order to subdivide this bar of length L into four elements of length $l_e = \frac{L}{4}$.

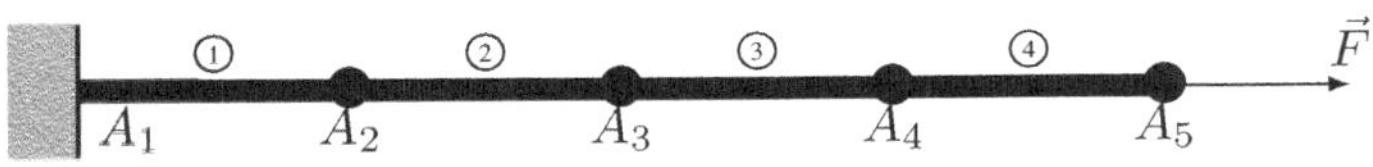

Figure A.6. *Discretization of a unidirectional bar under tension into four elements and five nodes*

The structure is subdivided into elements to form a mesh. The particular points at the intersections of these elements are called the nodes. Each node carries information which, as part of the mechanics of the solid, is its own displacement. The displacement $\boldsymbol{U}(x)$ solution of the discrete problem is entirely determined by the displacements at nodes.

A.3.1.2. *Construction of the elementary stiffness matrix and the external force vector*

A.3.1.2.1. Interpolation of the motion field

Writing the equilibrium of the system by the principle of virtual powers will involve calculating strains and stress for any x. Thus, the idea of the FEM is to pass from the continuous problem to a discretized problem whose displacements are interpolated only on those of the discretization nodes.

Thus, on each element defined between nodes i and j, the following interpolation of the displacement field is imposed:

$$u(x) = N_1(x)u_i + N_2(x)u_j \text{ with } \boldsymbol{u}_{el} = \begin{bmatrix} u_i \\ u_j \end{bmatrix} \qquad [A.46]$$

u_i and u_j are the nodal displacements of element el. They are collected in a vector of elementary displacements denoted as: $\boldsymbol{u}_{el}$

The functions $N_1(x)$ and $N_2(x)$ are generally polynomials and are called the interpolation functions. They make it possible to interpolate displacement on values at nodes and thus verify compatibility conditions of the type:

$$\begin{cases} N_1(0) = 1 \\ N_2(0) = 0 \\ N_1(l_e) = 0 \\ N_2(l_e) = 1 \end{cases} \text{which leads to: } \begin{cases} N_1(x) = 1 - \frac{x}{l_e} \\ N_2(x) = \frac{x}{l_e} \end{cases} \qquad [A.47]$$

Thus, in this simple example, each of the functions that have two conditions each to be complied with can be chosen as polynomials of degree 1. The discretized field $u(x)$ is then written as:

$$u(x) = \left(1 - \frac{x}{l_e}\right) u_i + \frac{x}{l_e} u_j \text{ or even } u(x) = \underbrace{\begin{bmatrix} 1 - \frac{x}{l_e} & \frac{x}{l_e} \end{bmatrix}}_{\boldsymbol{N}_{el}} \underbrace{\begin{bmatrix} u_i \\ u_j \end{bmatrix}}_{\boldsymbol{u}_{el}} \qquad [A.48]$$

This displacement field, which is continuous, now depends on only two unknowns u_i and u_j. These nodal displacements are therefore the only unknowns of the problem.

The matrix $\boldsymbol{N}_{el}$ is the matrix of shape functions and the vector $\boldsymbol{u}_{el}$ is the vector of nodal displacements.

A.3.1.2.2. Writing system equilibrium

Writing the system equilibrium by applying the principle of virtual powers therefore gives:

$$\forall u^\star(x),\ \mathcal{P}^\star_{int} = \mathcal{P}^\star_{ext} \qquad [A.49]$$

A.3.1.2.3. Writing the external power term (external force vector)

The power of external forces can be written simply as:

$$\mathcal{P}^\star_{ext} = F u^\star_5 + R u^\star_1 \text{ (where } R \text{ is the unknown support reaction)} \qquad [A.50]$$

NOTE A.9.– A simple way to not show support reactions (edge of the structure where the displacement is imposed) is to choose a virtual field that is kinematically admissible at 0. This means that where the real kinematics are imposed is chosen to be zero. This simply means that $u^\star_1 = 0$ and the calculation of the unknown kinematics from the structure of the support reactions (which can always be deduced later) is decoupled.

A.3.1.2.4. Writing the inner power term (stiffness matrix)

The power of inner forces can be written as:

$$\mathcal{P}^{\star}_{int} = \int_0^L \frac{du^{\star}(x)}{dx} N(x)dx \text{ or even: } \mathcal{P}^{\star}_{int} = \int_0^L \frac{du^{\star}(x)}{dx} ES \frac{du(x)}{dx} dx \qquad \text{[A.51]}$$

A key step is to calculate strains contained in this term of power of internal forces. This calculation is in practice not complicated since it is a question of deriving polynomial interpolation functions.

So, we can write:

$$\frac{du(x)}{dx} = \underbrace{\begin{bmatrix} -\frac{1}{l_e} & \frac{1}{l_e} \end{bmatrix}}_{\boldsymbol{B}_{el}} \underbrace{\begin{bmatrix} u_i \\ u_j \end{bmatrix}}_{\boldsymbol{u}_{el}} \qquad \text{[A.52]}$$

The same relationship can therefore be written for the virtual field and thus: $\frac{du^{\star}(x)}{dx} = \boldsymbol{B}_{el}\boldsymbol{u}^{\star}_{el}$.

These expressions are valid for each element. An additional step is therefore to breakdown the calculation of the integral on each element:

$$\mathcal{P}^{\star}_{int} = \int_0^L \frac{du^{\star}(x)}{dx} ES \frac{du(x)}{dx} dx = \sum_e \underbrace{\int_{l_e} \frac{du^{\star}(x)}{dx} ES \frac{du(x)}{dx} dx}_{\mathcal{P}^{el\star}_{int}} \qquad \text{[A.53]}$$

Thus, the calculation of power terms is done on each element as follows:

$$\mathcal{P}^{el\star}_{int} = {}^t\boldsymbol{u}^{\star}_{el} \underbrace{\int_0^L {}^t\boldsymbol{B}_{el} ES \boldsymbol{B}_{el} dx}_{\boldsymbol{k}_{el}} \, \boldsymbol{u}_{el} \qquad \text{[A.54]}$$

This last relationship is therefore an expression where the number of unknowns is now finite (these are the nodal displacements $\boldsymbol{u}_{el}$). Moreover, they are independent of the integral on the domain and can be removed from it.

The term $\boldsymbol{k}_{el}$ is homogeneous at a stiffness and provides the forces that should be imposed on the element when a displacement set $\boldsymbol{u}_{el}$ is applied to it. This is the stiffness matrix of the element.

An additional step will consist of the sum of the contributions of all the elements. By summing each elementary power term, we realize that we can always factor in the real nodal displacements (on the right) and the virtual displacements (on the left). This step can be seen as a step to assemble a stiffness matrix (written in the base of all the dof) from elementary matrices.

In practice, for the two-node bar element, the calculation of the elementary stiffness matrix is as follows:

$$\mathcal{P}_{int}^{el\star} = \begin{bmatrix} u_i^\star & u_j^\star \end{bmatrix} \underbrace{\frac{ES}{l_e} \begin{bmatrix} 1 & -1 \\ -1 & 1 \end{bmatrix}}_{\boldsymbol{k}_{el}} \begin{bmatrix} u_i \\ u_j \end{bmatrix} \qquad \text{[A.55]}$$

The principle of virtual powers applied over the whole structure makes it possible to provide the assembly of contributions. The factorization of this latter equation can be seen as the assembly of the four elementary stiffness matrices.

In the end, the system to be solved becomes (where $l_e = \frac{L}{4}$ for each element):

$$\underbrace{\frac{4ES}{L} \begin{bmatrix} 1 & -1 & 0 & 0 & 0 \\ -1 & 2 & -1 & 0 & 0 \\ 0 & -1 & 2 & -1 & 0 \\ 0 & 0 & -1 & 2 & -1 \\ 0 & 0 & 0 & -1 & 1 \end{bmatrix}}_{\boldsymbol{K}_g} \underbrace{\begin{bmatrix} u_1 \\ u_2 \\ u_3 \\ u_4 \\ u_5 \end{bmatrix}}_{\boldsymbol{U}_g} = \underbrace{\begin{bmatrix} R \\ 0 \\ 0 \\ 0 \\ F \end{bmatrix}}_{\boldsymbol{F}_g} \qquad \text{[A.56]}$$

We must be careful because the matrix $\boldsymbol{K}_g$ is not invertible at this stage. Indeed, in this problem we have not taken into account the conditions at the limits $u_1 = 0$ (which leads to the calculation of the unknown support reaction R in the right-hand term of the equation).

To properly calculate unknown displacements (from u_2 to u_5, it is therefore necessary to impose $u_1 = 0$, which amounts to crossing out the first column of the system. The first line of the system can be deduced from the other four to calculate the support reaction.

Thus, the system is resolved as follows:

– first, we solve:

$$\underbrace{\frac{ES}{l_e} \begin{bmatrix} 2 & -1 & 0 & 0 \\ -1 & 2 & -1 & 0 \\ 0 & -1 & 2 & -1 \\ 0 & 0 & -1 & 1 \end{bmatrix}}_{\boldsymbol{K}} \underbrace{\begin{bmatrix} u_2 \\ u_3 \\ u_4 \\ u_5 \end{bmatrix}}_{\boldsymbol{U}} = \underbrace{\begin{bmatrix} 0 \\ 0 \\ 0 \\ F \end{bmatrix}}_{\boldsymbol{F}} \qquad \text{[A.57]}$$

$\boldsymbol{K}$ is now invertible and free of stiff body movement. Thus, $\boldsymbol{U} = \boldsymbol{K}^{-1}\boldsymbol{F}$;

– which gives here:

$$u_2 = \frac{FL}{4ES}, u_3 = \frac{FL}{2ES}, u_4 = \frac{3FL}{4ES}, u_5 = \frac{FL}{ES} \qquad [A.58]$$

– then, to find the support reaction, we solve: $\frac{ES}{l_e}(u_1 - u_2) = R$.

A.3.2. *Planar element case*

In this section, we show how a triangular element with three nodes in the plane is formulated. The same type of demonstration can be done in 3D or considering elements with more nodes.

A.3.2.1. *Actual element and notion of reference element*

As seen in the example of the bar in section A.3.1, the FEM consists of first dividing a structure into elements carried by nodes and interpolating the movements on these nodes.

For 2D or 3D problems, interpolation functions are determined in the same way as in the 1D case, but the mathematical form of these interpolation functions as a function of (x, y, z) can prove difficult.

In order to simplify the writing of these interpolation functions, an intermediate step is carried out, which makes it possible to write these functions in a simpler manner. We therefore define a reference element, which is regular and defined in a space (ξ, η). Writing interpolation functions in this space leads to functions that are simple to manipulate.

In order to obtain these interpolation functions in the space of the actual element (x, y), it is therefore necessary to pass through a second step, which is a change of variable to pass from the triangle defined in (ξ, η) to (x, y). The image of the three nodes (m_1, m_2, m_3) is therefore the three nodes of the actual element.

The change of variable by the transformation τ can be expressed simply by defining functions $\bar{N}_a(\xi, \eta)$ such that $\bar{N}_a(\xi_b, \eta_b) = \delta_{ab}$:

$$\tau : \begin{bmatrix} \xi \\ \eta \end{bmatrix} \rightarrow \begin{bmatrix} x \\ y \end{bmatrix} = \sum_{a=1}^{3} \bar{N}_a(\xi, \eta) \begin{bmatrix} x_a \\ y_a \end{bmatrix} \qquad [A.59]$$

Thus, in the case of this triangular element with three nodes, to obtain these functions $\bar{N}_a$, it is necessary to respect $M_1 = \tau(m_1)$, $M_2 = \tau(m_2)$ and $M_3 = \tau(m_3)$. In other words:

$$\begin{cases} \bar{N}_1(0,0) = 1 \\ \bar{N}_1(0,1) = 0 \\ \bar{N}_1(1,0) = 0 \end{cases}, \quad \begin{cases} \bar{N}_2(0,0) = 0 \\ \bar{N}_2(0,1) = 1 \\ \bar{N}_2(1,0) = 0 \end{cases}, \quad \begin{cases} \bar{N}_3(0,0) = 0 \\ \bar{N}_3(0,1) = 0 \\ \bar{N}_3(1,0) = 1 \end{cases} \quad \text{[A.60]}$$

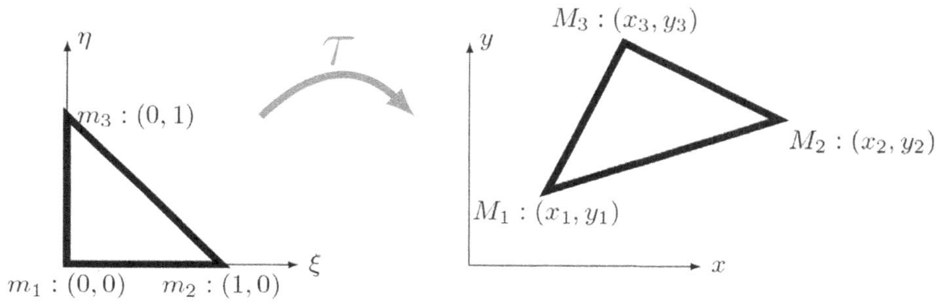

Figure A.7. *Representation of a parent element (regular) and the actual element and geometrical transformation function (variable change relationship)*

The following functions are suitable for respecting the images of nodes of the triangle between the reference element and the actual element:

$$\begin{cases} \bar{N}_1(\xi,\eta) = 1 - \xi - \eta \\ \bar{N}_2(\xi,\eta) = \xi \\ \bar{N}_3(\xi,\eta) = \eta \end{cases} \quad \text{[A.61]}$$

To conclude, the expression [A.66] is therefore written as:

$$\begin{cases} x = (1 - \xi - \eta)x_1 + \xi x_2 + \eta x_3 \\ y = (1 - \xi - \eta)y_1 + \xi y_2 + \eta y_3 \end{cases} \quad \text{[A.62]}$$

This expression is a variable change equation and has a Jacobian matrix:

$$\boldsymbol{J} = \begin{bmatrix} \dfrac{\partial x}{\partial \xi} & \dfrac{\partial x}{\partial \eta} \\ \dfrac{\partial y}{\partial \xi} & \dfrac{\partial y}{\partial \eta} \end{bmatrix} = \begin{bmatrix} x_2 - x_1 & x_3 - x_1 \\ y_2 - y_1 & y_3 - y_1 \end{bmatrix} \quad \text{[A.63]}$$

The inverse of the Jacobian matrix is given by:

$$J^{-1} = \frac{1}{det J}\begin{bmatrix} y_3 - y_1 & x_1 - x_3 \\ y_1 - y_2 & x_2 - x_1 \end{bmatrix} \quad [A.64]$$

with $det J = (x_2 - x_1)(y_3 - y_1) - (y_2 - y_1)(x_2 - x_1)$.

NOTE A.10.– The inverse of this Jacobian matrix contains by definition the terms:

$$J^{-1} = \begin{bmatrix} \dfrac{\partial \xi}{\partial x} & \dfrac{\partial \xi}{\partial y} \\ \dfrac{\partial \eta}{\partial x} & \dfrac{\partial \eta}{\partial y} \end{bmatrix} = \begin{bmatrix} \xi_{,x} & \xi_{,y} \\ \eta_{,x} & \eta_{,y} \end{bmatrix} \quad [A.65]$$

In practice, we therefore simply use the identification between equations [A.64] and [A.65] to calculate the derivatives of (ξ, η) as a function of (x, y).

These terms will later be useful for calculating various derivatives and integrals involved in the calculation of strains and the equilibrium of the element via the PVP*.

A.3.2.2. *Interpolation of displacements on the reference element*

When the relationship between real coordinates (x, y) and reference element coordinates (ξ, η) is known, it can be much simpler and interesting to define the interpolation functions of the displacements on the nodes of the parent element.

The conditions to be met for displacement are as follows:

$$U(\xi, \eta) = \begin{bmatrix} u \\ v \end{bmatrix} = \sum_{a=1}^{3} N_a(\xi, \eta) \begin{bmatrix} u_a \\ v_a \end{bmatrix} \quad [A.66]$$

Thus, in the case of this triangular element with three nodes, to obtain these functions N_a, it is necessary to respect:

$$\begin{cases} N_1(0,0) = 1 \\ N_1(0,1) = 0 \\ N_1(1,0) = 0 \end{cases}, \quad \begin{cases} N_2(0,0) = 0 \\ N_2(0,1) = 1 \\ N_2(1,0) = 0 \end{cases}, \quad \begin{cases} N_3(0,0) = 0 \\ N_3(0,1) = 0 \\ N_3(1,0) = 1 \end{cases} \quad [A.67]$$

The following functions meet these conditions:

$$\begin{cases} N_1 = 1 - \xi - \eta \\ N_2 = \xi \\ N_3 = \eta \end{cases} \quad [A.68]$$

NOTE A.11.– In this type of element, the N_a functions are the same as the geometric transformation functions $\bar{N}_a$ since they meet the same conditions. It is then said that the element is iso-parametric.

We can then define an interpolation matrix $\boldsymbol{N_{el}}$ such that $\boldsymbol{U} = \boldsymbol{N_{el}}\boldsymbol{u_{el}}$:

$$\boldsymbol{U} = \begin{bmatrix} u \\ v \end{bmatrix} = \underbrace{\begin{bmatrix} N_1 & 0 & N_2 & 0 & N_3 & 0 \\ 0 & N_1 & 0 & N_2 & 0 & N_3 \end{bmatrix}}_{\boldsymbol{N_{el}}} \begin{bmatrix} u_1 \\ v_1 \\ u_2 \\ v_2 \\ u_3 \\ v_3 \end{bmatrix} \quad \text{[A.69]}$$

The matrix $\boldsymbol{B_{el}}$ used to calculate strains at any point of the element can be defined as:

$$\boldsymbol{B_{el}} \text{ such that } \boldsymbol{\varepsilon} = \boldsymbol{B_{el}}\boldsymbol{u_{el}} = \begin{bmatrix} \varepsilon_{xx} \\ \varepsilon_{xy} \\ \sqrt{2}\varepsilon_{xy} \end{bmatrix} = \begin{bmatrix} \dfrac{\partial u}{\partial x} \\ \dfrac{\partial v}{\partial y} \\ \dfrac{\sqrt{2}}{2}\left(\dfrac{\partial u}{\partial y} + \dfrac{\partial v}{\partial x}\right) \end{bmatrix} \quad \text{[A.70]}$$

The calculation of the derivatives of functions is done with respect to the actual coordinates (x, y) taking care to take into account compound derivatives with respect to ξ and η. Thus:

$$\boldsymbol{\varepsilon} = \underbrace{\begin{bmatrix} \dfrac{\partial N_1}{\partial x} & 0 & \dfrac{\partial N_2}{\partial x} & 0 & \dfrac{\partial N_3}{\partial x} & 0 \\ 0 & \dfrac{\partial N_1}{\partial y} & 0 & \dfrac{\partial N_2}{\partial y} & 0 & \dfrac{\partial N_3}{\partial y} \\ \dfrac{\sqrt{2}}{2}\dfrac{\partial N_1}{\partial y} & \dfrac{\sqrt{2}}{2}\dfrac{\partial N_1}{\partial x} & \dfrac{\sqrt{2}}{2}\dfrac{\partial N_2}{\partial y} & \dfrac{\sqrt{2}}{2}\dfrac{\partial N_2}{\partial x} & \dfrac{\sqrt{2}}{2}\dfrac{\partial N_3}{\partial y} & \dfrac{\sqrt{2}}{2}\dfrac{\partial N_3}{\partial x} \end{bmatrix}}_{\boldsymbol{B_{el}}} \begin{bmatrix} u_i \\ v_i \\ u_j \\ v_j \\ u_k \\ v_k \end{bmatrix} \quad \text{[A.71]}$$

In calculating the above terms, it is important to take proper account of compound derivatives and therefore the terms of the Jacobian matrix calculated above:

$$\frac{\partial N_i}{\partial x^j} = \frac{\partial N_i}{\partial \xi^k}\frac{\partial \xi^k}{\partial x^j} \quad \text{[A.72]}$$

Thus, the matrix $\boldsymbol{B_e}$ can be expressed as follows:

$$\boldsymbol{B_{el}} = \frac{1}{\det(\boldsymbol{J})} \left[\begin{array}{ccc} y_2 - y_3 & 0 & y_3 - y_1 \\ 0 & x_3 - x_2 & 0 \\ \frac{\sqrt{2}}{2}(x_3 - x_2) & \frac{\sqrt{2}}{2}(y_2 - y_3) & \frac{\sqrt{2}}{2}(x_1 - x_3) \end{array} \right. \cdots$$
$$\cdots \left. \begin{array}{ccc} 0 & y_1 - y_2 & 0 \\ x_1 - x_3 & 0 & x_2 - x_1 \\ \frac{\sqrt{2}}{2}(y_3 - y_1) & \frac{\sqrt{2}}{2}(x_2 - x_1) & \frac{\sqrt{2}}{2}(y_1 - y_2) \end{array} \right] \quad \text{[A.73]}$$

NOTE A.12.– For the three-node triangle, the terms of the matrix $\boldsymbol{B_e}$ are independent of x and y since the interpolation functions are of degree 1. This is not the case for all elements. In general, we have a matrix $\boldsymbol{B_e}$ that depends on (x, y) (and therefore (ξ, η)).

Writing the equilibrium using the PVP*, and more particularly the power term of the internal forces, makes it possible to calculate the elementary stiffness matrix.

In linear elasticity, Hooke's law makes it possible connect the strain with the stress according to: $\boldsymbol{\sigma} = \boldsymbol{\Lambda}\boldsymbol{\varepsilon}$.

In 2D, in the case of planar stresses, this matrix $\boldsymbol{\Lambda}$ equals:

$$\boldsymbol{\Lambda} = \frac{E}{1 - \nu^2} \begin{bmatrix} 1 & \nu & 0 \\ \nu & 1 & 0 \\ 0 & 0 & 1 - \nu \end{bmatrix} \quad \text{[A.74]}$$

Thus, to simplify the calculation and since all the functions that appear in the matrix $\boldsymbol{B_e}$ are generally a function of (ξ, η), a variable change is made in the calculation of the integral:

$$\boldsymbol{k_{el}} = \int_{\Omega_e} {}^t\boldsymbol{B_{el}}\boldsymbol{\Lambda}\boldsymbol{B_{el}} d\Omega = \int_{\Omega_{parent}} {}^t\boldsymbol{B_{el}}\boldsymbol{\Lambda}\boldsymbol{B_{el}} |\det \boldsymbol{J}| d\Omega^{parent} \quad \text{[A.75]}$$

On the parent triangle, this integral is calculated as follows:

$$\boldsymbol{k_{el}} = \int_0^1 \left(\int_0^{1-\xi} {}^t\boldsymbol{B_{el}}\boldsymbol{\Lambda}\boldsymbol{B_{el}} |\det \boldsymbol{J}| d\eta \right) d\xi \quad \text{[A.76]}$$

$\boldsymbol{k_{el}}$ is a matrix 6×6 that connects the six displacements nodes with the six forces that would be required to move them by the quantity imposed.

NOTE A.13.– The formulation of a four node quadrangular element follows exactly the same steps by replacing the interpolation functions with:

$$\begin{cases} N_1(\xi,\eta) = \dfrac{1}{4}\,(1-\xi)\,(1-\eta) \\ N_2(\xi,\eta) = \dfrac{1}{4}\,(1+\xi)\,(1-\eta) \\ N_3(\xi,\eta) = \dfrac{1}{4}\,(1+\xi)\,(1+\eta) \\ N_4(\xi,\eta) = \dfrac{1}{4}\,(1-\xi)\,(1+\eta) \end{cases} \qquad [A.77]$$

and the matrix $\boldsymbol{N_{el}}$:

$$\boldsymbol{U} = \begin{bmatrix} u \\ v \end{bmatrix} = \underbrace{\begin{bmatrix} N_1 & 0 & N_2 & 0 & N_3 & 0 & N_4 & 0 \\ 0 & N_1 & 0 & N_2 & 0 & N_3 & 0 & N_4 \end{bmatrix}}_{\boldsymbol{N_{el}}} \begin{bmatrix} u_1 \\ v_1 \\ u_2 \\ v_2 \\ u_3 \\ v_3 \\ u_4 \\ v_4 \end{bmatrix} \qquad [A.78]$$

The matrix $\boldsymbol{B_{el}}$ will therefore be a matrix 3×8, which will ultimately give an elementary stiffness matrix of size 8×8.

A.3.3. *Beam elements case*

A.3.3.1. *Euler–Bernoulli elastic beams*

Here, we will show how the continuous problem of the Euler–Bernoulli beam defined in the plane and in its local coordinate system via the functions of shapes, and the construction of the stiffness matrix and the resulting force vector is discretized.

The nodal unknowns are as follows: ${}^t\boldsymbol{u_{el}} = [\boldsymbol{u}_i, \boldsymbol{u}_j]$, where ${}^t\boldsymbol{u}_i = [u_{1_i}, u_{2_i},\ u_{3_i}]$, the displacements of the node i expressed in the local coordinate system of the beam. For more readability, we will rename these dof as follows: ${}^t\boldsymbol{u}_i = [u_i, v_i, \theta_i]$, the axial, transversal and rotation displacements, respectively, of the node i.

A possible breakdown of the generalized displacements of all the points of the beam is given as:

$$\begin{cases} U_x(x) = u_i N_i(x) + u_j N_j(x) \\ U_y(x) = v_i H_1(x) + \theta_i H_2(x) + v_j H_3(x) + \theta_j H_4(x) \\ \theta_z(x) = v_i H_1'(x) + \theta_i H_2'(x) + v_j H_3'(x) + \theta_j H_4'(x) \end{cases} \qquad [A.79]$$

This interpolation of displacements makes it possible to verify the Euler–Bernoulli condition. The straight sections remain straight and perpendicular to the neutral fiber, which implies: $\theta_z = \frac{dU_y}{dx}$ and that the curvature $\kappa(x) = \frac{d\theta}{dx}$ is thus equal to $\kappa(x) = \frac{d^2U_y}{dx^2}$.

The conditions to be respected for these functions of interpolations on displacements at nodes make it possible to write as follows:

$$\begin{cases} N_1(x) = \left(1 - \frac{x}{l}\right) \\ N_2(x) = \frac{x}{l} \\ H_1(x) = \frac{1}{4}\left(2 - \frac{2x}{l}\right)^2\left(1 + \frac{2x}{l}\right) \\ H_2(x) = \frac{L}{8}\left(1 - (\frac{2x}{l} - 1)^2\right)\left(2 - \frac{2x}{l}\right) \\ H_3(x) = \frac{1}{4}\left(\frac{2x}{l}\right)^2\left(3 - \frac{2x}{l}\right) \\ H_4(x) = \frac{L}{8}\left((\frac{2x}{l} - 1)^2 - 1\right)\left(\frac{2x}{l} - 1\right) \end{cases} \quad \text{[A.80]}$$

This interpolation of displacements therefore makes it possible to write the generalized strain vector for the section in the case of Euler–Bernoulli kinematics:

$$\boldsymbol{e}(x) = \begin{bmatrix} \varepsilon(x) \\ \kappa(x) \end{bmatrix} = \begin{bmatrix} \frac{dU_x}{dx}(x) \\ \frac{d^2U_y}{dx^2}(x) \end{bmatrix}$$

$$\boldsymbol{e}(x) = \underbrace{\begin{bmatrix} N_1'(x) & 0 & 0 & N_2'(x) & 0 & 0 \\ 0 & H_1''(x) & H_2''(x) & 0 & H_3''(x) & H_4''(x) \end{bmatrix}}_{\boldsymbol{B}_{el}} \begin{bmatrix} u_i \\ v_i \\ \theta_i \\ u_j \\ v_j \\ \theta_j \end{bmatrix} \quad \text{[A.81]}$$

These generalized strain magnitudes at the sections can be connected by means of a constitutive law with generalized stress. These generalized stresses in beam sections are as follows: normal force $N(x)$, shear force $T(x)$ and bending moment $M(x)$).

To do this, it is necessary to calculate strains at any point of the section by the relationship $\varepsilon_{xx}(x, \mathbf{y}) = \varepsilon(c) - \mathbf{y}\kappa(x)$ (rigid body motion of the cross-section assumption). This then makes it possible to calculate stress at any point of the section using a Hooke's law $\sigma_{xx}(x, \mathbf{y}) = E\varepsilon_{xx}(x, \mathbf{y})$ (beam theory hypothesis: $\sigma_{yy} = \sigma_{zz} = 0$).

The integration of these stresses on the section makes it possible to return to the generalized stresses $\boldsymbol{s}(x)$ according to $N = \int_S \sigma_{xx}(x, \mathbf{y})dS$ and $M = \int_S -y\sigma_{xx}(x, \mathbf{y})dS$.

Thus, the constitutive law of the section in terms of generalized stresses and strains is deduced therefrom:

$$\underbrace{\begin{bmatrix} N(x) \\ M(x) \end{bmatrix}}_{\boldsymbol{s}(x)} = \underbrace{\begin{bmatrix} EA & 0 \\ 0 & EI \end{bmatrix}}_{\boldsymbol{\Lambda}} \underbrace{\begin{bmatrix} \varepsilon(x) \\ \kappa(x) \end{bmatrix}}_{\boldsymbol{e}(x)} \qquad [A.82]$$

According to the PVP* on each item, we can write:

$$\begin{aligned} & \forall \boldsymbol{V}^\star \text{ CA } 0, \int_\Gamma {}^t\boldsymbol{e}^\star.\boldsymbol{s}dx = \int_\Gamma {}^t\boldsymbol{V}^\star.\boldsymbol{f}dx + \sum_{x\in\partial\Omega} {}^t\boldsymbol{V}^\star\boldsymbol{t}(x) \\ & \Leftrightarrow \forall \boldsymbol{V}^\star \text{ CA } 0, \textstyle\sum_{el}\int_{\Gamma_{el}} {}^t\boldsymbol{e}^\star.\boldsymbol{s}dx = \sum_{el}\int_{\Gamma_{el}} {}^t\boldsymbol{V}^\star.\boldsymbol{f}dx + \displaystyle\sum_{x\in\partial\Omega} {}^t\boldsymbol{V}^\star\boldsymbol{t}(x) \\ & \Leftrightarrow \forall \boldsymbol{u}_{el}^\star \text{ CA } 0, \textstyle\sum_{el}\int_{\Gamma_{el}} {}^t\boldsymbol{u}_{el}^\star\,{}^t\boldsymbol{B}_{el}\boldsymbol{\Lambda}\boldsymbol{B}_{el}\boldsymbol{u}_{el}dx = \sum_{el}\int_{\Gamma_{el}} {}^t\boldsymbol{u}_{el}^\star\,{}^t\boldsymbol{N}_{el}\boldsymbol{f}dx \\ & + \sum_{x\in\partial\Omega} {}^t\boldsymbol{u}_{el}^\star\,{}^t\boldsymbol{N}_{el}\boldsymbol{t}(x) \\ & \Leftrightarrow \sum_{el} \underbrace{\left[\int_{\Gamma_{el}} {}^t\hat{\boldsymbol{B}}_{el}\boldsymbol{\Lambda}\hat{\boldsymbol{B}}_{el}dx\right]}_{\hat{\boldsymbol{k}}_{el}} \boldsymbol{U} = \underbrace{\sum_{el}\int_{\Gamma_{el}} {}^t\hat{\boldsymbol{N}}_{el}\boldsymbol{f}dx + \sum_{x\in\partial\Omega_e} {}^t\hat{\boldsymbol{N}}_{el}\boldsymbol{t}(x)}_{\hat{\boldsymbol{f}}_{el}} \end{aligned} \qquad [A.83]$$

NOTE A.14.– Caps refer to broken or extended matrices (written in the dof base of the problem). $\boldsymbol{U}$ is the vector that groups all the dof of the problem from $\boldsymbol{u}_{el}$.

The stiffness matrix (not expanded, i.e. reduced to the dof of the element) can therefore be calculated as follows:

$$\boldsymbol{k}_{el} = \int_0^l {}^t\boldsymbol{B}_{el}\boldsymbol{\Lambda}\boldsymbol{B}_{el}dx \qquad [A.84]$$

This expression therefore gives the following expression for the elementary stiffness matrix:

$$\boldsymbol{k}_{el} = \begin{bmatrix} \frac{EA}{l} & 0 & 0 & -\frac{EA}{l} & 0 & 0 \\ 0 & \frac{12EI}{l^3} & \frac{6EI}{l^2} & 0 & -\frac{12EI}{l^3} & \frac{6EI}{l^2} \\ 0 & \frac{6EI}{l^2} & \frac{4EI}{l} & 0 & -\frac{6EI}{l^2} & \frac{2EI}{l} \\ -\frac{EA}{l} & 0 & 0 & \frac{EA}{l} & 0 & 0 \\ 0 & -\frac{12EI}{l^3} & -\frac{6EI}{l^2} & 0 & \frac{12EI}{l^3} & -\frac{6EI}{l^2} \\ 0 & \frac{6EI}{l^2} & \frac{2EI}{l} & 0 & -\frac{6EI}{l^2} & \frac{4EI}{l} \end{bmatrix} \qquad [A.85]$$

A.3.3.2. *Elastic Timoshenko beams*

Timoshenko's kinematics no longer require the derivative of the transverse displacement with respect to x to be equal to the rotation of the section, so $\theta_z \neq \frac{dU_y}{dx}$.

In Timoshenko's kinematics, there is a distortion of the section $\gamma_y = \frac{dU_y}{dx} - \theta_z$, which characterizes the rotation of the section with respect to the perpendicular to the neutral fiber. This distortion physically represents the strain generated by the shear due to the shear force.

The finite element formulation of a Timoshenko beam is therefore a little different from that of Euler–Bernoulli in the choice of the interpolation of the kinematic field. In practice, the transverse displacement U_y and the rotation θ_z are interpolated independently. Several types of interpolations exist for Timoshenko beams. A comparison of the different formulations is detailed in Bitar et al. (2018).

In this work, this is the formulation of Friedman and Kosmatka (1993) which is used. The interpolation functions depend on the material parameters to avoid shear locking problems:

$$\underbrace{\begin{bmatrix} U_x(x) \\ U_y(x) \\ \theta_z(x) \end{bmatrix}}_{\boldsymbol{V}(x)} = \underbrace{\begin{bmatrix} N_1 & 0 & 0 & N_4 & 0 & 0 \\ 0 & N_8 & N_9 & 0 & N_{11} & N_{12} \\ 0 & N_{14} & N_{15} & 0 & N_{17} & N_{18} \end{bmatrix}}_{\boldsymbol{N}_{el}} \underbrace{\begin{bmatrix} u_i \\ v_i \\ \theta_i \\ u_j \\ v_j \\ \theta_j \end{bmatrix}}_{\boldsymbol{u}_{el}} \qquad \text{[A.86]}$$

Writing generalized strains $\boldsymbol{e}(x)$ (equation [A.87]) with the expression of the recalled form functions in Friedman and Kosmatka (1993) or Bitar et al. (2018) will lead to the expression of the stiffness matrix of equation [A.88]:

$$\boldsymbol{e}(x) = \begin{bmatrix} \varepsilon(x) \\ \gamma_y(x) \\ \kappa(x) \end{bmatrix} = \begin{bmatrix} \frac{dU_x}{dx}(x) \\ \frac{dU_y}{dx}(x) - \theta_z(x) \\ \frac{\theta_z}{dx}(x) \end{bmatrix} \qquad \text{[A.87]}$$

Any calculation done, the elementary stiffness matrix becomes:

$$\boldsymbol{k}_{el} = \begin{bmatrix} \frac{ES}{l} & 0 & 0 & -\frac{ES}{l} & 0 & 0 \\ 0 & \frac{12EI}{(1+\phi)l^3} & \frac{6EI}{(1+\phi)l^2} & 0 & -\frac{12EI}{(1+\phi)l^3} & \frac{6EI}{(1+\phi)l^2} \\ 0 & \frac{6EI}{(1+\phi)l^2} & \frac{(4+\phi)EI}{(1+\phi)l} & 0 & -\frac{6EI}{(1+\phi)l^2} & \frac{(2-\phi)EI}{(1+\phi)l} \\ -\frac{ES}{l} & 0 & 0 & \frac{ES}{l} & 0 & 0 \\ 0 & -\frac{12EI}{(1+\phi)l^3} & -\frac{6EI}{(1+\phi)l^2} & 0 & \frac{12EI}{(1+\phi)l^3} & -\frac{6EI}{(1+\phi)l^2} \\ 0 & \frac{6EI}{(1+\phi)l^2} & \frac{(2-\phi)EI}{(1+\phi)l} & 0 & -\frac{6EI}{(1+\phi)l^2} & \frac{(4+\phi)EI}{(1+\phi)l} \end{bmatrix} \qquad \text{[A.88]}$$

A.3.3.3. *Multifiber beam principle*

The multifiber beam concept makes it possible to decompose the fiber section whose uniaxial behavior can have a nonlinear behavior.

The generalized strains of the section are calculated from equation [A.81]. Based on these generalized strains, and because it is always assumed that the straight sections remain straight, it is possible to calculate local strains in each fiber $\varepsilon_{xx}(x, \mathbf{y})$ (in Euler–Bernoulli kinematics), and even an additional shear strain $\gamma_{xy}(x, \mathbf{y})$ in the case of Timoshenko kinematics (see Figure A.8(a)).

Different behaviors and therefore different materials can even be considered for each fiber (see Figure A.8(b)), which is very practical for modelling composite sections such as reinforced concrete.

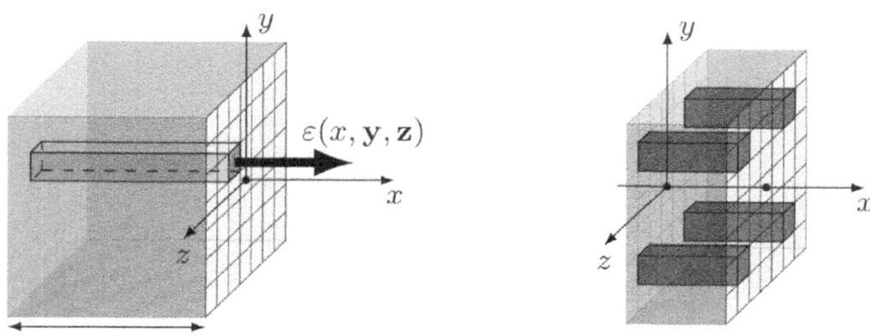

Figure A.8. *Illustration of the axial strain in each fiber in a section of beam of length dx (figure from Kotronis and Grange (2016)) and different materials representing the section (e.g. steel rebars in concrete)*

Thus, the calculation steps are as follows:

– from displacements at nodes: calculation of generalized strains $\mathbf{e}(x)$ and strain deformations in each fiber according to $\varepsilon_{xx}(x, \mathbf{y}) = \varepsilon(x) - \mathbf{y}\kappa(x)$ (rigid in-plane cross-section assumption);

– use of the behavioral law of each fiber: $\sigma_{xx}(x, \mathbf{y}) = f(\varepsilon_{xx}(x, \mathbf{y}))$;

– stress integration to return to generalized stresses $\boldsymbol{s}(x)$ according to $N = \int_S \sigma_{xx}(x, \mathbf{y}) dS$ and $M = \int_S -y\sigma_{xx}(x, \mathbf{y}) dS$;

– calculation of forces at nodes by projection using the matrix of derivatives of interpolation functions $\boldsymbol{B}_{el}$, according to $\boldsymbol{P} = {}^t[\boldsymbol{P}_1, \boldsymbol{P}_2] = \int_0^L {}^t\boldsymbol{B}_{el}\boldsymbol{s}(x)dx$.

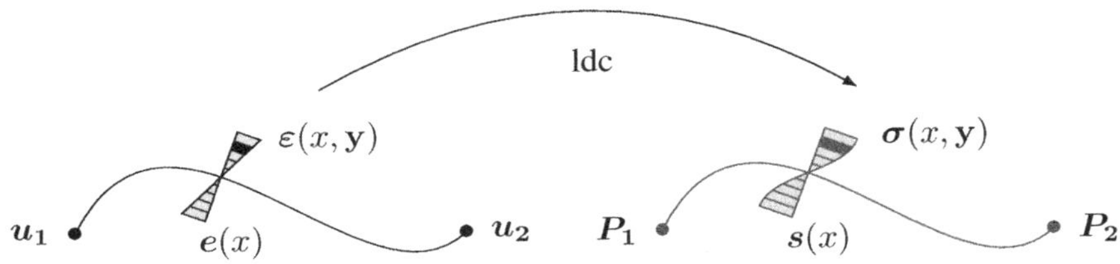

Figure A.9. *Calculation of strains, stresses and forces at nodes from displacements at nodes of the multifiber beam element (figure from Kotronis and Grange (2016)). For a color version of this figure, see www.iste.co.uk/mazars/damage.zip*

References

ACI (1995). Effect of restraint, volume change and reinforcement in cracking of mass concrete. Report, ACI Committee 207.

Aftes, R. (2012). Relations microstructure-fissuration-perméabilité dans les milieux granulaires cimentaires. PhD Thesis, Université de Montpellier II, Montpellier.

Ågårdh, L., Magnusson, J., Hansson, H. (1999). High strength concrete beams subjected to impact loading – An experimental study. FOA-R-99–01187–311–SE, Defense Research Establishment, Stockholm.

Association Eugène Freyssinet (2004). *Eugène Freyssinet – Une révolution dans l'art de construire*. Presses des Ponts et Chaussées, Paris.

Barre, F., Bisch, P., Chauvel, D., Cortade, J., Coste, J.-F., Dubois, J.-P., Erlicher, S., Gallitre, E., Labbé, P., Mazars, J. et al. (2016). *Control of Cracking in Reinforced Concrete Structures*. ISTE Ltd, London and John Wiley & Sons, New York.

Bascoul, A. and Maso, J.-C. (1974). Influence de la contrainte intermédiaire sur le comportement mécanique du béton en compression biaxiale. *Matériaux et constructions*, 42, 411–419.

Batoz, J. and Dhatt, G. (1990a). *Modélisation des structures par éléments finis 1. Solides élastique*. Les Presses de l'Université Laval, Quebec.

Batoz, J. and Dhatt, G. (1990b). *Modélisation des structures par éléments finis 2. Poutres et plaques*. Les Presses de l'Université Laval, Quebec.

Batoz, J. and Dhatt, G. (1990c). *Modélisation des structures par éléments finis 3. Coques*. Les Presses de l'Université Laval, Quebec.

Bazant, Z.P. (2002). Concrete fracture models: Testing and practice. *Engineering Fracture Mechanics*, 69, 165–205.

Bazant, Z.P. and Oh, B.H. (1983). Crack band theory for fracture of concrete. *Materials and Structures*, 16(3), 155–177.

Bitar, I., Grange, S., Kotronis, P., Benkemoun, N. (2018). A comparison of displacement based Timoshenko multi-fiber beams finite element formulations and elasto-plastic applications. *European Journal of Environmental and Civil Engineering*, 22(4), 464–490.

Bouhjiti, D. (2018). Analyse probabiliste de la fissuration et du confinement des grands ouvrages en béton armé et précontraint. PhD Thesis, Université Grenoble Alpes, Grenoble.

Bournazel, J.-P. (1995). Comportement thermomécanique des bétons compactés au rouleau sous l'effet d'automaturation : approche thermodynamique. *Materials and Structures*, 28, 318–323.

Braga, F., Gigliotti, R., Laterza, M., D'Amato, M., Kunnath, S. (2012). Modified steel bar model incorporating bondslip for seismic assessment of concrete structures. *Journal of Structural Engineering*, 138(11), 1342–1350.

Briffaut, M., Benboudjema, F., Torrenti, J.-M., Nahas, G. (2011). Numerical analysis of the thermal active restrained shrinkage ring test to study the early age behaviour of massive concrete structures. *Engineering Structures*, 33(4), 1390–1401.

Capdevielle, S., Grange, S., Dufour, F., Desprez, C. (2016). A multifiber beam model coupling torsional warping and damage for reinforced concrete structures. *European Journal of Environmental and Civil Engineering*, 20(8), 914–935.

CEOS.fr (2017). *Recommandations pour la maîtrise des phénomènes de fissuration*. Presses des Ponts et Chaussées, Paris.

Chen, W. and Forquin, P. (2019). Experimental and numerical study of the damage process in RC beam-column sub-assemblages during a progressive collapse scenario. *International Journal for Numerical and Analytical Methods in Geomechanics*, 43(9), 1704–1723.

CimBéton (2013). *Les bétons : formulation, fabrication et mise en œuvre. Fiches techniques, tome 2*. CimBéton, Paris.

Code_Aster (2020). [Online]. Available at: www.code_aster.org [Accessed January 2020].

Code Cast3M (2020). [Online]. Available at: http://www-cast3m.cea.fr/ [Accessed June 2020].

Corbin, M. and Garcia, M. (2016). International benchmark VeRCoRs 2015: Overview, synthesis and lessons learnt. Report, EDF, Lyon.

Coulomb, C.A. (1773). Essai sur une application des règles de maximis et minimis à quelques problèmes de statique relatifs à l'architecture. PhD Thesis, Académie royale des sciences, Paris. Annales de l'ITBTP, 478, 137–156.

Crambuer, R., Richard, B., Ile, F., Ragueneau, F. (2013). Experimental characterization and modeling of energy dissipation in reinforced concrete beams subjected to cyclic loading. *Engineering Structures*, 56(0), 919–934.

Cusson, D. and Pautre, P. (1994). High-strength concrete columns confined by rectangular ties. *Journal of Structural Engineering*, 120(3), 783–804.

Davenne, L., Saouridis, C., Piau, J.-M. (1989). Un code de calcul pour la prévision du comportement des structures endommageables en béton. *Annales de l'ITBTP*, 478, 137–156.

De Biasio, M., Grange, S., Dufour, F. (2014). Modeling structural-wall buildings through an enriched ERC method and a new damage-based concrete constitutive law: The SMART2013 case study. In *SMART 2013 Workshop*, Paris-Saclay.

De Schutter, G. (1999). Degree of hydration based Kelvin model for the basic creep of early age concrete. *Materials and Structures*, 32, 260–265.

Desprez, C., Mazars, J., Kotronis, P., Paultre, P. (2013). Damage model for FRP-confined concrete columns under cyclic loading. *Engineering Structures*, 48, 519–531.

EN1992-2, Eurocode 2 (2004). Calcul des structures en béton. European norm.

Engström, B. (2011). Restraint cracking of reinforced concrete structures. Thesis, Chalmers University of Technology, Gothenburg.

Erzar, B. and Forquin, P. (2011). Experiments and mesoscopic modelling of dynamic testing of concrete. *Mechanics of Materials*, 43, 505–527.

Erzar, B. and Forquin, P. (2014). Analysis and modelling of the cohesion strength of concrete at high strain-rates. *International Journal of Solids and Structures*, 51(14), 2559–2574.

Fellenius, W. (1936). Calculation of the stability of earth dams. In *2nd Congress on Large Dams*, Washington, DC.

Friedman, Z. and Kosmatka, J.B. (1993). An improved two-node Timoshenko beam finite element. *Computers & Structures*, 47(3), 473–481.

Gabet, T. (2006). Comportement triaxial du béton sous fortes contraintes : influence du trajet déchargement. PhD Thesis, Université Joseph-Fourier, Grenoble.

Gabet, T., Malecot, Y., Daudeville, L. (2008). Triaxial behavior of concrete under high stresses: Influence of the loading path on compaction and limit states. *Cement and Concrete Research*, 38(3), 403–412.

Gallagher, R.H. (1976). *Introduction aux éléments finis*. Pluralis, Paris.

Gatuingt, F., Snozzi, L., Molinari, J.F. (2013). Numerical determination of the tensile response and the dissipated fracture energy of concrete: Role of the mesostructure and influence of the loading rate. *International Journal for Numerical and Analytical Methods in Geomechanics*, 37(18), 3112–3130.

Giry, C., Dufour, F., Mazars, J. (2011). Stress-based nonlocal damage model. *International Journal of Solids and Structures*, 48(25/26), 3431–3443.

Grange, S. (2017). ATL4S–A tool and language for simplified structural solution strategy. International Report, GEOMAS/INSA Lyon Laboratory.

Halm, D. and Dragon, A. (1998). An anisotropic model of damage and frictional sliding for brittle materials. *European Journal of Mechanics – A/Solids*, 17, 439–460.

Heilmann, H.G., Hilsdorf, H.H., Finsterwalder, K. (1969). Festigkeit und verformung von beton unter Zugspannungen. In *Deutscher Ausschuss für Stahl-beton*. W. Ernst & Sohn, Berlin.

Hilal, A.A., Thom, N.H., Dawson, A.R. (2012). Pore structure and permeation characteristics of foamed concrete. *Journal of Advanced Concrete Technology*, 12, 535–544.

Hillerborg, A., Modeer, M., Peterssonn, P.E. (1976). Analysis of crack formation and growth in concrete by means of fracture mechanics and finite elements. *Cement & Concrete Research*, 6, 773–782.

Hopkinson, B. (1914). A method of measuring the pressure produced in the detonation of high explosives or by the impact of bullets. *Philosophical Transactions of the Royal Society A*, A213, 437–456.

Hrennikoff, A. (1941). Solution of problems of elasticity by the frame-work method. *ASME Journal of Applied Mechanics*, 8, 619–715.

Jaafari, C. (2020). Effect of early age drying shrinkage on the seismic response of RC structures. PhD Thesis, Laboratoire GEOMAS/INSA Lyon, Lyon.

Jardin, A., Leblond, J.-B., Berghhezan, D. (2009). Par-delà la loi de Miner pour la modélisation de la fatigue des élastomères. In *19e congrès français de mécanique*, CFM, Marseille.

Jason, L., Huerta, G., Pijaudier-Cabot, S., Ghavamian, S. (2006). An elastic plastic damage formulation for concrete: Application to elementary tests and comparison with an isotropic model. *Computer Methods in Applied Mechanics and Engineering*, 195, 7077–7092.

Jirásek, M. (2004). Non-local damage mechanics with application to concrete. *French Journal on Civil Engineering*, 8, 683–707.

Jungwirth, J. and Muttoni, A. (2004). Structural use of ultra-high performance fiber reinforced concrete. In *FCR in Switzerland: Research, Applications and Perspectives, Conference: FRC – From Theory to Practice*, Bergamo.

Kachanov, L.M. (1958). On time to rupture in creep conditions (in Russian). *Izviestia Akademii Nauk SSSR, Otdelenie Tekhnicheskikh Nauk*, 8, 26–31.

Kolani, B., Buffo-Lacarriere, L., Sellier, A. (2012). Exploitation des résultats du test THM des structures RG. Report, CEOS.fr R12CE0090 and R13CE0096.

Kotronis, P. and Grange, S. (2016). Méthodes simplifiées pour le calcul non linéaire de structures de génie civil. *Techniques de l'ingénieur. Méthodes de calcul et conception*, c6002 [Online]. Available at: https://doi.org/10.51257/a-v1-c6002.

Kotronis, P. and Mazars, J. (2005). Simplified modelling strategies to simulate the dynamic behavior of R/C walls. *Journal of Earthquake Engineering*, 9(2), 285–306.

Kotronis, P., Mazars, J., Davenne, L. (2003). The equivalent reinforced concrete model for simulating the behavior of shear walls under dynamic loading. *Engineering Fracture Mechanics*, 70(7/8), 1085–1097.

Kupfer, H.B. and Gerstle, K.H. (1973). Behavior of concrete under biaxial stresses. *Journal of the Engineering Mechanics Division*, 99(4), 853–866.

Kupfer, H.B., Hilsdorf, H.K., Rüsch, H. (1969). Behavior of concrete under biaxial stresses. *ACI Journal*, 66(66/62), 656–666.

Labbé, P., Pegon, P., Molina, J., Gallois, C., Chauvel, D. (2015). The SAFE experimental research on the frequency dependence of shear wall seismic design margins. *Journal of Earthquake Engineering*, 101–128.

Laborderie, C., Mazars, J., Pijaudier-Cabot, G. (1994). Response of plain and reinforced concrete structures under cyclic loadings. In *Damage Mechanics Model for Reinforced Concrete Structures Under Cyclic Loading*, Gerstle, W., Bazant, Z.P. (eds). ACI Special Publication, Indianapolis, IN, 134, 147–172.

Lavinia, S., Benboudjema, F., Torrenti, J.M., Bissonnette, B. (2010). Prediction of elastic properties of cement pastes at early ages. *Computational Materials Science*, 47(3), 775–784.

Lee, J. and Fenves, G.L. (1998). Plastic-damage model for cyclic loading of concrete structures. *Journal of Engineering Mechanics*, 124, 892.

Lemaitre, J. and Chaboche, J.-L. (1978). Aspect phénoménologique de la rupture par endommagement. *Journal de mécanique appliquée*, 2, 317–363.

Lemaitre, J. and Chaboche J.-L. (1990). *Mechanics of Solid Materials*. Cambridge University Press, Cambridge.

Lemaitre, J. and Desmorat, J. (2005). *Engineering Damage Mechanics*. Springer-Verlag, Berlin.

Li, Z.X. and Mróz, Z. (1994). Viscoplastic model combining damage and smeared crack for softening of concrete. *Acta Mechanica Solida Sinica*, 7(2), 125–136.

Lubliner, J., Oliver, J., Oller, S., Onate, E. (1989). A plastic-damage model for concrete. *International Journal of Solids and Structures*, 25(3), 299–326.

Mang, C. (2015). Modélisation de la liaison acier-béton dans le calcul des structures en béton armé. PhD Thesis, Université Paris Ouest, Paris.

Mayer, U. and Eligehausen, R. (1998). Bond behaviour of ribbed bars at inelastic steel strains. In *2nd International PhD Symposium in Civil Engineering*, Balàzs, G.L. (ed.). Technical University of Budapest, Budapest.

Mazars, J. (1984). Application de la mécanique de l'endommagement au comportement non linéaire et à la rupture du béton de structure. PhD Thesis, Université Paris 6, Paris.

Mazars, J. (1986). A description of micro and macroscale damage of concrete structure. *Engineering Fracture Mechanics*, 25, 729–737.

Mazars, J. and Bournazel, J.-P. (1998). Modelling of damage processes due to volumic variations for maturing and matured concrete. In *Concrete from Material to Structure*, Malier, Y. and Bournazel, J.P. (eds). Rilem, Paris.

Mazars, J. and Coste, J.-F. (2014). CEOS.fr national project: Behaviour and assessment of special structures in relation with shrinkage and creep. *The French Technology of Concrete CEB-Fib*, Mumbai.

Mazars, J. and Grange, S. (2015). Simplified modeling of reinforced concrete structures for engineering issues. *Computers & Concrete*, 16, 683–701.

Mazars, J. and Grange, S. (2017). Simplified strategies based on damage mechanics for concrete under dynamic loading. *Philosophical Transactions – A Royal Society*, A375 [Online]. Available at: http://dx.doi.org/10.1098/rsta.2016.0170.

Mazars, J. and Pijaudier-Cabot, G. (1989). Continuum damage theory: Application to concrete. *Journal of Engineering Mechanics*, 115(2), 345–365.

Mazars, J., Ramtani, S., Berthaud, Y. (1989). An experimental procedure to delocalize tensile failure and to identify the unilateral effect of distributed damage. In *Cracking and Damage*, Mazars, J., Bazant, Z.P. (eds). Elsevier, London.

Mazars, J., Berthaud, Y., Ramtani, S. (1990). The unilateral behaviour of damaged concrete. *Engineering Fracture Mechanic*, 35(4/5), 629–635.

Mazars, J., Pijaudier-Cabot, G., Saouridis, C. (1991). Size effect and continuous damage in cementitious materials. *International Journal of Fracture*, 51, 159–173.

Mazars, J., Grange, S., Briffaut, M. (2018). Simplified modeling strategy for the thermo-mechanical analysis of massive reinforced concrete structures at an early age. *Applied Sciences*, 8, 448.

Mazars, J., Hamon, F., Grange, S. (2015). A new 3D damage model for concrete under monotonic, cyclic and dynamic loading. *Materials and Structures*, 48, 3779–3793.

MC2010 (2013). *fib Model Code for Concrete Structures 2010*. Ernst & Sohn, Berlin.

Merceron-Vicat, M. (1999). *Louis Vicat, sa vie et ses travaux*. Lonrai.

Miner, M. (1945). Cumulative damage in fatigue. *Journal of Applied Mechanics*, 67, A159–A164.

Mivelaz, P. (1996). Étanchéité des structures en béton armé : fuite au travers d'un élément fissuré. PhD Thesis, EPF Lausanne, Lausanne.

Murakami, S. (2014). *Continuum Damage Mechanics*. Springer, Dordrecht.

Muralidhara Rao, T. and Gunneswara Rao, T.D. (2009). Fracture parameters of high strength concrete – An experimental study. *Journal of Structural Engineering*, 35(6), 397–403.

Norton, F.N. (1929). *The Creep of Steel at High Temperature*. McGraw-Hill, New York.

Ollivier, J.-P., Torrenti, J.M., Carcassès, M. (2012). *Propriétés physiques du béton et de ses constituants*. Hermès Sciences Lavoisier, Paris.

Pijaudier-Cabot, G. and Bazant, Z. (1987). Non-local damage theory. *Journal of Engineering Mechanics*, 133, 1512–1533.

Pijaudier-Cabot, G., Mazars, J., Pulikowski, J. (1991). Steel-concrete bond analysis with non-local continuous damage. *Journal of Structural Engineering ASCE*, 117(3), 862–882.

Planas, J., Elices, M., Guinea, G.V. (1992). Measurement of the fracture energy using three-point bend tests. *Materials and Structures*, 25, 305–312.

Pliskin, L. (1995). Béton. *Volume 4. Encyclopedia Universalis*, Paris.

Pontiroli, C., Rouquand, A., Mazars, J. (2010). Predicting concrete behavior from quasi-static loading to hypervelocity impact. An overview of the PRM model. *European Journal of Environmental and Civil Engineering*, 14(6/7), 703–727.

Rabotnov, Y.N. (1969). *Creep Problem in Structural Members*. North-Holland Publishing Co, Amsterdam/London.

Ragueneau, F., Lebon, G., Delaplace, A. (2010). Analyse expérimentale du comportement cyclique de poutres en béton armé. LMT Report, Cachan.

Ramtani, S. (1990). Contribution à la modélisation du comportement multiaxial du béton endommagé avec description du caractère unilatéral. PhD Thesis, Université Paris 6, Paris.

Rendulic, L. (1935). Ein Beitrag zur Bestimmung der Gleitsicherheit. *Der Bauingenieur*, 19–20, 230–233.

Richard, B., Ragueneau, F., Lucas, A., Crémona, C. (2011). A multi-fiber approach for modelling corroded reinforced concrete structures. *European Journal of Applied Mechanics A/Solids*, 30, 950–961.

Richard, B., Fontan, M., Mazars, J. (2014). SMART 2013: Overview, synthesis and lessons learnt from the International Benchmark. In *SMART 2013 Workshop*, Paris-Saclay.

Richard, B., Voldoire, F., Fontan, M., Mazars, J., Chaudat, T., Abouri, S., Bonfils, N. (2018). SMART 2013: Lessons learned from the international benchmark about the seismic margin assessment of nuclear RC. *Engineering Structures*, 161, 217–222.

RILEM Technical Committee-89 (1990). Determination of fracture parameters of plain concrete using three-point bends tests. RILEM Report, Paris.

Ritchie, R.O. and Launey, M.E. (2013). Crack growth in brittle and ductile solids. In *Encyclopedia of Tribology*, Jane Wang, Q., Wah Chung, Y. (eds). Springer Science, Dordrecht.

Rospars, C. and Chauvel, D. (2014). CEOS.fr experimental programme and reference specimen tests results. *Journal of European Environmental and Civil Engineering*, 18(7/8), 738–753.

Rouquand, A., Pontiroli, C., Mazars, J. (2007). Concrete structures under severe loading: A strategy to model the response for a large range of dynamic loads. In *FraMCoS-6 Proceedings*, Catania.

Salençon, J. (1983). *Calcul à la rupture et analyse limite*. Presses de l'ENPC, Paris.

Salençon, J. (1996). *Mécanique du continu. 1. Concepts généraux*. Universités francophones/ Ellipses, Paris.

Saouridis, C. and Mazars, J. (1989). A multiscale approach to distributed damage and its usefulness for capturing structural size effect. In *Cracking and Damage*, Mazars, J., Bazant, Z.P. (eds). Elsevier Applied Science, London.

Saouridis, C. and Mazars, J. (1992). Prediction of the failure and size effect in concrete via a bi-scale damage approach. *Engineering Computations*, 9, 329–344.

Sciumè, G., Benboudjema, F., De Sa, C., Pesavento, F., Berthaud, Y., Schrefler, B. (2012). A multi-physics model for concrete at early age applied to repairs problems. *Engineering Structures*, 57, 374–387.

Sellier, A. and Buffo-Lacarrière, L. (2011). Towards a simple and unified modelling of basic creep, shrinkage and drying creep of concrete. *European Journal of Environmental and Civil Engineering*, 13(10), 1161–1182.

Simo, J. and Hughes, T. (1998). *Computational Inelasticity*, Springer-Verlag, New York.

Smadi, M., Slate, F.O., Nilson, A.H. (1985). High-, medium-, and low-strength concretes subject to sustained overloads-strains, strengths, and failure mechanisms. *ACI Journals*, 82(5), 657–664.

SMART Website (2013). Seismic design and best-estimate methods assessment for reinforced concrete buildings subjected to torsion and non-linear effects [Online]. Available at: https://www.institut-seism.fr/en/projects/smart/ [Accessed June 2020].

Tang, S., Yao, Y., Andrade, C., Li, Z.J. (2015). Recent durability studies on concrete structure. *Cement and Concrete Research*, 143–154.

Terrien, M. (1980). Émission acoustique et comportement mécanique post-critique d'un béton sollicité en traction. *Bulletin des Ponts et Chaussées*, 105, 65–72.

Timoshenko, S.P. and Young, D.H. (1945). *Theory of Structures*. McGraw-Hill Book Co, New York.

Ulm, F.J. and Coussy, O. (1998). Couplings in early-age concrete: From material modeling to structural design. *International Journal of Solids and Structures*, 35(31/32), 4295–4311.

Verbeck, G.J. and Helmuth, R.H. (1968). Structure and physical properties of cement paste. In *Proceedings of the 5th International Symposium Chemistry of Cement*, Tokyo.

VeRCoRs Presentation (2018). Dissemination of EDF VeRCoRs project [Online]. Available at: https://www.tu1404.eu/news/dissemination-of-edf-vercors-project.html.

Wang, X. and Liu, X. (2004). Modeling bond strength of corroded reinforcement without stirrups. *Cement and Concrete Research*, 34, 1331–1339.

Wassereau, T. (2016). Caractérisation de matériaux composites par problème inverse vibratoire. PhD Thesis, Université du Maine, Le Mans.

Webb, J.P. (1990). Imposing linear constraints in finite-element analysis. *Applied Numerical Methods*, 6(6), 471–475.

Zhao, G. (2019). Experimental and numerical investigation of the progressive collapse resistance of reinforced concrete beam-column sub-assemblages. PhD Thesis, Université Grenoble Alpes, Grenoble.

Zienkiewicz, O., Taylor, R., Zhu, J.Z. (2013). *The Finite Element Method: Its Basis and Fundamentals*, 7th edition. Elsevier Ltd, Oxford.

Index

D

E, F

G, H, I

L

M

P, R

S

T

U, V, W

Other titles from

ISTE

in

Civil Engineering and Geomechanics

2020

SALENÇON Jean
Elastoplastic Modeling

2019

KOTRONIS Panagiotis
Risk Evaluation and Climate Change Adaptation of Civil Engineering Infrastructures and Buildings: Project RI-ADAPTCLIM

LAMBERT David Edward, PASILIAO Crystal L., ERZAR Benjamin, REVIL-BAUDARD Benoit, CAZACU Oana
Dynamic Damage and Fragmentation

PERROT Arnaud
3D Printing of Concrete: State of the Art and Challenges of the Digital Construction Revolution

SALENÇON Jean
Viscoelastic Modeling for Structural Analysis

ZEMBRI-MARY Geneviève
Project Risks: Actions Around Uncertainty in Urban Planning and Infrastructure Development

2018

FROSSARD Etienne
Granular Geomaterials Dissipative Mechanics: Theory and Applications in Civil Engineering

KHALFALLAH Salah
Structural Analysis 1: Statically Determinate Structures
Structural Analysis 2: Statically Indeterminate Structures

VERBRUGGE Jean-Claude, SCHROEDER Christian
Geotechnical Correlations for Soils and Rocks

2017

LAUZIN Xavier
Civil Engineering Structures According to the Eurocodes

PUECH Alain, GARNIER Jacques
Design of Piles Under Cyclic Loading: SOLCYP Recommendations

SELLIER Alain, GRIMAL Étienne, MULTON Stéphane, BOURDAROT Éric
Swelling Concrete in Dams and Hydraulic Structures: DSC 2017

2016

BARRE Francis, BISCH Philippe, CHAUVEL Danièle *et al.*
Control of Cracking in Reinforced Concrete Structures: Research Project CEOS.fr

PIJAUDIER-CABOT Gilles, LA BORDERIE Christian, REESS Thierry, CHEN Wen, MAUREL Olivier, REY-BETHBEDER Franck, DE FERRON Antoine
Electrohydraulic Fracturing of Rocks

TORRENTI Jean-Michel, LA TORRE Francesca
Materials and Infrastructures 1
Materials and Infrastructures 2

2015

AÏT-MOKHTAR Abdelkarim, MILLET Olivier
Structure Design and Degradation Mechanisms in Coastal Environments

MONNET Jacques
In Situ Tests in Geotechnical Engineering

2014

DAÏAN Jean-François
Equilibrium and Transfer in Porous Media – 3-volume series
Equilibrium States – Volume 1
Transfer Laws – Volume 2
Applications, Isothermal Transport, Coupled Transfers – Volume 3

2013

AMZIANE Sofiane, ARNAUD Laurent
Bio-aggregate-based Building Materials: Applications to Hemp Concretes

BONELLI Stéphane
Erosion in Geomechanics Applied to Dams and Levees

CASANDJIAN Charles, CHALLAMEL Noël, LANOS Christophe, HELLESLAND Jostein
Reinforced Concrete Beams, Columns and Frames: Mechanics and Design

GUÉGUEN Philippe
Seismic Vulnerability of Structures

HELLESLAND Jostein, CHALLAMEL Noël, CASANDJIAN Charles, LANOS Christophe
Reinforced Concrete Beams, Columns and Frames: Section and Slender Member Analysis

LALOUI Lyesse, DI DONNA Alice
Energy Geostructures: Innovation in Underground Engineering

LEGCHENKO Anatoly
Magnetic Resonance Imaging for Groundwater

2012

BONELLI Stéphane
Erosion of Geomaterials

JACOB Bernard *et al.*
ICWIM6 – Proceedings of the International Conference on Weigh-In-Motion

OLLIVIER Jean-Pierre, TORRENTI Jean-Marc, CARCASSES Myriam
Physical Properties of Concrete and Concrete Constituents

PIJAUDIER-CABOT Gilles, PEREIRA Jean-Michel
Geomechanics in CO_2 Storage Facilities

2011

BAROTH Julien, BREYSSE Denys, SCHOEFS Franck
Construction Reliability: Safety, Variability and Sustainability

CREMONA Christian
Structural Performance: Probability-based Assessment

HICHER Pierre-Yves
Multiscales Geomechanics: From Soil to Engineering Projects

IONESCU Ioan R. *et al.*
Plasticity of Crystalline Materials: from Dislocations to Continuum

LOUKILI Ahmed
Self Compacting Concrete

MOUTON Yves
Organic Materials for Sustainable Construction

NICOT François, LAMBERT Stéphane
Rockfall Engineering

PENSÉ-LHÉRITIER Anne-Marie
Formulation

PIJAUDIER-CABOT Gilles, DUFOUR Frédéric
Damage Mechanics of Cementitious Materials and Structures

RADJAI Farhang, DUBOIS Frédéric
Discrete-element Modeling of Granular Materials

RESPLENDINO Jacques, TOUTLEMONDE François
Designing and Building with UHPFRC

2010

ALSHIBLI A. Khalid
Advances in Computed Tomography for Geomechanics

BUZAUD Eric, IONESCU Ioan R., VOYIADJIS Georges
Materials under Extreme Loadings / Application to Penetration and Impact

LALOUI Lyesse
Mechanics of Unsaturated Geomechanics

NOVA Roberto
Soil Mechanics

SCHREFLER Bernard, DELAGE Pierre
Environmental Geomechanics

TORRENTI Jean-Michel, REYNOUARD Jean-Marie, PIJAUDIER-CABOT Gilles
Mechanical Behavior of Concrete

2009

AURIAULT Jean-Louis, BOUTIN Claude, GEINDREAU Christian
Homogenization of Coupled Phenomena in Heterogenous Media

CAMBOU Bernard, JEAN Michel, RADJAI Fahrang
Micromechanics of Granular Materials

MAZARS Jacky, MILLARD Alain
Dynamic Behavior of Concrete and Seismic Engineering

NICOT François, WAN Richard
Micromechanics of Failure in Granular Geomechanics

2008

BETBEDER-MATIBET Jacques
Seismic Engineering

CAZACU Oana
Multiscale Modeling of Heterogenous Materials

HICHER Pierre-Yves, SHAO Jian-Fu
Soil and Rock Elastoplasticity

JACOB Bernard *et al.*
HVTT 10

JACOB Bernard *et al.*
ICWIM 5

SHAO Jian-Fu, BURLION Nicolas
GeoProc2008

2006

BALAGEAS Daniel, FRITZEN Claus-Peter, GÜEMES Alfredo
Structural Health Monitoring

DESRUES Jacques *et al.*
Advances in X-ray Tomography for Geomaterials

FSTT
Microtunneling and Horizontal Drilling

MOUTON Yves
Organic Materials in Civil Engineering

2005

PIJAUDIER-CABOT Gilles, GÉRARD Bruno, ACKER Paul
Creep Shrinkage and Durability of Concrete and Concrete Structures CONCREEP – 7

Printed and bound by CPI Group (UK) Ltd, Croydon, CR0 4YY

16/09/2026

14963083-0004